Theory and Application of Secondary Electron Emission

This book examines the physical mechanisms of secondary electrons and recent developments, focusing on the theory and applications of secondary electron emission and suppression technology.

With the development of the modern information industry, space microwave technology, new materials, devices, and high-precision processing methods, secondary electron emission and its applications face new scientific and engineering challenges and opportunities. This book discusses theoretical models, measurements, data results, surface processes, and dielectric charging analysis of secondary electron emission. It covers key areas such as particle physics theory, simulation algorithms, magneto-hydrodynamics, surface processes, and surface characterization. The secondary electron yield and energy spectrum data are disclosed and analyzed, benefiting fields such as space microwave multipactor, scanning electron microscope (SEM), particle accelerator, and microwave plasma. Several typical application scenarios are also presented. Using the secondary electron emission data provided, this book shows how the secondary electron multiplication effect can be suppressed and exploited to improve device performance.

This title will be a valuable reference for researchers and students interested in the physical mechanisms of secondary electrons, SEM, accelerators, electrostatic discharge, and space multipactor.

Wanzhao Cui is a PhD supervisor and Chief Professor at the Northwestern Polytechnical University, researcher of the China Academy of Space Technology (CAST), and academic leader of satellite communication at China Aerospace Science and Technology Corporation (CASC). He is also an IEEE senior member. His main research interests include secondary electron emission effects, space microwave technology, and satellite communication technology.

Guobao Feng is a senior engineer at the China Academy of Space Technology and the youth director of Shaanxi Vacuum Society. His main research interests include theoretical simulation of secondary electron emission and space irradiation effects.

Space Science, Technology and Application Series

Series Editors - Weimin Bao, Yulin Deng

Simulation Method of Multipactor and Its Application in Satellite Microwave Components
Wanzhao Cui, Yun Li, Hongtai Zhang, Jing Yang

Passive Intermodulation: Concepts and Technology
Edited by Wanzhao Cui, Jun Li, Huan Wei, Xiang Chen

Spacecraft System Design
Edited by Zhang Qingjun and Liu Jie

Space Terahertz Remote Sensing Technology
Weidong Hu, Yuming Bai, and Xin Lv

Fundamentals of Ultrasonic Testing
Chunguang Xu and Weibin Li

Theory and Application of Secondary Electron Emission
Edited by Wanzhao Cui and Guobao Feng

For more information about this series, please visit: https://www.routledge.com/Space-Science-Technology-and-Application-Series/book-series/SSTA

Theory and Application of Secondary Electron Emission

Edited by
Wanzhao Cui and Guobao Feng

CRC Press
Taylor & Francis Group
Boca Raton London New York

CRC Press is an imprint of the
Taylor & Francis Group, an **informa** business

北京理工大学出版社
BEIJING INSTITUTE OF TECHNOLOGY PRESS

Designed cover image: Cover image created by Beijing Institute of Technology Press Co., Ltd

First edition published in English 2026
by CRC Press
2385 NW Executive Center Drive, Suite 320, Boca Raton FL 33431

and by CRC Press
4 Park Square, Milton Park, Abingdon, Oxon, OX14 4RN

CRC Press is an imprint of Taylor & Francis Group, LLC

English Version by permission of Beijing Institute of Technology Press

ISBN: 978-1-032-66708-9 (hbk)
ISBN: 978-1-032-91366-7 (pbk)
ISBN: 978-1-003-56286-3 (ebk)

DOI: 10.1201/9781003562863

Typeset in Minion
by codeMantra

Contents

Figures

Tables

Foreword

Looking back on the development of China's space industry over the past 40 years, from the first artificial satellite "Dongfanghong-1" to the successful manned flight of "Shenzhou-5," to the successful return of "Chang'e-5" with the moon soil, to the successful global network of "Beidou-3," and then to the successful development of the "Tianwen-1" exploration project, people who strive for China's space industry have always adhered to the traditional space spirit, the spirit of "Two Bombs, One Satellite" and the spirit of manned spaceflight. In the face of a series of scientific and technological problems with a weak industrial base at that time, China's space industry was fearless, overcame one difficulty after another, and built into today's aerospace power with proud achievements and amazing speed, ranking among the forefront of the world.

In the process of human exploration of the universe and the stars, it is destined to face endless unknowns and risks. In the process of moving into the vastness of space, China's space program also faces various challenges. In the development process, China's space payload faces similar challenges, where the secondary electron multiplication effect under high-power conditions is a serious bottleneck that limits the development of high-performance microwave payloads. The cascading secondary electron emission process of accelerated seed electrons between microwave pole plates can eventually evolve into strong pulse discharges within the components, damaging spacecraft microwave components. Therefore, for a payload, every microwave active and passive component needs to be ground tested to ensure they can withstand the designed power levels. However, for a considerable number of devices, mitigating secondary electron multiplication at high power has become a major test before they leave the factory. For the new era of space missions, high-power and integration requirements make spacecraft payload components face a more severe test of secondary electron multiplication, and more and more attention is being focused on the secondary electron emission problem itself.

As early as the end of the 19th century, scientists discovered that when electrons were irradiated on the surface of a metal sample, there would be secondary electron emission in excess of the original number of electrons. Since then, there has been significant progress in secondary electron-related research and applications. The secondary electrons emitted from the surface are used in many scanning electron microscopes to investigate the surface microscopic morphology. In addition, secondary electrons are also used in applications such as electron multiplier tubes, electron beam exposure, and particle accelerators. On the one hand, high secondary electron yield (SEY) materials are used to make electron multiplier tubes to amplify electrical signals; on the other hand, the electron cloud effect caused by secondary electron multiplication in high-energy particle accelerators severely limits the particle beam quality. In high-power microwave components in space, the secondary electron emission poses significant challenges due to the induced multiplication discharge effect. The research on secondary electrons has been well established by our researchers, but the reports on secondary electron emission are primarily limited to academic publications. As frontline workers in space science and technology, the authors of this book have been researching the high-power microwave secondary electron multiplication effect in space for more than 10 years. They have written this book based on their solid theoretical foundation and rich practical experience in engineering. This book provides a comprehensive and detailed explanation of all aspects, from microscopic particle collision theory to macroscopic surface control technology, and will be an invaluable resource for scientific research and engineering applications.

Every aspect of aerospace engineering must adhere to a rigorous and robust work style, and solving many space-related challenges requires a deep dive into fundamental disciplines to address the bottlenecks that hinder technological breakthroughs in aerospace. I am glad that the authors of this book, as frontline researchers in aerospace engineering, have written this book to consolidate the research in basic sciences, which is of high reference value both for the promotion of basic scientific research and for the analysis of related engineering problems, thus contributing to the scientific and technological development and progress of China's aerospace industry.

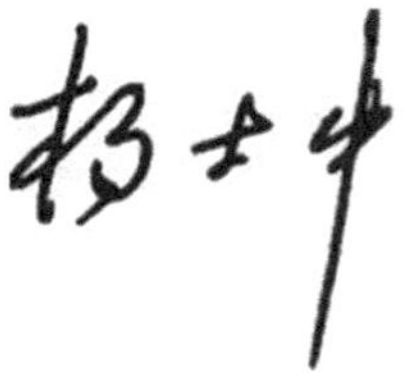

Yang Shizhong

Academician of CAE Member

Preface

SINCE THE EARLY 20TH-CENTURY discovery of the phenomenon of additional secondary electron emission under electron bombardment, the essential mechanisms have been largely elucidated through more than a century of research. With the development of the modern information industry, new materials and high-precision processing methods, as well as other scientific technologies, have brought new challenges to a physical phenomenon that was thought to be solved in the last century. For example, microwave payloads in the natural vacuum of the space orbit environment and high-power microwave fields that drive seed electrons in the cavity can lead to a cascade of secondary electron emission multiplication, potentially causing high-power discharges and operational anomalies within the device. In China's Scattered Neutron Source/Fast Cycle Synchrotron, the electron cloud effect formed by the secondary electron emission phenomenon directly affects the stability of the particle accelerator operating state and severely limits the beam quality of the ring cyclotron. In addition, for electron multiplier tubes applied in ultrafast diagnostics, it is also necessary to find new plating materials to achieve more considerable secondary electron emission. In the traditional electron microscopy and electron beam exposure fields, the image charge drift, "ghosting," and scattering range diffusion, which are directly related to secondary electrons, are still important issues that affect the image and processing accuracy. At the same time, the emergence of new materials and surface processing technology innovation for the secondary electron emission of the physical phenomenon of research has brought a new direction of technical research. Although the era of vacuum electron tubes has long gone, the development of space microwave technology, large scientific devices, and high-precision processing technology has opened a new chapter for the classical physical phenomenon of secondary electron emission.

Given the long history of research on the phenomenon of secondary electron emission, the classical theories of secondary electrons have evolved through models such as Joy's, Vaughan's, Furman's, and the Monte Carlo method. Physicists such as D.C. Joy, J. Cazaux, H.J. Fitting, and D.R. Penn have made significant contributions to the scientific understanding of secondary electron emission.

This book focuses on the theoretical models of secondary electron emission and their applications, with several typical cases discussed in various chapters. It covers theoretical models of secondary electron emission, measurement techniques, data results, surface processes, and dielectric charging analysis, and introduces typical application scenarios. It covers a range of fields including particle physics theory, simulation algorithms, magneto-hydrodynamics, surface processes, and surface characterization.

This book was written under the guidance of several experts and scholars. We would like to thank Professor Cao Meng of Xi'an Jiaotong University, Hu Tiancun, Wang Xinbo, Wang Rui, Xie Guibai, Yang Jing, He Yun, Bai Chunjiang, Chen Xiang, Wang Qi, Wei Huan, Bai He, and Chen Zeyu for their valuable comments on this book. Part of the research work in this book was supported by the Natural Science Foundation projects (No. 61901360, 62101434).

Despite our best efforts, it is inevitable that there will be some shortcomings in the book. Readers are welcome to provide feedback.

Contributors

Yahui Cai
School of Microelectronics
Xi'an Jiaotong University
Xi'an, China

Wanzhao Cui
National Key Laboratory of Science and Technology on Space Microwave
China Academy of Space Technology
Xi'an, China

Hongjie Dong
School of Physics and Optoelectronic Engineering
Nanjing University of Information Science and Technology
Nanjing, China

Guobao Feng
National Key Laboratory of Science and Technology on Space Microwave
China Academy of Space Technology
Xi'an, China

Yongning He
Faculty of Electronic and Information Engineering
School of Microelectronics, Xi'an Jiaotong University
Xi'an, China

Shulin Liu
Institute of High Energy Physics
Chinese Academy of Sciences
Beijing, China

Yifan Liu
School of Physics and Optoelectronic Engineering
Nanjing University of Information Science and Technology
Nanjing, China

Guanghui Miao
National Key Laboratory of Science and Technology on Space Microwave
China Academy of Space Technology
Xi'an, China

Fang Wang
Key Laboratory for Physical Electronics and Devices of the Ministry of Education
Xi'an Jiaotong University
Xi'an, China

Aigen Xie
School of Physics and Optoelectronic Engineering
Nanjing University of Information Science and Technology
Nanjing, China

Baojun Yan
Institute of High Energy Physics
Chinese Academy of Sciences
Beijing, China

Ming Ye
College of Information and Control Engineering
Xi'an University of Architecture and Technology
Xi'an, China

CHAPTER 1

Introduction

Wanzhao Cui and Guobao Feng

1.1 BASIC CONCEPT

As a special effect that affects the reliable and safe operation of spacecraft payloads in orbit, the occurrence of secondary electron emission and charging effects can lead to serious consequences. In recent years, with the development of space technology, the secondary electron emission induced by electron bombardment and the charging effects of spacecraft dielectric materials have received great attention, which are the main topics covered in this book.

Secondary electron emission refers to the phenomenon where electrons or other particles with certain energy emit electrons from the surface of solid materials when they are irradiated. When the pressure is at 1×10^{-3} Pa or lower, it is easy to produce resonance discharge through secondary electron emission under high power conditions, which is known as the multipactor effect. The occurrence process of bilateral multipactor is as follows: under the acceleration of the electric field between the plates, initial electrons move from the lower plate to the upper plate and collide with the surface of the upper plate. Due to secondary electron emission, the upper plate may emit multiple electrons (the specific emission conditions depend on the material and incident electron parameters). When these electrons are emitted, the electric field between the plates changes direction, allowing these electrons to move from the top plate to the bottom plate and collide with the surface of the bottom plate under the action of the electric field, thereby exciting more secondary electrons. If certain conditions are met, this process will continue, resulting in an exponential increase in the number of electrons between the plates, creating a secondary electron multiplier effect, or multipactor effect.

The mechanism of secondary electron generation can be briefly summarized as follows: when electrons are incident on the material surface with a certain energy, they scatter several times with the atoms or molecules inside the material (Figure 1.1), and part of the electrons are bounced back directly from the surface atoms by elastic scattering, forming elastic backscattered electrons (BSE). The primary electrons that enter the material interior may inelastically scatter with the material atoms and excite inner secondary electrons, which are

DOI: 10.1201/9781003562863-1

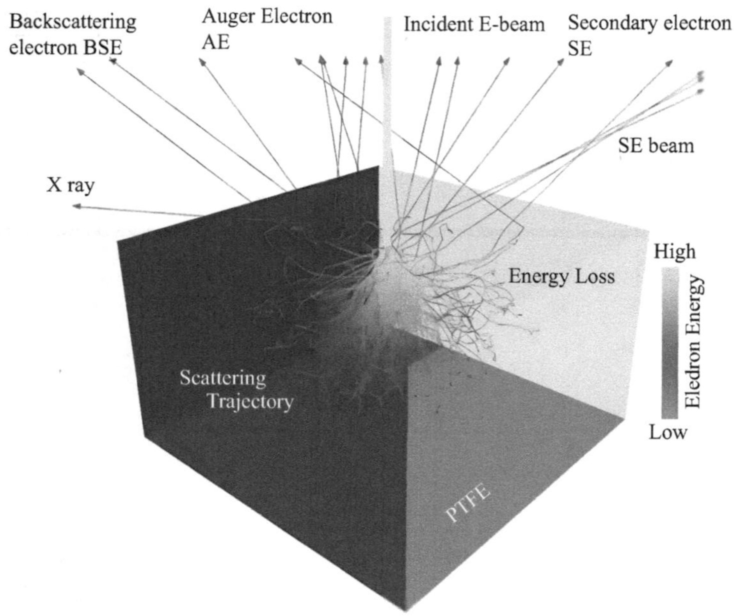

FIGURE 1.1 Scattering process of interaction between electron and material.

mainly formed by the incident electrons ionizing the sample atoms in the conduction band, valence band, or a small amount of inner shell electrons escaping from the sample. Some of the inner secondary electrons move toward the surface, overcome the work function, and are emitted, forming intrinsic secondary electrons (SE). Some of the primary electrons change their trajectories and lose energy through multiple scattering until they escape from the surface to form inelastic backscattered electrons or stay inside the sample after consuming all their energy. The elastic and inelastic backscattered electrons are usually referred to as backscattered electrons. To distinguish between these two types of electrons, their classification can theoretically be judged according to the type of scattering. In experiments, the type of secondary electrons is usually judged by their energy, with secondary electrons having energy less than 50 eV regarded as intrinsic secondary electrons and those with energy greater than 50 eV as backscattered electrons. To measure the ability of secondary electron emission, secondary electron yield (SEY) characterization is typically used. SEY is defined as the ratio of the number of secondary electrons, including both intrinsic secondary electrons and backscattered electrons, emitted from the material surface to the number of electrons incident on the material surface.

In the space environment, in addition to escaping from the sample surface in the form of secondary electrons, the electrons irradiated into the dielectric material accumulate in the form of deposited charges within the sample or in the superficial layers, thus causing charging phenomena in the spacecraft. The process of electron charge effect after incident dielectric is divided into two main steps: first, electrons scatter and ionize with atoms, generating a large number of electron pairs and hole pairs, forming a localized charge-dense region with plasma characteristics. Then, the local plasma causes electrons and holes to form a certain space charge distribution in the surface and deep layers of the material by transport and trapping, and at the same time affects the space charge properties of the material by changing the secondary electron emission coefficient through the generated surface electric field. The charging of a dielectric sample irradiated by an electron beam is caused by the imbalance between the electron beam current incident on the sample and the secondary electron current and leakage current leaving the sample. The charging process is influenced by both the sample conditions and the incident electron beam conditions. When the electron beam energy is greater than the second critical energy that makes the secondary electron emission coefficient equal to 1, the electron beam current is greater than the secondary electron current leaving the sample surface, and the sample is negatively charged; otherwise, it is positively charged.

The charging and discharging phenomena of spacecraft dielectric materials are the most common causes of abnormal operation and failure of spacecraft. The charging and discharging process will exacerbate and damage the surface pollution of the spacecraft, and the high-intensity current pulse generated during discharge can cause abnormal operation or even damage to electronic devices. The phenomenon of charging and discharging of spacecraft materials is not only related to the space environment but also to the physical properties of the spacecraft materials themselves. Polymer materials, which are used extensively in spacecraft, have good dielectric properties, and when operating in space, charged particles from space are stored on their surface and deep within them, forming a charging phenomenon. The secondary electron emissivity, dielectric strength, volume conductivity, and dielectric properties of polymers are closely related to their charging phenomena. Among these physical properties, the volume conductivity and dielectric properties of polymers are important parameters for evaluating the charge-discharge behavior of polymers. Moreover, with the effective control of the high and low-temperature effects of the space environment in recent years, the charge-discharge effect caused by space irradiation has become an important issue hindering the long-term normal operation of microstrip antennas in orbit.

1.2 THE STATE OF THE ART

1.2.1 Research of Secondary Electron Emission Characteristics

To suppress the secondary electron multiplication effect in microwave devices, methods that reduce the secondary electron emission coefficient through surface treatment, including the creation of surface grooves and trap structures, ion sputtering cleaning, and heating treatment, have received extensive attention and undergone in-depth research.

Since the 1970s, the European Space Agency (ESA) has continuously funded projects dedicated to researching the suppression of secondary electron multiplication effects [1]. In the initial phase of the research, various surface treatment techniques were mainly proposed to effectively suppress secondary electron emission [2]. Building on this, the second project focused on studying porous metal surfaces and the environmental stability of the secondary electron emission suppression characteristics. The third project concentrated on the issue of electrical loss characteristics [1]. At the MULCOPIM conference in 2011, a theory was presented that surface roughness structures on the micro-nanometer scale can significantly increase the component threshold. However, it also faced two major challenges: the problem of increased insertion loss due to the roughened surface and the environmental stability of component performance due to the adsorption process of water molecules [1,3]. In fact, there are still some results on the suppression of secondary electron emission in the field of gas pedals. A more notable example is that Pivi and Wang et al. of the Stanford University group studied the effect of regular surface topography on the secondary electron emission coefficient using the Monte Carlo method and Furman probability model. They used the method of carving grooves on the material surface to reduce the secondary electron emission coefficient [4,5]. In 2010, Mostajeran et al. in Iran studied the motion of electrons in the RF field, and simulation results showed that the adsorption of electrons inside the hemispherical depression in the 2D model led to a significant reduction in the secondary electron emission coefficient [6]. In 1973, Henrich et al. at Lincoln Laboratory performed ion cleaning of Si materials with Ar ions and found that the secondary electron emission coefficient was significantly reduced from 1.5 to 0.9 [7]. In 2000, Baglin et al. at CERN used Ar ion glow discharge cleaning to suppress secondary electron emission, and the peak secondary electron emission coefficient of Cu material was reduced from 2.4 to 1.3 and Al from 3 to about 1 [8]. In 2006, Le Pimpec et al. at the Swiss National Laboratory, in collaboration with Kirby et al. at the Stanford Linear Accelerator Center (SLAC), used different doses of positive N_2 ions to bombard TiZrV film samples to suppress secondary electron emission in the storage ring, and the measured secondary electron emission coefficient decreased with increasing ion dose [9]. In 2000, Bojko et al. of CERN heated Cu samples to 350°C, and the secondary electron emission coefficient decreased from 2.5 to 1.4, mainly due to the desorption of water molecules adsorbed in the surface layer during heating. The results of Ar ion cleaning were also compared, and it was found that the secondary electron emission coefficients of ion cleaning were smaller than those of heating above 300°C, indicating that ion sputtering cleaning has a more pronounced effect on suppressing the secondary electron emission coefficients [10].

In the study of theoretical analysis and numerical simulation of secondary electron emission and multiplication, the analytical derivation of the multipactor threshold of parallel plate structures based on the electron motion equation and resonance conditions was the main method for analyzing secondary electron multiplication theory at that time, as computers were not widespread in the early days [11–13]. With the development of computer technology, the study of numerical simulation of multipactor effects using computers has made significant progress in recent years. Representative numerical simulation methods include particle-in-cell (PIC) [14], single-electron tracking [15], and statistical theory [16,17]. These numerical simulation methods closely approximate the actual physical processes, not only accounting for details about the secondary electron emission process but also, in combination with sophisticated electromagnetic field calculation tools, analyzing the multipactor of microwave devices including coaxial, diaphragm, and other complex structures. The theoretical model of secondary electron emission is described in detail in Chapter 2.

Many researchers in China have also used various methods to suppress secondary electron emission. To suppress the hazard of secondary electron emission from the collector electrode in space traveling wave tubes, Beijing Vacuum Electronic Technology Co., Ltd. analyzed its suppression effect on secondary electron emission by using Mo atomic ion sputtering deposition and reduced the secondary electron emission coefficient of Cu to below 1 through a special process, significantly improving the efficiency of the traveling wave tube [18–21]. Additionally, Chang Chao et al. of the Northwest Institute of Nuclear Technology, in their study of high-power dielectric window discharge, proposed different surface configurations to suppress the one-sided multiplication process of electrons by analyzing the evolution of secondary electron multiplication, thus effectively improving the discharge threshold [22]. Cui Wanzhao's team at the Xi'an Branch of the Chinese Academy of Space Technology conducted an in-depth experimental study of secondary electron emission during their study of high-power multipactor and built the first high-precision secondary electron test platform in China. This platform can measure the secondary electron emission yield and energy spectrum characteristics in situ under different conditions of temperature and ion cleaning, reaching the international leading level [23,24]. In 2012, the team used an electrophoretic deposition method to deposit carbon nanofilms on silver-plated substrates and showed that the uniform and dense carbon nanofilms with a distinct trapping structure can effectively suppress secondary electron emission and improve the multipactor threshold, providing a new method for fabricating high-power microwave components [25]. From the theoretical analysis of the secondary electron trapping effect of micro-trap structures to the realization method of cylindrical hole trap surfaces, a team led by He Yongning at Xi'an Jiaotong University has conducted a principle experimental study on the suppression of secondary electron emission from spacecraft microwave components using micro-trap structures. They were able to effectively control the increase in insertion loss under large porosity conditions [26–30]. For example, a regular micro-trap structure was realized on the surface of a silver-plated aluminum alloy specimen by a semiconductor lithography process with in situ Ar ion cleaning, and the peak secondary electron emission coefficient was suppressed by 37.6%. The prepared micro-trap structure can improve the multipactor threshold of microwave devices by about 4 dB and can reduce the insertion loss to 0.3 dB. Since 2009, Zhang Haibo's research team at Xi'an Jiaotong University has proposed a new multi-band model for multiple interactions of secondary electrons with metal surfaces based on Monte Carlo simulations of electron-metal interactions and found that rectangular slots and circular pits can effectively suppress secondary electron emission [31]. In recent years, the team has worked closely with the Xi'an Branch of the Chinese Academy of Space Technology to conduct experimental research on the secondary electron emission properties of metals, and the secondary electron emission coefficient of Cu materials could be reduced from 2 to about 1.1 by using Ar ion sputtering cleaning on metal materials [32], and the secondary electron emission coefficient could be reduced from 2 to about 1.6 by heating silver-plated materials commonly used in spacecraft to about 300°C [33].

Although the study of the multipactor effect caused by secondary electron emission has been carried out for more than half a century, due to the complexity of its physical nature, the problem has not been well solved. This includes the theoretical description of its formation process, the saturation mechanism, or the prediction of the multipactor threshold of high-power microwave devices, which are still under development. With the development of high-power microwave charging technology in China, the demand for experimental verification of secondary electron emission is becoming increasingly urgent. Therefore, the experimental research on secondary electron emission of spacecraft dielectric materials is an important part of the independent reliability test of China's spacecraft and is of great significance for improving the on-orbit reliability of China's high-power microwave payload spacecraft.

1.2.2 Secondary Electron Dynamic Emission

The phenomenon of secondary electron emission was discovered by Campbell in 1899 and reported by the German scientists Austin and Starke in 1902 [34]. Essentially, for dielectric materials, if the number of secondary electrons escaping the surface potential barrier is not equal to the number of electrons incident on the material's surface, the interior of the material will exhibit the phenomenon of charged samples due to the accumulation of charges resulting from this imbalance of incoming and outgoing charges. Therefore, the secondary electron emission yield (SEY) is directly related to the amount of charge carried by the dielectric material when it is irradiated by electrons and the charged state it presents [35].

When electrons of a certain energy are incident on the sample, the electrons first undergo a complex series of processes, including electron scattering and charge transport within the material [36]. After the electrons hit the sample material, elastic and inelastic collision processes occur, and the electrons in the surface layer that overcome the work function to leave the sample form outgoing secondary electrons, while the charges deposited within the sample cause it to exhibit charged properties.

As electrons travel inside the material, they inevitably collide with atoms or molecules. These collisions cause the electrons to change their trajectory according to certain laws, a process known as scattering. The scattering process of electrons can be divided into elastic scattering and inelastic scattering, depending on whether energy is lost or not, as shown in Figure 1.2. Elastic scattering is the scattering of electrons by the Coulomb potential of the nucleus and the electron cloud outside the nucleus. Since the mass of the nucleus is much larger than the mass of the electron (more than three orders of magnitude larger), the energy change of both the atom and the incident electron before and after scattering

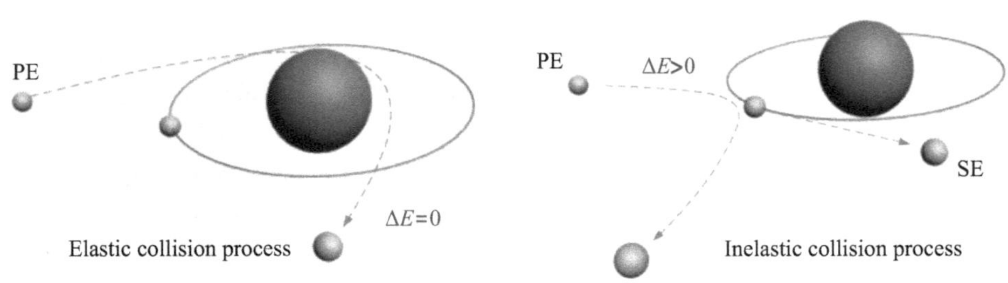

FIGURE 1.2 Elastic and inelastic collisions of incident electrons in collisions with materials.

is negligible, only the direction of motion of the incident electron changes. In the inelastic scattering process, the incident electron interacts with the outer electrons, and not only changes its direction of motion before and after scattering but also loses some energy. This energy is transferred to the outer electrons, causing them to escape from the nucleus and become excited, forming inner secondary electrons. Some of the inner secondary electrons will move toward the surface, overcome the work function, and escape, forming intrinsic secondary electrons, also called true secondary electrons. Some of the primary electrons change their trajectory and lose energy through multiple internal scatterings. They either escape from the surface as inelastic backscattered electrons or lose all their energy and remain within the sample, being absorbed by the material.

To obtain more accurate secondary electron emission yields, the Xi'an Branch of the Chinese Academy of Space Technology (CAST) has developed an accurate secondary electron emission test platform. The platform can measure and study the SEY and SES of materials at different electron incident energies and angles in an ultra-high vacuum environment with an in situ surface treatment device.

The research platform is composed of a test platform and two sets of control cabinets, where the test platform mainly includes three parts: vacuum system, test system, and sample processing system. Figure 1.3 shows the structure principle of the secondary electron emission characteristic research platform. The main vacuum chamber can achieve an ultimate vacuum of 10^{-9} Pa, equipped with two electron guns operating at 10–5,000 eV and 5–30 keV, and the electron beam current ranges from 50 pA to 2 nA. The main test

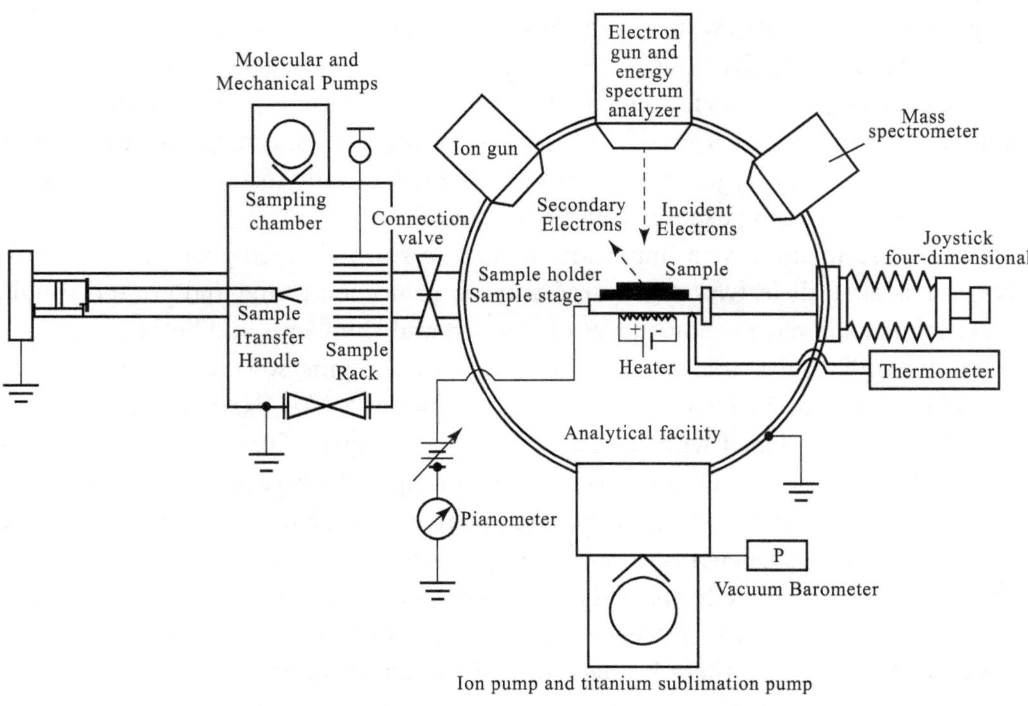

FIGURE 1.3 Structure principle of secondary electron emission characteristics measurement platform.

chamber also incorporates multiple functions such as sample heating (up to 400°C), Ar^+ ion cleaning, SEM imaging, and XPS surface composition analysis, and has high test accuracy and stability.

For dielectric samples, the imbalance between the continuous electron irradiation and secondary electron emission makes the samples charged, and the charged state of the surface in turn affects the secondary electron emission, so that the secondary electron emission exhibits dynamic changes. R. Renoud et al. of the College of Science and Technology, University of Nantes, France, in studying the secondary electron emission characteristics of dielectric materials under continuous electron irradiation, found that when the incident electron energy is high (2.5–10 keV), the surface of dielectric materials exhibits negatively charged characteristics. With continued irradiation, the secondary electron yield gradually increases from an initial state of less than 1 to near 1, and finally reaches a stable equilibrium. When the incident electron energy is low (100–1,000 eV), the corresponding secondary electron yield gradually decreases from the initial state of greater than 1 to 1 during the continuous increase of the irradiation dose and finally reaches stability. In the case of higher energy electron incidence, the outgoing electrons are less than the incoming electrons, and the material is negatively charged, while the negative surface potential will repel the incoming electrons and reduce the incoming electron energy, thus increasing the secondary electron yield until the secondary electron yield $SEY \approx 1$ reaches equilibrium. In the case of low incident electron energy, the initial secondary electron yield is greater than 1, and the surface is positively charged, when the outgoing secondary electrons return to the surface under the action of the local positive electric field, resulting in the actual escape of the outgoing secondary electrons decreases, and finally reaches the charged equilibrium. In general, surface positive charging only requires a surface potential of a few tens of eV to pull back the outgoing secondary electrons to reach equilibrium, while for negative charging, when the incident electron energy is very large, the surface potential can usually reach several keV, so the surface negative charging usually shows more significant.

Numerical simulation is an important method to study charged phenomena in space electron irradiated dielectrics. Compared with on-orbit monitoring and ground simulation experiments, it has the advantages of low cost and simplicity, and helps to study the charged dynamic properties. Commercial spacecraft charging software has been developed and put into practice by major space agencies, such as NASA's NASCAP-2K [37], open source SPIS [38] developed with the support of the European Space Agency (ESA) and others in 2002, and MUSCAT [39] developed by the Japan Aerospace Exploration Agency (JAXA) in 2004. This commercial software is powerful enough to model spacecraft structures in the space plasma environment and to perform macroscopic system-level analysis of spacecraft charging. The focus of the studies has been on spacecraft surface charging and potentials. However, the influence of surface conditions on secondary emissions has not been considered, and there is a lack of exploration of the microscopic mechanisms of space charge accumulation [38,39]. This results in several limitations in studying the dynamic characteristics of the charging process.

REFERENCES

1. Nistor V C, Aguilera L, Montero I, et al. Strategies for anti-multipactor coatings of suppressed secondary emission and low insertion losses for high power RF components in satellite systems. *7th International Workshop on Multipactor, Corona and Passive Intermodulation in Space RF Hardware*, Valencia, Spain. 2011: 25–32.
2. Wolk D, Damaschke J, Vicente C. AO-4025 ITT ESA–Surface treatment and coating for the reduction of multipactor and Passive Intermodulation (PIM) effects in RF components. *Proceedings of the 4th International Workshop on Multipactor, Corona and Passive Intermodulation in Space RF Hardware (MULCOPIM2003)*, Valencia, Spain. 2003: 8–11.
3. Aguilera L, Montero I, Dávila M E. ESA survey on secondary emission yield of industry materials and their impact on multipactor threshold. *ESA MULCOPIM*, Valencia, Spain, 2011.
4. Wang L, Bane K, Chen C, et al. Suppression of secondary electron emission using triangular grooved surface in the ILC dipole and wiggler magnets. *2007 IEEE Particle Accelerator Conference (PAC)*. IEEE, Albuquerque, NM, 2007: 4234–4236.
5. Pivi M, King F K, Kirby R E, et al. Sharp reduction of the secondary electron emission yield from grooved surfaces. *Journal of Applied Physics*, 2008, 104(10), 104904.
6. Mostajeran M, Rachti M L. Importance of number of gap crossings on secondary emission in the simulation of two-sided multipactor. *Journal of Instrumentation*, 2010, 5(8), P08003.
7. Henrich V E. Fast, accurate secondary-electron yield measurements at low primary energies. *Review of Scientific Instruments*, 1973, 44(4), 456–462.
8. Baglin V, Bojko J, Gröbner O, et al. The secondary electron yield of technical materials and its variation with surface treatments. *Proceedings of EPAC 2000*, Vienna, 2000: 26–30.
9. Le Pimpec F, Kirby RE, King FK, et al. The effect of gas ion bombardment on the secondary electron yield of TiN, TiCN and TiZrV coatings for suppressing collective electron effects in storage rings. *Nuclear Instruments and Methods in Physics Research Section A*, 2006, 564(1), 44–50.
10. Bojko I, Hilleret N, Scheuerlein C. Influence of air exposures and thermal treatments on the secondary electron yield of copper. *Journal of Vacuum Science & Technology A: Vacuum, Surfaces, and Films*, 2000, 18(3), 972–979.
11. Hatch A J, Williams H B. The secondary electron resonance mechanism of low-pressure high-frequency gas breakdown. *Journal of Applied Physics*, 1954, 25(4), 417–423. doi:10.1063/1.1721656.
12. Hatch A J, Williams H B. Multipacting modes of high-frequency gaseous breakdown. *Physical Review*, 1958, 112(3), 681–685. doi:10.1103/PhysRev.112.681.
13. Vaughan J R M. Multipactor. *IEEE Transactions on Electron Devices*, 1988, 35(7), 1172–1180.
14. Li Y, Cui W Z, Zhang N, et al. Three-dimensional simulation method of multipactor in microwave components for high-power space application. *Chinese Physics B*, 2014, 23(4), 048402.
15. de Lara J, Pérez F, Alfonseca M, et al. Multipactor prediction for on-board spacecraft RF equipment with the MEST software tool. *IEEE Transactions on Plasma Science*, 2006, 34(2), 476–484.
16. Rasch J, Johansson J F. Non-resonant multipactor—A statistical model. *Physics of Plasmas*, 2012, 19(12), 123505.
17. 林舒, 李永东, 曹猛, et al. 基于统计理论的Au与Ag两种材料微放电敏感区域的计算. *真空电子技术*, 2012, (3), 1–3, 22. https://d.wanfangdata.com.cn/periodical/ChpNaW5lclBlcmlvZGljYWxDSEkyMDIzMDMyMRIPemtkempzMjAxMjAzMDAxGghldGlpZDN6dQ%3D%3D.
18. Ding M Q, Huang M G, Feng J J, et al. Ion surface modification for space TWT multistage depressed collectors. *Applied Surface Science*, 2008, 255(5), 2196–2199.
19. 丁明清, 黄明光, 冯进军, et al. 离子束表面处理抑制空间行波管多级降压收集级二次电子发射的研究. *真空科学与技术学报*, 2009, (3), 247–250.

20. 白国栋, 丁明清, 赵青平, et al. 抑制行波管多级降压收集极二次电子发射的工艺研究. *真空电子技术*, 2009, (5), 22–25. https://d.wanfangdata.com.cn/periodical/zkdzjs200905006.
21. 刘腊梅, 赵世柯. 工艺参数对Ar+改性无氧铜表面形貌及二次电子发射的影响. *真空科学与技术学报*, 2012, (10), 861–866. https://d.wanfangdata.com.cn/periodical/zkkx201210002.
22. Chang C, Huang H J, Liu G Z, et al. The effect of grooved surface on dielectric multipactor. *Journal of Applied Physics*, 2009, 105(12), 123305.
23. 崔万照, 杨晶, 张娜. 空间金属材料的二次电子发射系数测量研究. *空间电子技术*, 2013, 10(2), 75–78. https://d.wanfangdata.com.cn/periodical/kjdzjs201302014.
24. 张娜, 曹猛, 崔万照, et al. 高性能多功能超高真空金属二次电子发射特性测试平台. *真空科学与技术学报*, 2014, (5), 554–558. https://d.wanfangdata.com.cn/periodical/zkkx201405022.
25. 杨晶, 崔万照, 贺永宁. 碳纳米涂层材料的二次电子发射系数及工艺研究. *现代电子技术*, 2012, (20), 143–145. https://d.wanfangdata.com.cn/periodical/xddzjs201220043.
26. Ye M, He Y N, Hu S G. Simulation of secondary electron emission suppression of porous metal surface. *7th International Workshop on Multipactor, Corona and Passive Intermodulation in Space RF Hardware*, Valencia, Spain, 2011.
27. 贺永宁, 叶鸣, 胡少光. 铝合金镀银表面粗糙化处理方法及其SEY抑制机理研究. 昆明, 2012. DOI: 10.16490/j.cnki.issn.1001-3660.2018.05.001
28. Ye M, He Y N, Hu S G, et al. Suppression of secondary electron yield by micro-porous array structure. *Journal of Applied Physics*, 2013, 113(7), 074904.
29. Ye M, He Y N, Hu S G, et al. Investigation into anomalous total secondary electron yield for micro-porous Ag surface under oblique incidence conditions. *Journal of Applied Physics*, 2013, 114(10), 104905.
30. 叶鸣, 贺永宁, 王瑞, et al. 基于微陷阱结构的金属二次电子发射系数抑制研究. *物理学报*, 2014, (14), 354–362. DOI: 10.7498/aps.63.147901
31. 许石多. 粗糙金属表面二次电子发射特性的数值仿真研究. 西安交通大学, 2012.
32. Zhang H B, Hu X C, Wang R, et al. Note: Measuring effects of Ar-ion cleaning on the secondary electron yield of copper due to electron impact. *Review of Scientific Instruments*, 2012, 83(6), 066105.
33. Zhang H B, Hu X C, Cao M, et al. The quantitative effect of thermal treatment on the secondary electron yield from air-exposed silver surface. *Vacuum*, 2014, 102, 12–15.
34. Jbara O, Fakhfakh S, Belhaj M, et al. Charging effects of PET under electron beam irradiation in a SEM. *Journal of Physics D: Applied Physics*, 2008, 41(24), 245504.
35. Nagasawa K, Honjoh M, Miyake H, et al. Charge accumulation in various electron-beam-irradiated polymers. *IEEJ Transactions on Electrical and Electronic Engineering*, 2010, 5, 410–415.
36. Renoud R, Mady F, Attard C, et al. Secondary electron emission of an insulating target induced by a well-focused electron beam—Monte Carlo simulation study. *Physica Status Solidi A*, 201, 2119–2133.
37. Mandell M J, Davis V A, Cooke D L, et al. Nascap-2k spacecraft charging code overview. *IEEE Transactions on Plasma Science*, 2006, 34, 2084–2093.
38. Roussel J F, Rogier F, Dufour G, et al. SPIS open-source code: Methods, capabilities, achievements, and prospects. *IEEE Transactions on Plasma Science*, 2008, 36, 2360–2368.
39. Muranaka T, Hosoda S, Kim J H, et al. Development of multi-utility spacecraft charging analysis tool (MUSCAT). *IEEE Transactions on Plasma Science*, 2008, 36, 2336–2349.

CHAPTER 2

Secondary Electron Emission Model

Guobao Feng, Wanzhao Cui, Fang Wang,
and Guanghui Miao

In order to quantitatively describe the secondary electron emission characteristics of electrons emitted from the material surface after interacting with the material, a theoretical model is required to obtain the secondary electron yield emitted from the material surface. Since the incident electrons of different energies change the yield of secondary electrons emitted from the material surface, it is necessary to obtain the secondary electron yield curves with respect to the incident electron energy [1]. Currently, the main approaches to calculate the secondary electron yield include macroscopic parameter fitting models (e.g., Vaughan model [2,3], Furman model [4,5], and Joy–Everhart model [1,6,7]) and microscopic process tracking models (e.g., Monte Carlo model [8–10] and semi-physical model [11]). This chapter will give a brief introduction to Vaughan model, Furman model, and Monte Carlo model, as well as semi-physical model.

2.1 VAUGHAN MODEL

At the end of the 20th century, Vaughan proposed a new curve function for fitting experimental data based on the work of his predecessors Lye and Dekker. In the low incident energy part, the Gibbons curve, a general curve shape, was used. This is because the Gibbons curve is generally considered to be general; the researchers believe that it is a good expression of the quadratic emission coefficient for most surfaces at any incidence angle. In contrast to the Lye and Dekker curves, for the high incident energy part, Vaughan collected a large number of experimental results and normalized their respective V_{max} and δ_{max} to obtain a similar curve, as shown in Figure 2.1.

DOI: 10.1201/9781003562863-2

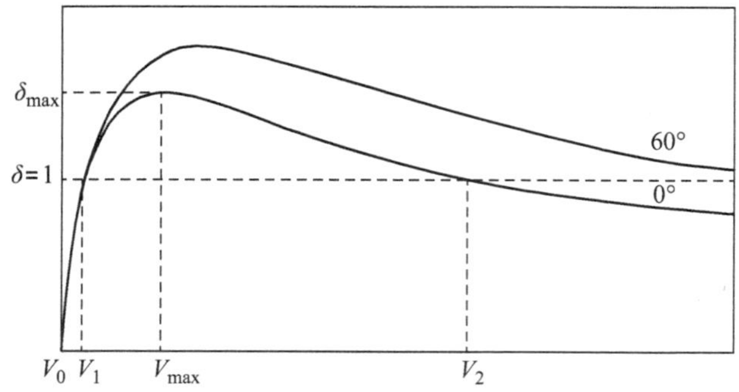

FIGURE 2.1 Typical secondary electron emission curves at 0° and 60° incidence.

The corresponding model SEY is calculated as

$$\frac{\delta}{\delta_{max}} = \left(ve^{1-v}\right)^k, v = \frac{V_i - V_0}{V_{max} - V_0}, \tag{2.1}$$

where V_{max} is the secondary electron emission curve SEY maximum point based on the previous formulas of Bruining and Muller; V_0 is the incident voltage where the secondary electron emission coefficient is greater than 0, and its reference actual value is set to 12.5 V. We have

$$k = k_1 = 0.62,\ v < 1 \tag{2.2}$$

$$k = k_2 = 0.25,\ v > 1, \tag{2.3}$$

where k is the fitting parameter. Using the above k, the Vaughan equation fits the Gibbons curve very well. At low incident energies, the Vaughan equation fits the "generic curve shape" better than the Lye and Dekker equations, and at high incident energies, it fits the experimentally tested SEY curve better. It is worth noting that the discontinuity of k at $v = 1$ is irrelevant when considering the SEY value in practical terms, but to ensure the rigor of the model, the discontinuity of k can be adjusted by the following equation, namely

$$k = \frac{k_1 + k_2}{2} - \frac{k_1 - k_2}{\pi}\arctan(\pi \ln v), \tag{2.4}$$

where $k_{1,2}$ can be determined by accurately measuring V_1 and V_2, and we have

$$k_{1,2} = \frac{\ln \delta_{max}}{v_{1,2} - \ln v_{1,2} - 1},\ v_{1,2} = \frac{V_{1,2} - V_0}{V_{max} - V_0} \tag{2.5}$$

And then, in 1993, based on the new accurate data collected by Shih and Hor, Vaughan made appropriate modifications to the original model, and the final formula for the quadratic emission coefficient was:

$$v = \frac{V_i - V_0}{V_{max} - V_0} \cdot \left(\frac{1e}{1e}\right) = \frac{E_i - E_0}{E_{max} - E_0}, \tag{2.6}$$

where E_i is the impact energy of the particle; E_0 is the impact energy of the particle when SEY > 0. And

$$\frac{\delta(\theta)}{\delta_{max}(\theta)} = \left(ve^{1-v}\right)^k, \ v \le 3.6 \tag{2.7}$$

$$\frac{\delta(\theta)}{\delta_{max}(\theta)} = \frac{1.125}{v^{0.35}}, \ v \ge 3.6, \tag{2.8}$$

where k is taken as

$$k = k_1 = 0.56, \ \text{when } v < 1 \tag{2.9}$$

$$k = k_1 = 0.25, \ \text{when } 1 < v \le 3.6 \tag{2.10}$$

The maximum value of SEY calculated is:

$$E_{max}(\theta) = E_{max}(\theta)\left(1 + \frac{k_{sE}\theta^2}{2\pi}\right) \tag{2.11}$$

$$\delta_{max}(\theta) = \delta_{max}(\theta)\left(1 + \frac{k_{s\theta}\theta^2}{2\pi}\right) \tag{2.12}$$

Up to this point, the Vaughan model not only fully relies on the research results of its predecessors but also continues to modify its own model according to the research results of contemporaries, which is extremely simplified in form and has reached a certain level of accuracy. Therefore, the Vaughan model was once favored by researchers, and even now, there are many researchers who use it. However, the Vaughan model considers less physical content and is not very close to the physical reality, and there are still some errors in some aspects of the study.

2.2 FURMAN MODEL

The Furman model, also known as the F–P model, is a model of secondary electron emission first proposed by Furman and Pivi in 2002. The Furman model is unique in its detailed theoretical analysis based on the physical mechanism and the establishment of the corresponding secondary electron emission model based on the different classifications of secondary electrons. The Furman model is unique in that it is based on a detailed theoretical analysis of the physical mechanism, and the corresponding secondary electron emission model is established according to the different classifications of secondary electrons. In addition, the probability distribution is added to the model, which is closer to the real process of the stochastic characteristics of secondary electron emission. Therefore, the Furman model is the best model that agrees with the experimental results. However, its applicability is relatively poor due to its complex expressions and numerous adjustable parameters.

2.2.1 Elastic Backscattered Electrons

The elastic backscattered electron has its own secondary electron emission coefficient and energy spectrum model. According to the concept of Furman model, the secondary electron emission coefficient is consistent with the result of area integration of the secondary electron emission energy spectrum region at the corresponding energy:

$$\delta_e(E_0,\theta_0)=\left\{P_{1,e}(\infty)+\left[P_{1,e\max}-P_{1,e}(\infty)\right]\exp\left(-\frac{\left|E_0-E_{e\max}\right|^p}{pW^p}\right)\right\}\left[1+e_1\left(1-\cos^{e_2}\theta_0\right)\right] \tag{2.13}$$

$$f_{1,e}=\theta(E)\theta(E_0-E)\delta_e(E_0,\theta_0)\frac{2\exp\left[-(E-E_0)^2/2\sigma_e^2\right]}{\sqrt{2\pi}\sigma_e\operatorname{erf}\left(E_0/\sqrt{2}\sigma_e\right)}, \tag{2.14}$$

where $P_{1,e}(\infty)$ is the elastic backscattering coefficient when the electron is incident vertically and the incident electron energy tends to infinity; $P_{1,emax}$ is the maximum value of the elastic backscattering coefficient when the electron is incident vertically; E_{emax} is the incident electron energy corresponding to the maximum value of the elastic backscattering coefficient when the electron is incident vertically; W and p are fitting parameters that together determine the elastic backscattering when the electron is incident vertically. The function of $\theta(x)$ is to ensure that the value in the parentheses is non-negative, so as to determine the range of the variable; erf (x) function is the error function. The whole expression of the elastic backscattered electron energy spectrum reaches its maximum at $E=E_0$.

2.2.2 Inelastic Backscattered Electrons

From the above analysis, the probability of inelastic backscattering is relatively small and the span is relatively large:

$$\delta_r(E_0,\theta_0)=\left\{P_{1,r}(\infty)+\left[1-\exp\left(-\left(E_0/E_r\right)^r\right)\right]\right\}\cdot\left[1+r_1\left(1-\cos^{r_2}\theta_0\right)\right] \tag{2.15}$$

$$f_{1,r}=\theta(E)\theta(E_0-E)\delta_r(E_0,\theta_0)\frac{(q+1)}{E_0}\left(\frac{E}{E_0}\right)^q, \tag{2.16}$$

where $P_{1,r}(\infty)$ is the inelastic backscattering coefficient when the electron is incident vertically and the incident energy tends to infinity; E_r and r determine the trend of the inelastic backscattered electron emission coefficient at vertical incidence; r_1 and r_2 determine the effect of the angle on the inelastic backscattered electron emission coefficient; the function $\theta(x)$ serves to ensure that the value in the parentheses is non-negative, thus determining the range of the variable; for a given value of q, the expression of the total inelastic backscattered electron energy spectrum is an exponential function.

2.2.3 True Secondary Electron

True secondary electrons are more complex than elastic and inelastic backscattered electrons. The randomness of the number of true secondary electrons and the probability difference of different numbers of electrons emitted determine the complexity of the structure of the true secondary electron emission spectrum:

$$\delta_{\mathrm{ts}}(E_0,\theta_0)=\delta_{\mathrm{ts\,max}}(0)\cdot\frac{\dfrac{s\cdot E_0\cdot\left[1+t_1\left(1-\cos^{t_2}\theta_0\right)\right]}{E_{\mathrm{ts\,max}}(0)\cdot\left[1+t_3\left(1-\cos^{t_4}\theta_0\right)\right]}}{s-1+\left\{\dfrac{E_0}{E_{\mathrm{ts\,max}}(0)\cdot\left[1+t_3\left(1-\cos^{t_4}\theta_0\right)\right]}\right\}^{s}} \tag{2.17}$$

$$\frac{\mathrm{d}\delta_{\mathrm{ts}}}{\mathrm{d}E}=\sum_{\mathrm{n}=1}^{\infty}\frac{nP_{\mathrm{n,ts}}(E_0)(E/\varepsilon_{\mathrm{n}})^{p_{\mathrm{n}}-1}e^{-E/\varepsilon_{\mathrm{n}}}}{\varepsilon_{\mathrm{n}}\Gamma(p_{\mathrm{n}})P(np_{\mathrm{n}},E_0/\varepsilon_{\mathrm{n}})}\times P\left[(n-1)p_{\mathrm{n}},(E_0-E)/\varepsilon_{\mathrm{n}}\right] \tag{2.18}$$

A further simplification can be obtained when $E_0 >> E$ is satisfied,

$$\frac{\mathrm{d}\delta_{\mathrm{ts}}}{\mathrm{d}E}=\sum_{\mathrm{n}=1}^{\infty}nP_{\mathrm{n,ts}}(E_0)\frac{(E/\varepsilon_{\mathrm{n}})^{p_{\mathrm{n}}-1}e^{-E/\varepsilon_{\mathrm{n}}}}{\varepsilon_{\mathrm{n}}\Gamma(p_{\mathrm{n}})}, \tag{2.19}$$

where δ_{tsmax} (0) is the maximum value of the true secondary electron emission coefficient that can be reached at vertical incidence; E_{tsmax} (0) is the incident electron energy corresponding to the maximum value of the true secondary electron emission coefficient at vertical incidence; s determines the trend of the true secondary electron emission coefficient; t_1, t_2, t_3, t_4 together determine the effect of the change in the angle of the incidence on the true secondary electron emission coefficient. $P_{\mathrm{n,ts}}$ (x) is the probability distribution function; $\Gamma(x)$ is a typical gamma function; $P(z,x)$ is an incomplete gamma function; n is the number of secondary electrons produced when a single electron is incident on the material surface, and n takes a value generally less than 10; ε_{n} and p_{n} are the fitting parameters.

Compared with elastic and inelastic backscattering energy spectra, the energy spectrum model of true quadratic electrons considers a probabilistic model with a complex structure and more parameters.

2.2.4 Secondary Electron Emission Spectrum

$$\frac{\mathrm{d}\delta}{\mathrm{d}E}=f_{1,\mathrm{e}}+f_{1,\mathrm{r}}+\frac{\mathrm{d}\delta_{\mathrm{ts}}}{\mathrm{d}E} \tag{2.20}$$

The main advantage of the Furman model is that it discusses three types of secondary electrons separately based on the physical mechanism and establishes the corresponding expressions, which is a great breakthrough. In addition, the model introduces a "probabilistic model" in the true secondary electron part, which is closer to the physical nature of the randomness of secondary electron emission. Therefore, the Furman model is a major theoretical advance, and the factors considered are more comprehensive. In particular, the analysis of single-electron and multi-electron events, i.e., single-electron event, means that the number of emitted secondary electrons is one, and the secondary electrons can only be one of the three types of elastic backscattered electrons, inelastic backscattered electrons, and true secondary electrons, while multi-electron events can only occur in the case of true secondary electron emission. However, the expressions of the Furman model are complicated and there are many unknown parameters. In addition, the expressions of the Furman model contain fewer parameters about the physical mechanism of secondary electron emission, and there is no unified parameter adjustment standard for fitting the secondary electron emission spectrum for different materials, different incident electron energies, and different incident angles. Accurate simulations for specific materials and incident energies may give better results but are not conducive to accurate simulations of general significance. Therefore, the practical application value of the Furman model is relatively limited.

2.3 MONTE CARLO NUMERICAL MODEL

The quantitative description of the secondary electron emission characteristics can be obtained by following the electron motion process and eventually by statistical methods, in addition to the result-oriented parametric fitting model. When electrons are incident on a sample at a certain energy, firstly, a complex series of processes including electron scattering and charge transport occur with the material inside the sample. In this section, the simulation of electron scattering and transport processes is performed using the Monte Carlo method (referred to as MC method). Given the adaptability of the Monte Carlo method to computer simulation of many random events, this MC simulation method has been widely used in theoretical research and analysis in many fields such as condensed matter physics, applied physics, theoretical physics, chemistry, and nonlinear phenomena, and has become a standard tool in scientific research. After hitting the sample material by means of electrons, elastic and inelastic collision processes occur, and the electrons in the surface layer leave the sample, overcoming the work function to form outgoing secondary electrons. In addition, the charge deposited inside the sample undergoes transport processes. Since the incident electrons studied in this book are all below 10 keV, the Mott scattering model is used for the elastic scattering process, while the Perm dielectric function model and the fast secondary electron model are used for the inelastic scattering (Figure 2.2).

2.3.1 Scattering Model for Incident Electrons

In the process of traveling inside a material, electrons will inevitably collide with atoms or molecules, and this collision will cause the electrons to change their trajectory according to certain laws, and this process is called scattering. The process of electron

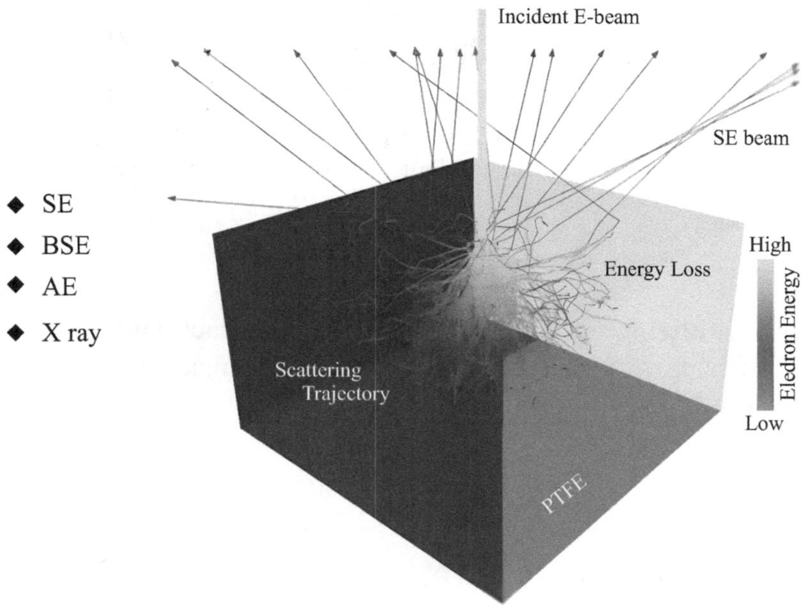

FIGURE 2.2 Process of electron-material interaction.

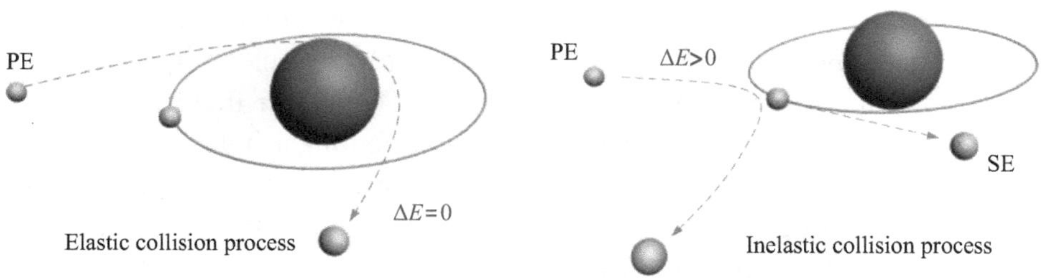

FIGURE 2.3 Elastic and inelastic collision processes between electrons and atoms.

scattering can be divided into elastic scattering and inelastic scattering, depending on whether energy is lost or not. Elastic scattering is the scattering of electrons by the Coulomb potential of the nucleus and the electron cloud outside the nucleus. Since the mass of the nucleus is much larger than the mass of the electron (more than three orders of magnitude larger), the energy change of both the atom and the incident electron before and after scattering is negligible; only the direction of motion of the incident electron changes. In inelastic scattering, however, the incident electrons not only change their direction of motion before and after scattering but also lose a certain amount of energy (Figure 2.3).

To analyze the scattering process of electrons, it is first necessary to calculate the scattering cross section of the electrons. Depending on the type of scattering, the scattering cross section of electrons is equally divided into elastic and inelastic scattering cross sections. The elastic scattering cross section of the incident electron is the integral of the differential elastic scattering cross section of the incident electron over all directions, while the inelastic scattering cross section of the incident electron is the integral of the differential inelastic scattering cross section of the incident electron over all directions and various energy losses. Among them, the differential elastic scattering cross section is the probability $d\sigma_e/d\Omega$ of the elastic scattering of the incident electron in a unit solid angle in a certain direction, and the total elastic scattering cross section is the integral of the differential elastic scattering cross section of the incident electron in all directions,

$$\sigma_e = \int \frac{d\sigma_e}{d\Omega} d\Omega, \tag{2.21}$$

and the inelastic scattering cross section σ_{in} of the atom is a double integral of the differential inelastic scattering $\sigma'_{in}(\Omega, E_f)$ of the atom over all solid angles and all incident electron final state energies E_f, i.e.,

$$\sigma_{in} = \iint \sigma'_{in}(\Omega, E_f) d\Omega dE_f \tag{2.22}$$

2.3.1.1 *Elastic Scattering Model*

Differential scattering cross sections can be used to describe the elastic scattering process for incident electrons of different energies. For high-energy incident electrons (>10 keV), the differential scattering cross section can be obtained by solving the Schrödinger equation under the Born approximation when describing the electron elastic scattering model, i.e., the Rutherford elastic scattering model. For lower incident electron energies (greater than 10 keV), the Born approximation no longer holds, and the exact differential elastic scattering cross section should be derived from the relativistic Dirac equation, i.e., the mathematical expression is obtained using the scattering fractional wave method. Therefore, for the incident electron energy below 10 keV considered in this book, the Mott scattering model is used to calculate the differential elastic scattering cross section. Mott elastic scattering is a spin-related electron elastic scattering theory, where the initial plane wave of the incident electron has its own phase shift in the transmission process, and the corresponding scattering amplitude is the summation of the fractional waves, so that the Mott differential elastic scattering cross section can be expressed as

$$\frac{d\sigma_e}{d\Omega} = |f(\theta)|^2 + |g(\theta)|^2, \tag{2.23}$$

where σ_e is the elastic scattering cross section cm^2/atom; Ω is the stereo angle deg; $f(\theta)$ and $g(\theta)$ are the incident and scattered wave splitting functions, which can be obtained by the wave splitting method,

$$f(\theta) = \frac{1}{2ik} \sum \{(l+1)[\exp(2i\eta_l) - 1] + l[\exp(2i\eta_{-l-1}) - 1]\} P_l(\cos\theta) \tag{2.24}$$

$$g(\theta) = \frac{1}{2ik} \sum [-\exp(2i\eta_l) + \exp(2i\eta_{-l-1})] P_l^1(\cos\theta), \tag{2.25}$$

where, the index of summation sign can add $l = 0, 1, 2, \ldots, \infty$ in the number of the partial waves; $P_l(\cos\theta)$ and $P_l^{(1)}(\cos\theta)$ are the Legendre function and the first-order joint Legendre function, respectively. The phase shift η_l, η_{-l-1} of each partial wave is obtained from the radial solution of the Dirac equation for the central potential field.

The total scattering cross section of Mott's elastic scattering can be obtained by integrating the Mott differential scattering cross section in each direction,

$$\sigma_e = 2\pi \int_0^{\pi} \frac{d\sigma_e}{d\Omega} \sin\theta \, d\theta, \text{ Unit: cm}^2/\text{atom} \tag{2.26}$$

After finding the scattering cross section at the corresponding incident electron energy, the elastic scattering angle can be obtained by a random number R_{Mott} between 0 and 1, i.e.,

$$R_{\text{Mott}} = \frac{\int_0^{\theta_M} \frac{d\sigma_e}{d\Omega} \sin\theta \, d\theta}{\int_0^{\pi} \frac{d\sigma_e}{d\Omega} \sin\theta \, d\theta} \tag{2.27}$$

For a multi-element compound material, it is first necessary to determine with which atom the elastic collision occurs. Again, a random number R_{MottAtom} must be used to determine with which atom the elastic scattering occurs. For a compound containing N elements, the probability P_i that an electron will elastically scatter with the ith element is

$$P_i = \frac{C_i \sigma_i}{\sum_{i=1}^{N} C_i \sigma_i}, \tag{2.28}$$

and when the random number R_{MottAtom} satisfies

$$\sum_{i=1}^{k-1} P_i < R_{\text{MottAtom}} < \sum_{i=1}^{k} P_i, \tag{2.29}$$

The electron then collides elastically with the kth atom (when $k = 1$, the left equation is 0).

2.3.1.2 Inelastic Scattering Model

In addition to elastic scattering with no energy loss, inelastic scattering processes with energy loss occur between electrons and sample materials. To improve the accuracy of the model, different models are used for inelastic scattering models at different electron energy bands. The fast quadratic electron model is used to model the inelastic scattering process for electrons with energy higher than 3 keV, while the Penn dielectric function model is more suitable to deal with the energy lower than 3 keV.

a. Fast secondary electron model

   For the fast quadratic electron model, the differential scattering cross section $d\sigma_{in}/d\Omega$ considering quantum mechanics during the inelastic scattering of electrons from the sample is

$$\frac{\mathrm{d}\sigma_{\text{in}}}{\mathrm{d}\Omega}=\frac{\pi e^4}{E^2}\left(\frac{1}{\Omega^2}+\frac{1}{(1-\Omega)^2}+\left(\frac{\tau}{\tau+1}\right)^2-\frac{2\tau+1}{(\tau+1)^2\Omega(1-\Omega)}\right), \tag{2.30}$$

where E is the electron energy (keV); Ω is the normalized energy loss factor, i.e., the ratio of the lost energy to the energy of the electron before scattering; τ is the ratio of the kinetic energy of the electron to the rest mass energy.

Considering the energy range of the electrons (≤10 keV) at $\tau=1$, the differential cross section of Eq. (2.30) can be simplified to

$$\frac{\mathrm{d}\sigma_{\text{in}}}{\mathrm{d}\Omega}=\frac{\pi e^4}{E^2}\left(\frac{1}{\Omega^2}+\frac{1}{(1-\Omega)^2}-\frac{1}{\Omega(1-\Omega)}\right) \tag{2.31}$$

Considering that the number of electrons lost in each inelastic scattering process cannot exceed half, i.e., $0<\Omega\leq 0.5$, the total inelastic scattering cross section can be expressed as

$$\sigma_{\text{in}}=\int_{\Omega_c}^{0.5}\frac{\mathrm{d}\sigma_{\text{in}}}{\mathrm{d}\Omega}\mathrm{d}\Omega=\frac{\pi e^4}{E^2}\left\{\frac{1}{\Omega_c}-\frac{1}{1-\Omega_c}+\ln\left(\frac{\Omega_c}{1-\Omega_c}\right)\right\} \tag{2.32}$$

Similarly, the scattering angle and the possible excited secondary electrons of the inelastic scattering process can be obtained by a random number R_{QSE},

$$R_{\text{QSE}}=\frac{\int_{\Omega_c}^{\Omega_{\text{QSE}}}\frac{\mathrm{d}\sigma_{\text{in}}}{\mathrm{d}\Omega}\mathrm{d}\Omega}{\sigma_{\text{in}}}, \tag{2.33}$$

By determining the loss factor Ω_{QSE}, the scattering angle of the incident electron is found to be

$$\sin^2\theta=\frac{2\Omega_{\text{QSE}}}{2+\tau-\tau\Omega_{\text{QSE}}}, \tag{2.34}$$

The scattering angle of the secondary electron is

$$\sin^2\phi=\frac{2\left(1-\Omega_{\text{QSE}}\right)}{2+\tau\Omega_{\text{QSE}}}, \tag{2.35}$$

For the energy loss in the inelastic scattering process, the Bethe energy loss formula modified by Joy and Luo is used for the fast quadratic electron model. The energy loss of electrons per unit length is given by

$$\left(\frac{\mathrm{d}E}{\mathrm{d}S}\right)_{\text{Bethe}}=78{,}500\frac{\rho Z}{AE}\ln\left(\frac{1.166(E+kJ)}{J}\right), \text{unit: keV/cm}, \tag{2.36}$$

where

Z is the atomic number of the material;

A is the atomic mass (g/mol);

ρ is the density of the material (g/cm^3);

J is the average ionization energy of the material (keV);

k is the ionization energy correction factor related to the atomic number Z.

For compound materials, the above atomic number Z, atomic mass A, and average ionization energy J are weighted averages of the various elements.

b. Dielectric function models

For low-energy incident electrons (<3 keV), the dielectric function model based on optical constants proposed by Penn is more applicable to the problem of inelastic scattering of a medium under low-energy electron beam irradiation. Since the inelastic scattering of electrons in a medium is mainly related to the valence electron excitation, the dielectric function theory can be used to obtain its inelastic scattering cross section. However, the dielectric functions associated with momentum transfer and energy loss are often difficult to obtain in practice. Penn proposes that by extending the experimental optical constants to the unknown momentum transfer region, the corresponding dielectric functions can be found. In this study, the Penn dielectric function model is used to calculate the inelastic scattering cross section for low-energy incident electrons (<3 keV). The corresponding inelastic differential scattering cross section is given by

$$\frac{\mathrm{d}\sigma_{\text{in}}(E,\hbar\omega)}{\mathrm{d}(\hbar\omega)}=\frac{me^2}{2\pi\hbar^2 NE}\operatorname{Im}\left[-\frac{1}{\varepsilon(0,\hbar\omega)}\right]S\left(\frac{\hbar\omega}{E}\right), \tag{2.37}$$

where m is the electron mass; e is the primitive electron charge; N is the molecular number density of the material; E is the incident electron energy; and $\hbar\omega$ is the energy loss. The inelastic total scattering cross section can be obtained from the differential cross-sectional area fraction.

In the Penn model, electron–phonon interactions are considered in addition to electron-electron interactions. The probability of electron–phonon interaction increases rapidly at incident electron energies below several material band gaps. According to Frohlich's theory, for an electron with energy E, its interaction with lattice vibrations produces a longitudinal optical phonon of energy $\hbar\omega$ (according to energy conservation, the energy loss of the electron is also $\hbar\omega$) with the inverse of the mean free path,

$$\lambda_{\text{in-phonon}}^{-1}=\frac{1}{a_0}\left|\frac{\frac{1}{\left(e^{\hbar\omega/kT}-1\right)}+1}{2}\right|\cdot\left|\frac{1}{\varepsilon(\infty)}-\frac{1}{\varepsilon(0)}\right|\cdot\frac{\hbar\omega}{E}\ln\left\{\frac{\left[1+\left(1-\hbar\omega/E\right)^{1/2}\right]}{\left[1-\left(1-\hbar\omega/E\right)^{1/2}\right]}\right\}, \tag{2.38}$$

where a_0 is the Bohr radius; k is the Boltzmann constant; $\varepsilon(0)$ and $\varepsilon(\infty)$ are the static and high frequency dielectric constants of the material, respectively. The energy loss of electron–phonon interaction is taken as $W_{\text{ph}} = \hbar\omega = 0.1\,\text{eV}$.

In addition, the Penn model considers the polariton effect of low-energy electrons that will be subjected to the induced polarization field generated in their vicinity. According to the theory of Ganachaud et al., the inverse of the mean free path of the low-energy electron trapped by the ion lattice due to the polariton effect is

$$\lambda_{\text{in-polaron}}^{-1}(E) = Ce^{-\gamma E}, \tag{2.39}$$

where both C and γ are constants related to the material properties. According to the hypothesis of Ganachaud et al., when the low-energy electrons produce polaritons, the residual energy of the electrons can be almost neglected, which is equivalent to being trapped.

Since the electron–electron, phonon, and polariton interactions are considered in the Penn dielectric function model, it is first necessary to determine which inelastic scattering interaction has occurred. The inverse of the total mean free path of inelastic scattering under the Penn model is used here to determine that

$$\lambda_{\text{in}}^{-1} = (\lambda_{\text{in-electron}})^{-1} + (\lambda_{\text{in-phonon}})^{-1} + (\lambda_{\text{in-polaron}})^{-1}, \tag{2.40}$$

and find the probability of occurrence of each type of scattering, i.e., $\lambda_{\text{in}}/\lambda_{\text{in-electron}}$, $\lambda_{\text{in}}/\lambda_{\text{in-phonon}}$, and $\lambda_{\text{in}}/\lambda_{\text{in-polaron}}$, and then determine which scattering process occurs by a random number $R_{\text{in-Penn}}$.

After determining which type of scattering occurs, the electron energy loss during inelastic scattering and the scattering angle need to be obtained.

When inelastic electron-electron scattering occurs, similar to the fast quadratic electron model, a random number $R_{\text{in-We}}$ is still used here to calculate the energy loss W of the electron as well as the scattering angle θ,

$$R_{\text{in-We}} = \frac{1}{\sigma_{\text{in}}(E)} \int_0^W \frac{d\sigma_{\text{in}}(E,\hbar\omega)}{d\hbar\omega} d\hbar\omega \tag{2.41}$$

After the energy loss W of electrons is determined, the scattering angle of the incident electrons and the scattering angle of the generated secondary electrons can be calculated separately, while the corresponding azimuthal angles are all randomly and uniformly distributed using $(0 \sim 2\pi)$.

When inelastic scattering of electrons and phonons occurs, the electron energy loss $W = \hbar\omega$ and, according to the theory of Llacer et al., the scattering angle can also be obtained from a random number $R_{\text{in-Wp}}$,

$$\cos\theta = \left(\frac{E+E'}{2\sqrt{EE'}}\right)\left[1 - \left(\frac{E+E'+2\sqrt{EE'}}{E+E'-2\sqrt{EE'}}\right)^{\mu_5}\right] + \left(\frac{E+E'+2\sqrt{EE'}}{E+E'-2\sqrt{EE'}}\right)^{\mu_5}, \tag{2.42}$$

where E and E' are the energies before and after electron–phonon interaction, respectively.

When a polariton effect is produced, the electrons are believed to be bound to the interior of the material and no longer follow their trajectory.

2.3.2 Inner Electron Escape from the Surface Potential Barrier

The outgoing secondary electrons can finally be formed when the inner electrons escape from the surface potential barrier. Here, the inner electrons can include both the original incident electrons and the excited inner secondary electrons, which we must consider uniformly as they cross the potential barrier. When the electrons reach the inner surface, they need to overcome the surface barrier before they can be emitted into the vacuum.

When calculating the transmission probability of electrons, it is necessary to consider both the energy of the electron E_{in} and the shape of the potential barrier $U(x)$. For a given surface barrier distribution $U(x)$, the electron density of states inside the metal and outside the barrier can be obtained by solving Schrödinger's equation, and the transmission coefficient $T(E_{in})$ can be obtained from the ratio of the electron density of states on both sides of the barrier. For a certain potential distribution $U(x)$, the electron wave function ψ satisfies Schrödinger's equation, i.e.

$$\frac{\partial^2 \psi(x)}{\partial x^2} + (E_{\text{in}} - U(x))\psi(x) = 0 \tag{2.43}$$

In the vacuum region inside the metal and outside the surface potential barrier, $U(x)$ is constant; for a simple rectangular potential barrier, the transmission coefficient T is given by

$$T(E+U) = \begin{cases} 0, & E<0 \\ \dfrac{4\sqrt{1-(U+E_{\text{F}})/(E+E_{\text{F}}+U)}}{\left(1+\sqrt{1-(U+E_{\text{F}})/(E+E_{\text{F}}+U)}\right)^2}, & E>0 \end{cases} \tag{2.44}$$

In the calculation of electron emission using the MC method, when the projection coefficient of electrons of a certain energy at the surface potential barrier is obtained, it can be determined whether electrons can be ejected from the surface to form secondary electrons by generating a random number (Figure 2.4).

2.3.3 Simulation Process of Electron Scattering

For the scattering process of a large number of electrons with the sample, the MC method is used to simulate the energy of the scattering process by numerical methods. The specific simulation procedure is as follows:

a. The electron step is calculated based on the energy of the incident electron and the coordinates of the position at the next moment are obtained;

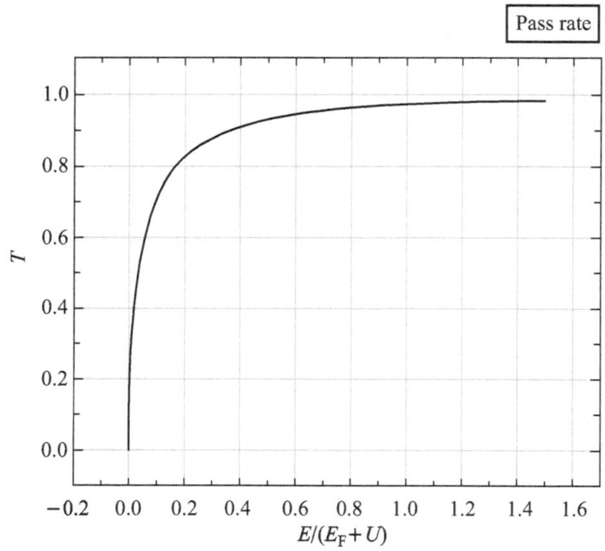

FIGURE 2.4 Variation of transmission coefficient with electron energy.

b. According to the reciprocal λ_e^{-1} of the elastic free path and the reciprocal λ_{in}^{-1} of the inelastic free path and the proportion of the reciprocal $\lambda_m^{-1} = \lambda_e^{-1} + \lambda_{in}^{-1}$ of the total mean free path, that is, λ_m/λ_e and λ_m/λ_{in}, a random number $R_{e\text{-}in}$ of $0 \sim 1$ is used to determine whether elastic scattering occurs.

c. For elastic scattering, the angle after scattering is calculated using the Mott elastic scattering model. When inelastic scattering occurs, the inelastic scattering model is first selected according to the magnitude of the incident electron energy: the fast quadratic electron model is used if it is larger than 3 keV; the Penn dielectric function model is used if it is smaller than 3 keV. For the fast quadratic electron model, a secondary electron is produced if the energy lost by scattering is greater than the band gap of the dielectric material and a hole is left at the collision site. In contrast, for the Penn model, if the inelastic scattering model is electron-electron scattering, a secondary electron is produced, again leaving a hole.

d. For the scattered electron and the newly generated secondary electron, a coordinate transformation is performed to determine whether it will leave the sample in the next step. If it leaves the sample, this electron tracking ends and the outgoing emission is recorded; if it is still in the sample, it goes to the first step and continues to track its next scattering.

When all the electrons have been tracked, the charge distribution of all the internal deposits and the number of outgoing electrons are recorded.

Based on the above Monte Carlo simulation method, the calculated secondary electron emission yields are compared with the results of relevant experimental and theoretical studies using PTFE as an example, as shown in Figure 2.5.

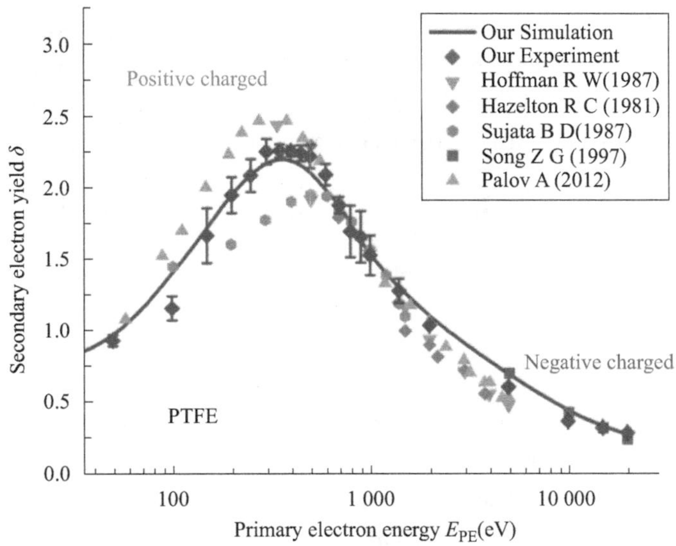

FIGURE 2.5 PTFE secondary electron yield calculated based on Monte Carlo simulation method.

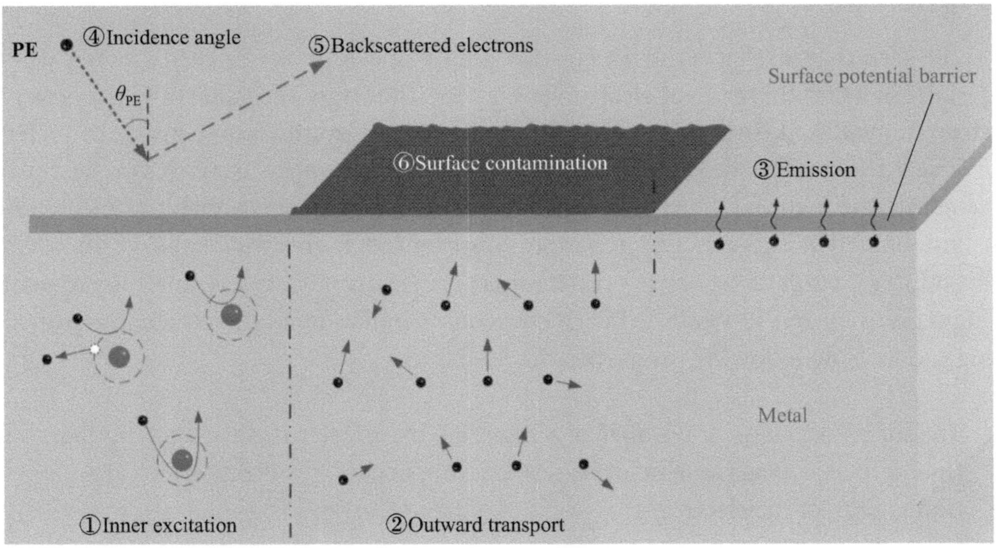

FIGURE 2.6 Schematic diagram of the secondary electron semi-physical model simulation.

2.4 SEMI-PHYSICAL MODEL

In addition to the Monte Carlo simulation method, which tracks all the processes of the inner electron, a semi-physical model of the secondary electron emission can be used to perform the calculation, which is simpler in comparison. The process of electron motion after incidence includes the excitation process of inner electrons, the motion of inner secondary electrons, the surface potential barrier on inner electron escape, and the angular correction and backscattering correction (Figure 2.6).

2.4.1 Inner Secondary Electron Excitation

When electrons of a certain energy are incident and scattered in a metallic material, inner secondary electrons are excited in the range of their incident depth, so the electron range of incident electrons of a certain energy in the material is a parameter closely related to the secondary electron emission.

For amorphous materials, the electron range $R(E_{pe})$ and the incident electron energy satisfy

$$R(E_{pe}) = \frac{L_k}{\rho_r}\left(\frac{E_{pe}}{E_k}\right)^{\alpha}, \tag{2.45}$$

where E_{pe} is the energy of the incident electron; E_k takes the value of 1 keV; L_k is the Lane-Zaffarano constant, which takes the value of 76nm; ρ_r is the density of the material relative to water. The densities of commonly used aerospace metal materials are given in Table 2.1.

Equation (2.45) can also be expressed as

$$\ln\left(\frac{R(E_{pe})}{L_k}\right) = \alpha \ln\left(\frac{E_{pe}}{E_k}\right) - \ln \rho_r \tag{2.46}$$

It can be seen that α reflects the penetration ability of electrons of a certain energy in the material. For a certain incident electron energy E_{pe}, the larger the value of α, the larger the electron range $R(E_{pe})$, and the stronger the ability of the electron to penetrate the material. Conversely, the smaller the value of α, the weaker the ability of electrons of a certain energy to penetrate the material. Lane and Zaffarano concluded that α is a material-independent constant, namely 1. This conclusion is more consistent for higher incident electron energies (>several keV), while the error is greater when the electron energy is small. By analyzing the data given by NIST, combined with computer simulation studies of the electron scattering process, the following properties can be found

a. The higher the atomic number of a material, the greater the average angular deflection of the electrons at which elastic scattering occurs, resulting in a weaker penetration energy of the electrons.

b. The higher the atomic number of the material, the higher the number of electrons outside the nucleus. The greater the probability of energy loss due to energy exchange between the incident electrons and the electrons outside the nucleus of the material, which also weakens the electron penetration ability.

TABLE 2.1 Density of Common Metal Materials

Material Name	Atomic Number (Z)	Density/(g/cm3)
Gold (Au)	79	19.30
Silver (Ag)	47	10.49
Copper (Cu)	29	8.96
Aluminum (A1)	13	2.70

Through quantitative analysis, it is found that the coefficient α in the electron range is material dependent, and for materials such as aluminum, gold, silver, and copper the coefficient and the material atomic number are obtained to satisfy the following relationship:

$$\alpha = 1.26 + 0.46\exp\left(-Z/19.92\right) \quad (2.47)$$

The relationship between the coefficient α and the atomic number is shown in Figure 2.5.

A series of inelastic scattering of incident electrons will occur in the material. Therefore, inner secondary electrons of a certain energy are excited in the electron range. Strictly speaking, the number of inner secondary electrons will vary at different positions in the range of incident electrons, that is, the density of inner secondary electrons will have a certain distribution in the depth direction. However, in the escape depth range, which is closely related to the exit of secondary electron, the distribution of the inner secondary electrons is nearly uniform, so the following approximation can be made, i.e.,

$$n(z) = \frac{N\left(E_{\text{pe}}\right)}{R\left(E_{\text{pe}}\right)}, \quad (2.48)$$

where $R\left(E_{\text{pe}}\right)$ is the range of the incident electrons; $N\left(E_{\text{pe}}\right)$ is the total number of inner secondary electrons (Figure 2.7).

2.4.2 Motion of Inner Secondary Electrons

Inner secondary electrons excited by incident electrons collide with free electrons and lattice atoms and lose energy as they move inside the material. After a certain distance, if the inner secondary electrons cannot reach the surface of the material, they may be trapped by the material atoms due to energy loss and stay in the material, so that only a part of the inner secondary electrons can move to the surface of the material. Bruining and Wittry et al. simplify the direction of movement of the inner secondary electron to two cases,

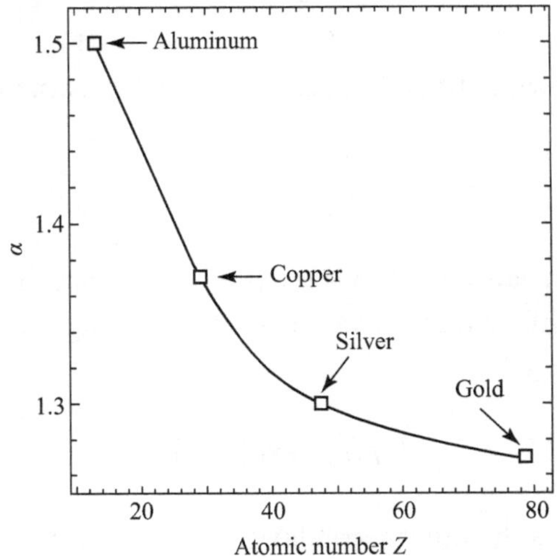

FIGURE 2.7 Relationship between coefficient α and atomic number.

upward and downward, and they believe that the probability of the inner secondary electron moving to the surface of the material at depth z is

$$p(z)=\frac{1}{2}\exp\left(-\frac{z}{\lambda_{\mathrm{eff}}}\right), \tag{2.49}$$

where λ_{eff} is the effective escape depth of the inner secondary electron.

The actual direction of motion of the inner secondary electrons is not simply downward or upward but may move in all directions. The probability that an electron moving in a particular direction will be absorbed in its motion during the outgoing stroke is set to be $p_{\mathrm{abs}}(\mathrm{d}s)$, and the value of this probability is proportional to the stroke ds,

$$p_{\mathrm{abs}}(\mathrm{d}s)=\gamma \mathrm{d}s, \tag{2.50}$$

where γ is the proportionality constant and represents the absorption coefficient per unit stroke. According to Eq. (2.50), the probability that the inner secondary electrons are not absorbed during the stroke of ds is

$$f(\mathrm{d}s)=1-p_{\mathrm{abs}}(\mathrm{d}s)=1-\gamma \mathrm{d}s \tag{2.51}$$

If the probability that an electron is not absorbed after a distance of s is $f(s)$, and the probability that it is still not absorbed after a distance of $s+\mathrm{d}s$ is $f(s+\mathrm{d}s)$, then according to the knowledge of probability we can get

$$f(s+\mathrm{d}s)=f(s)f(\mathrm{d}s)=f(s)(1-\gamma \mathrm{d}s) \tag{2.52}$$

or

$$\frac{f(s+\mathrm{d}s)-f(s)}{\mathrm{d}s}=-\gamma f(s) \tag{2.53}$$

That is, we have the following differential equation for $f(s)$, namely

$$\frac{\mathrm{d}f(s)}{\mathrm{d}s}=-\gamma f(s) \tag{2.54}$$

Obviously, the physical meaning of $f(0)$ is the probability that the electron is not absorbed at the initial moment, which takes the value of 1. Combining the differential equation and the initial conditions, we get

$$f(s)=\exp(-\gamma s) \tag{2.55}$$

The direction of motion of the inner secondary electrons must be considered. As shown in Figure 2.8, for an inner secondary electron with a depth of z and an angle φ between the direction of motion and the direction of the outer surface of the material, the length of the path taken by its motion to the surface is $z/\cos\varphi$.

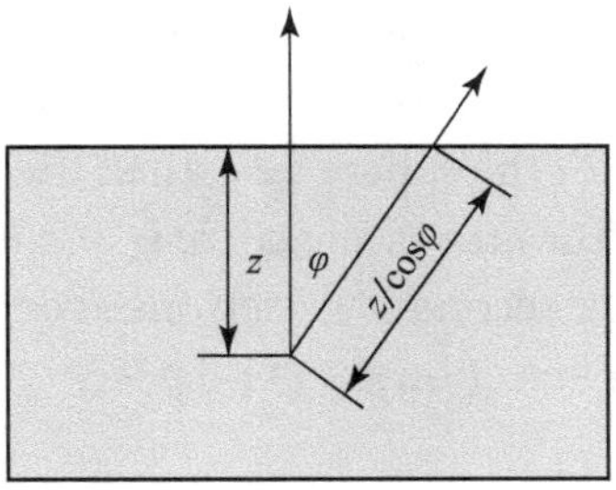

FIGURE 2.8 The distance traveled to the surface by the inner secondary electrons with different directions of motion.

TABLE 2.2 Mean Free Path of Electron Motion of Common Aerospace Metal Materials

Materials	**Atomic Number (Z)**	**Free Path of Electron Motion (λ/nm)**
Gold (Au)	79	1.6
Silver (Ag)	47	1.4
Copper (Cu)	29	1.2
Aluminum (Al)	13	3.8

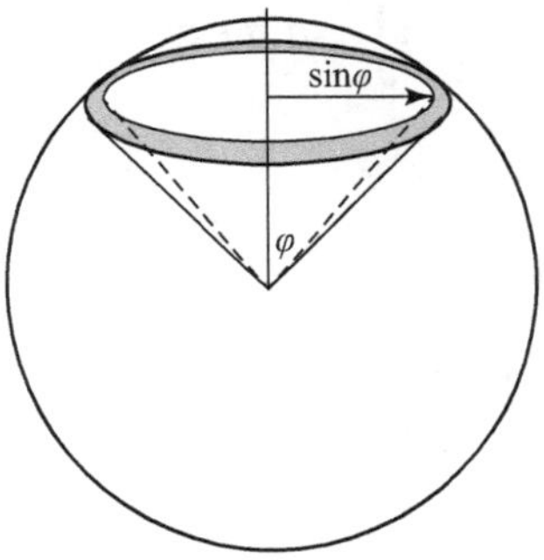

FIGURE 2.9 Schematic diagram of the proportion of inner secondary electrons moving in the φ direction.

According to the law of exponential decay of the emission probability, the probability of the inner secondary electron motion to the material surface in the direction φ is given by

$$p(z,\varphi)=\exp\left(-\frac{z}{\lambda\cos\varphi}\right), \tag{2.56}$$

where λ is the mean free path of electron motion in the material, and the commonly used aerospace metal materials take the values shown in Table 2.2.

As the inner secondary electrons move in the material, they also undergo non-scattering phenomena with atoms and generate new inner secondary electrons, a process called cascade scattering. For non-crystalline materials, the material is isotropic and the direction of motion of the inner secondary electrons in the material after cascade scattering can be considered isotropic. As shown in Figure 2.9, the ratio of the inner secondary electrons whose direction of motion is in $\varphi \sim \varphi + \mathrm{d}\varphi$ to all inner secondary electrons in all directions is the ratio of the gray shaded circle to the total area of the sphere in the Figure. Without loss of generality, let the radius of the sphere be 1, then the radius of the circle is $\sin\varphi$, and the width of the circle is $\mathrm{d}\varphi$, so the area of the circle is $2\pi\sin\varphi\mathrm{d}\varphi$, and the area of the whole sphere is 4π. The density of inner secondary electrons at depth z is $n(z)$, and the number of inner secondary electrons with direction $\varphi \sim \varphi + \mathrm{d}\varphi$ is

$$n(z,\varphi)\mathrm{d}\varphi = \frac{1}{2}n(z)\sin\varphi\mathrm{d}\varphi \tag{2.57}$$

The probability of these inner secondary electron motions reaching the surface is obtained from Eq. (2.57), so that the number of inner secondary electron motions to the surface with depths at $z \sim z + \mathrm{d}z$ and directions of motion at $\varphi \sim \varphi + \mathrm{d}\varphi$ can be obtained as

$$n(z,\varphi)p(z,\varphi)\mathrm{d}z\mathrm{d}\varphi = \frac{1}{2}n(z)\sin\varphi\exp\left(-\frac{z}{\lambda\cos\varphi}\right)\mathrm{d}z\mathrm{d}\varphi \tag{2.58}$$

The number of inner secondary electrons moving to the surface in each direction at the depth of $z \sim z + \mathrm{d}z$ is

$$n_s(z)\mathrm{d}z = \frac{1}{2}n(z)\int_0^{\frac{\pi}{2}}\sin\varphi\exp\left(-\frac{z}{\lambda\cos\varphi}\right)\mathrm{d}\varphi\mathrm{d}z \tag{2.59}$$

The angle is integrated first, that is

$$\begin{aligned}\int_0^{\frac{\pi}{2}}\sin\varphi\exp\left(-\frac{z}{\lambda\cos\varphi}\right)\mathrm{d}\varphi &= -\int_0^{\frac{\pi}{2}}\exp\left(-\frac{z}{\lambda\cos\varphi}\right)\mathrm{d}(\cos\varphi)\\ &= \int_0^1\exp\left(-\frac{z}{\lambda y}\right)\mathrm{d}y \\ &= \exp\left(-\frac{z}{\lambda}\right) - z\Gamma\left(0,\frac{z}{\lambda}\right),\end{aligned} \tag{2.60}$$

where Γ denotes the gamma function $\Gamma(a,z) = \int_z^{\infty} t^{a-1}e^{-t}\,\mathrm{d}t$. Obviously, only the inner secondary electrons moving toward the surface can be emitted through the material surface, so that the integration interval for the angle φ is $[0 \quad \pi/2]$.

The contribution of the inner secondary electrons to the surface secondary electrons with depth $z \sim z + \mathrm{d}z$ is obtained by substituting the obtained density of the inner secondary electrons, i.e.,

$$n_s(z)\mathrm{d}z = \frac{N(E_{pe})}{2R(E_{pe})}\left(\exp\left(-\frac{z}{\lambda}\right) - z\Gamma\left(0,\frac{z}{\lambda}\right)\right)\mathrm{d}z \tag{2.61}$$

The total number of inner secondary electrons at the surface is the sum of all inner secondary electrons moving to the surface within the range of the incident electrons,

$$\begin{aligned} N_s &= \int_0^{R(E_{pe})} n_s(z)\mathrm{d}z \\ &= \frac{N(E_{pe})}{2R(E_{pe})}\int_0^{R(E_{pe})}\left(\exp\left(-\frac{z}{\lambda}\right) - z\Gamma\left(0,\frac{z}{\lambda}\right)\right)\mathrm{d}z \\ &= \frac{N(E_{pe})\lambda}{4R(E_{pe})}F_\Gamma\left(\frac{R(E_{pe})}{\lambda}\right), \end{aligned} \tag{2.62}$$

where

$$F_\Gamma(x) = 2\int_0^x \left(\exp(-x') - z\Gamma(0,x')\right)\mathrm{d}x' \tag{2.63}$$

It cannot be given in terms of a primitive function, but its result can be approximated as

$$F_\Gamma(x) \approx 1 - \exp(-1.6x), \tag{2.64}$$

Thus, the number of inner secondary electrons at the surface is obtained as

$$N_s = \frac{N(E_{pe})\lambda}{4R(E_{pe})}\left(1 - \exp\left(-1.6\frac{R(E_{pe})}{\lambda}\right)\right) \tag{2.65}$$

The emission of inner secondary electrons at the surface is closely related to the energy of the inner secondary electrons. The energy distribution of the inner secondary electrons has a certain energy distribution. Joy et al. considered that the energy of the inner secondary electrons is all average ionization energy. Streitwolf et al. give the energy distribution of the inner secondary electrons as satisfying the following equation

$$S(E_s) \propto \frac{1}{(E_s - E_F)^2} \tag{2.66}$$

This conclusion has an essential physical flaw: if the calculated energy of all electrons larger than E_F, if the integral does not converge, i.e., the number of inner secondary electrons tends to infinity, which is obviously contradictory to the physical reality. At the same time, this conclusion only considers the process of excitation of the inner secondary electrons by the incident electrons and ignores the cascade scattering process of the inner secondary electrons.

Considering the cascade scattering process of inner secondary electrons, the energy of inner secondary electrons is considered to satisfy the exponential distribution, i.e.

$$S_R(E_s) = A\exp\left(-\frac{E_s - E_F}{E_v}\right), \tag{2.67}$$

where E_s is the energy of the inner secondary electron; E_v is the expected value of the inner secondary electron energy; E_F is the Fermi energy level of the material. $S_R(E_s)dE_s$ represents the probability that the energy of the inner secondary electron is between $E_s + dE_s$. The coefficient A satisfies the following normalization condition,

$$\int_{E_F}^{E_{pe}} S_R(E_s)\,dE_s = 1, \tag{2.68}$$

then we have

$$A = \frac{1}{E_v\left(1 - \exp\left(-\frac{E_{pe} - E_F}{E_v}\right)\right)} \tag{2.69}$$

According to the conservation of energy, the sum of all inner secondary electron energies should be equal to the incident electron energy. The number of inner secondary electrons with energy between E_s and $E_s + dE_s$ is $N(E_{pe})S(E_s)dE_s$. The starting point of the energy of the inner secondary electrons is chosen at the bottom of the conduction band, that is, the energy gained by the inner secondary electrons is the part of E_s greater than the material Fermi energy level E_F. Therefore, the energy gained by the above inner secondary electrons is $N(E_{pe})S(E_s)(E_s - E_F)dE_s$. The energy gained by all the inner secondary electrons should be equal to the total energy of the incident electrons, so we have

$$N(E_{pe})\int_{E_F}^{E_{pe}} S_R(E_s)(E_s - E_F)\,dE_s = E_{pe} \tag{2.70}$$

The maximum energy of the secondary electron can be the incident electron energy E_{Pe}, so the upper limit of the integration is chosen as E_{Pe}, then we have

$$N(E_{pe}) = \frac{E_{pe}}{E_v + \frac{\exp\left(-\frac{E_{pe} - E_F}{E_v}\right)(E_{pe} - E_F)}{1 - \exp\left(-\frac{E_{pe} - E_F}{E_v}\right)}} \tag{2.71}$$

2.4.3 Surface Potential Barrier on Inner Electron Escape

The inner secondary electrons moving to the surface need to cross the surface barrier as they exit from the surface, and the probability depends on the height of the surface barrier and the kinetic energy of the secondary electrons. The probability of crossing a potential barrier of height U for an inner secondary electron with energy E_s is

$$p_e(E_s) = \begin{cases} 0 & E_s \le U \\ \dfrac{4\sqrt{1 - U/E_s}}{\left(1 + \sqrt{1 - U/E_s}\right)^2} & E_s > U \end{cases} \tag{2.72}$$

This function can be simplified by the following approximate expression, namely

$$f_0(E_s) = \begin{cases} 0 & E_s \le U \\ \dfrac{29(E_s/U-1)}{1+29(E_s/U-1)} & E_s > U \end{cases} \tag{2.73}$$

The average probability of secondary electron emission at the surface is

$$\begin{aligned} P_s &= \int_0^{E_{pe}} S_R(E_s)p_e(E_s)\mathrm{d}E_s \\ &= 4A\int_U^{E_{pe}} \exp\left(-\frac{E_s - E_F}{E_v}\right)\cdot\frac{\sqrt{1-U/E_s}}{\left(1+\sqrt{1-U/E_s}\right)^2}\mathrm{d}E_s \end{aligned} \tag{2.74}$$

Combining the given number of secondary electrons at the surface, the secondary electron yield is obtained as

$$\delta = N_s P_s = \frac{N(E_{pe})P_s\lambda}{4R(E_{pe})}\left(1-\exp\left(-1.6\frac{R(E_{pe})}{\lambda}\right)\right) \tag{2.75}$$

In the case of $E_{pe} \ge E_F$, $N(E_{pe})$ can be approximated as

$$N(E_{pe}) = \frac{E_{pe}}{E_v} \tag{2.76}$$

Finally, we have

$$\delta = N_s P_s = \frac{E_{pe}P_s\lambda}{4E_v R(E_{pe})}\cdot\left(1-\exp\left(-1.6\frac{R(E_{pe})}{\lambda}\right)\right) \tag{2.77}$$

The secondary electron emission coefficient is related to the incident electron energy, so that the incident electron energy at the maximum emission coefficient is E_m, and the electron range corresponding to this incident electron energy is R_m, then we have

$$\left.\frac{\partial\delta}{\partial E_{pe}}\right|_{E_{pe}=E_m} = 0 \tag{2.78}$$

We obtain that R_m satisfies the equation

$$1.6\frac{R_m}{\lambda} = \left(1-\frac{1}{\alpha}\right)\left(\exp\left(1.6\frac{R_m}{\lambda}\right)-1\right) \tag{2.79}$$

Let $R_m = R'\lambda$, we have R' satisfies the equation

$$1.6R' = \left(1-\frac{1}{\alpha}\right)\left(\exp(1.6R')-1\right) \tag{2.80}$$

TABLE 2.3 Values of R' for Different Materials

Metal	Atomic Number (Z)	R'
Al	13	1.19
Cu	29	1.39
Ag	47	1.53
Au	79	1.61

When α is a certain value, R' can be obtained according to the above equation. The values of R' for different materials are given in Table 2.3.

After obtaining the value of R', where $R_m = R(E_m) = \frac{L_k}{\rho_r}\left(\frac{E_m}{E_k}\right)^{\alpha}$, E_m can be obtained according to the following equation, i.e.,

$$E_m = E_k \left(\frac{R'\rho_r \lambda}{L_k}\right)^{1/\alpha} \tag{2.81}$$

It's easy to get

$$\frac{R(E_{pe})}{R_m} = \left(\frac{E_{pe}}{E_m}\right)^{\alpha} \tag{2.82}$$

$$\frac{R(E_{pe})}{\lambda} = \left(\frac{E_{pe}}{E_m}\right)^{\alpha} \cdot \frac{R_m}{\lambda} = R'\left(\frac{E_{pe}}{E_m}\right)^{\alpha} \tag{2.83}$$

So, we have

After substitution, we get

$$\delta = \frac{\delta_m}{1-\exp(-1.6R')}\left(\frac{E_{pe}}{E_m}\right)^{1-\alpha}\left(1-\exp\left(-1.6R'\left(\frac{E_{pe}}{E_m}\right)^{\alpha}\right)\right), \tag{2.84}$$

where

$$\delta_m = \frac{E_m P_s}{4E_v R'}\left(1-\exp(-1.6R')\right) \tag{2.85}$$

2.4.4 Angular Correction

Many studies have shown that the secondary electron emission coefficient increases significantly with the increase of the incidence angle. In the case of perpendicular incidence, let the range of the incident electrons of a certain energy be $R(E_{pe})$, then the depth of incidence is $R\cos\theta$ when the electrons are injected into the material obliquely with the angle system θ normal to the metal surface, so the distribution function of the electrons in the depth direction should be modified as

$$n(z,\theta_{in}) = \frac{N(E_{pe})}{R(E_{pe})\cos\theta} \tag{2.86}$$

Thus, the contribution of the inner secondary electrons in the range of $z \sim z+\mathrm{d}z$ to the surface secondary electrons is

$$n_s(z)\mathrm{d}z = \frac{N(E_{pe})}{2R(E_{pe})\cos\theta}\left(\exp\left(-\frac{z}{\lambda}\right) - z\Gamma\left(0,\frac{z}{\lambda}\right)\right)\mathrm{d}z \tag{2.87}$$

The upper limit of integration of the number of secondary electrons on the surface calculated in turn should be corrected to $R\cos\theta$, and we have

$$\begin{aligned} N_s(\theta) &= \int_0^{R(E_{pe})\cos\theta} n_s(z)\mathrm{d}z \\ &= \frac{N(E_{pe})}{2R(E_{pe})\cos\theta}\int_0^{R(E_{pe})\cos\theta}\left(\exp\left(-\frac{z}{\lambda}\right) - z\Gamma\left(0,\frac{z}{\lambda}\right)\right)\mathrm{d}z \\ &= \frac{N(E_{pe})\lambda}{4R(E_{pe})\cos\theta}F_\Gamma\left(\frac{R(E_{pe})\cos\theta}{\lambda}\right) \\ &\approx \frac{N(E_{pe})\lambda}{4R(E_{pe})\cos\theta}\left(1-\exp\left(-1.6\frac{R(E_{pe})\cos\theta}{\lambda}\right)\right) \end{aligned} \tag{2.88}$$

Let $\lambda_\theta = \lambda/\cos\theta_{in}$, then the formula for the secondary electron emission coefficient at oblique incidence only needs to replace λ with λ_θ in form.

If expressed in terms of the maximum yield and its corresponding incident electron energy, we get

$$E_m(\theta_{in}) = \left(\frac{R'\rho\lambda_\theta}{B}\right)^{1/\alpha} = \left(\frac{R'\rho\lambda}{B\cos\theta_{in}}\right)^{1/\alpha} = E_m(0)\left(\frac{1}{\cos\theta_{in}}\right)^{1/\alpha} \tag{2.89}$$

The corresponding maximum yield is

$$\delta_m(\theta_{in}) = \delta_m\left(\frac{1}{\cos\theta_{in}}\right)^{1/\alpha} \tag{2.90}$$

Thus, we can obtain

$$\delta = \frac{\delta_m(\theta)}{1-\exp(-1.6R')}\left(\frac{E_{pe}}{E_m(\theta)}\right)^{1-\alpha}\left(1-\exp\left(-1.6R'\left(\frac{E_{pe}}{E_m(\theta)}\right)^{\alpha}\right)\right) \tag{2.91}$$

It is worth noting that in the above study of the electron motion process in the material, it was assumed that the trajectory of the incident electrons is a straight line along the incidence direction. The results obtained from this assumption do not deviate significantly from the actual ones when the incidence angle is small. However, when the angle of incidence exceeds 60°, a large deviation occurs. Therefore, the above equation for the incident angle is only suitable for the case where the incident angle is within 60°. In the actual multipactor process, the electrons move back and forth between the two metal walls and interact with the electromagnetic field to produce a multiplication. Obliquely incident electrons quickly leave the area where the multipactor occurs due to lateral motion, so they have little effect on the multipactor. Therefore, the analysis of multipactor mainly requires the study of the secondary electron emission characteristics within 60° of the angle of incidence.

2.4.5 Backscattering Correction

The backscattered electrons contain two parts: one is the low-energy electrons directly reflected by the surface potential barriers, and the other is the electrons elastically scattered by the surface atoms. The share of low-energy backscattered electrons directly reflected by the surface potential barrier is

$$\eta_1(E_{pe}) = \eta_1(0)\exp\left(-\frac{E_{pe}}{E'_{BS}}\right), \tag{2.92}$$

where $\eta_1(0)$ is the low-energy backscattering coefficient, which takes the value of 0 for an ideal clean surface and about 0.6 for a material with adsorption on the surface; the value of E'_{BS} is taken as 12.5 eV.

The portion of electrons bounced back by surface atoms at vertical incidence is given by Cazaux as

$$\eta_2(E_{pe},0) = \eta_2(\infty)\left(1-\exp\left(-\frac{E_{pe}}{E_{BS}}\right)\right), \tag{2.93}$$

where $\eta_2(\infty)$ is the high-energy backscattering coefficient, whose value is material dependent; the value of E_{BS} is also material dependent.

In 2002, Furman gave a probabilistic statistical model of the secondary electron yield, where the backscatter yield is related to the angle of incidence θ by the equation

$$\eta_2(E_{pe},\theta) = \eta_2(E_{pe},0)\left(1+e_1(1-\cos^{e_2}\theta)\right), \tag{2.94}$$

where e_1 and e_2 are the correction parameters, and for commonly used metal materials, e_1 and e_2 are assumed to be 0.8 and 0.4, respectively. Substituting them into the equation, a modified expression for the oblique incidence backscattering yield is obtained.

The total elastic backscattering coefficient is

$$\eta = \eta_1 + \eta_2 \tag{2.95}$$

If the incident electrons undergo elastic backscattering at the surface, they cannot enter the material and cannot excite the inner secondary electrons, so the incident electrons contributing to the intrinsic secondary electrons account for $(1-\eta)$ of all incident electrons, and the total secondary electron emission coefficient is

$$\delta_{\text{total}} = \eta + (1-\eta)\delta \tag{2.96}$$

2.4.6 Simulation Example

Based on the conclusions obtained from the above analysis, the results of the model calculations were compared with the experimental results. Since a certain oxide layer or adsorbed layer exists on the surface of the actual material, to eliminate as much as possible, the possible influence of surface oxidation and adsorption on the surface potential monopoly of the material, Ar ions were used to clean the surface of the sample before testing the secondary electron emission coefficients.

Firstly, the secondary electron emission coefficients of Ag-plated materials were compared. Ar ion cleaning can effectively remove the oxide layer and adsorbed materials from the surface of Ag materials, so the values of ideal Ag materials were used for the material parameters in the model, and the specific values are shown in Table 2.4.

A comparison of the model calculations with the experimental results for different incidence angles of *d* is shown in Figure 2.10.

TABLE 2.4 Parameters of the Model for Cleaning Ag Materials

Parameter	U/eV	α	R'	E_m/eV	δ_m	$\eta_2(\infty)$	E_{BS}/eV
Value	4.53	1.3	1.5	391	2.06	0.55	29.1

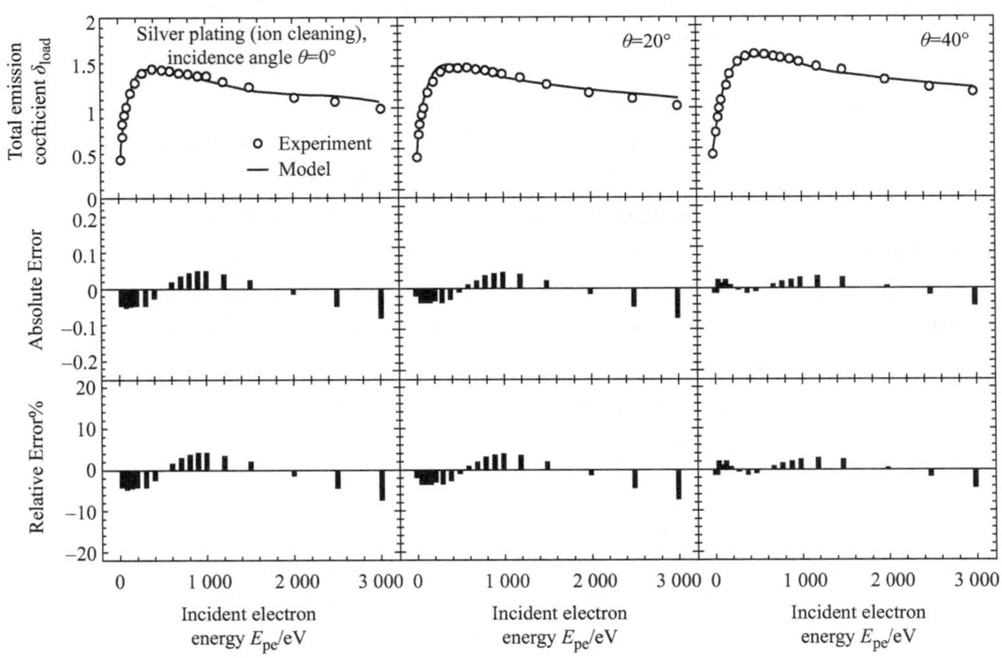

FIGURE 2.10 Comparison of model and experimental results of total secondary electron emission coefficient for cleaning Ag.

TABLE 2.5 Parameters of the Model for Uncleaning Ag Materials

Parameter	U/eV	α	R'	E_m/eV	δ_m	$\eta_2(\infty)$	E_{BS}/eV
Value	2.74	1.4	1.34	243	3.46	0.53	22.8

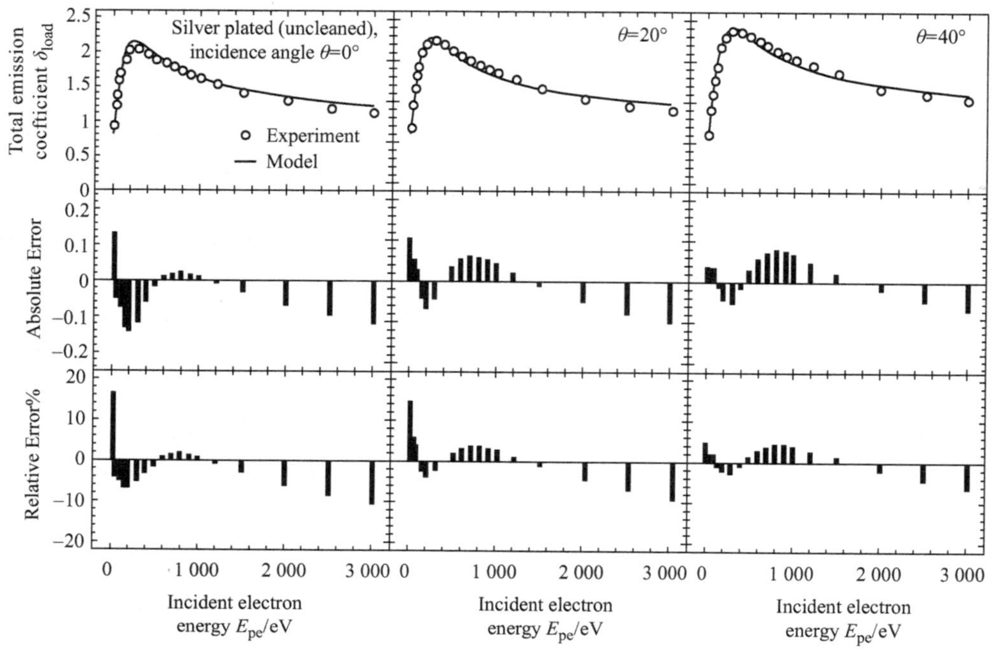

FIGURE 2.11 Comparison of the modeled total secondary electron emission coefficient of uncleaned Ag with experimental results.

For the uncleaned sample, there will be an oxide layer on the Ag surface, so the surface potential is assumed to be close to the work function of its oxide. Also, since the silver-plated material has more defects than those of pure silver, the average free range of electron motion is correspondingly smaller than that of pure silver, which is taken as 0.72 nm. The values of each parameter of the model are shown in Table 2.5.

The comparison between the model calculation results and the experimental results for different incidence angles is shown in Figure 2.11. For materials without surface treatment, there is a high probability that oxidation and adsorption on the surface and the surface potential barrier height will change. For Ag, oxidation and adsorption will result in lower potential barriers and higher SEY values.

REFERENCES

1. Feng G B, Li Y, Li X J, et al. Characteristics of secondary electron emission from few layer graphene on silicon (111) surface. *Chinese Physics B*, 2022, 31(10): 107901.
2. Joy D C. A database on electron-solid interactions. *Scanning*, 1995, 17(5): 270–275.
3. Penn D R. Electron mean-free-path calculations using a model dielectric function. *Physical Review B*, 1987, 35(2): 482.

4. Feng G B, Cui W Z, Zhang N, et al. A synthetic semi-empirical physical model of secondary electron yield of metals under E-beam irradiation. *Chinese Physics B*, 2017, 26(9): 097901.
5. Furman M A, Pivi M T F. Erratum: Probabilistic model for the simulation of secondary electron emission [Phys. Rev. ST Accel. Beams 5, 124404 (2002)]. *Physical Review Special Topics—Accelerators and Beams*, 2013, 16(6): 069901.
6. Furman M A, Pivi M T F. Probabilistic model for the simulation of secondary electron emission. *Physical Review Special Topics—Accelerators and Beams*, 2002, 5(12): 124404.
7. Vaughan J R M. A new formula for secondary emission yield. *IEEE Transactions on Electron Devices*, 1989, 36(9): 1963–1967.
8. Vaughan R M. Secondary emission formulas. *IEEE Transactions on Electron Devices*, 1993, 40(4): 830.
9. Lye R G, Dekker A J. Theory of secondary emission. *Physical Review*, 1957, 107(4): 977.
10. Mao S F, Li Y G, Zeng R G, et al. Electron inelastic scattering and secondary electron emission calculated without the single pole approximation. *Journal of Applied Physics*, 2008, 104(11): 114907.1–10.
11. Lin Y, Joy D C. A new examination of secondary electron yield data. *Surface and Interface Analysis: An International Journal Devoted to the Development and Application of Techniques for the Analysis of Surfaces, Interfaces and Thin Films*, 2005, 37(11): 895–900.

CHAPTER 3

Secondary Electron Emission Measurement and Database

Wanzhao Cui, Guobao Feng, Fang Wang, and Guanghui Miao

3.1 SECONDARY ELECTRON EMISSION MEASUREMENT

The measurement of the SEY needs to be performed in a vacuum environment. The SEY can be obtained by counting the number of electrons emitted by electron bombardment of the sample surface. Due to the charge accumulation effect, metallic secondary electron measurement and dielectric secondary electron measurement are required for conductive and non-conductive materials, respectively. In this section, a representative secondary electron emission device is selected separately for introduction.

3.1.1 Secondary Electron Emission Measurement of Metallic Materials

The measurement of the secondary electron emission coefficient of metallic materials is relatively simple compared to dielectric materials, and has developed from the triode electron gun method and tetrode electron gun method to the most used electron gun method at present. Using the electron gun as a stable electron source, the focused electron beam hits the sample to be measured, and the secondary electron emission coefficient can be obtained by testing the collected secondary electron current. The electron gun method can also analyze the relationship between the secondary electron emission coefficient and the angle of incidence. The electron gun method has significantly improved the reliability and accuracy of the measurement.

The following is an example of the secondary electron emission coefficient test system for metals (Figure 3.1) developed by the National Key Laboratory of Space Microwaves to introduce the test method of secondary electron emission coefficient for metallic materials in detail. The test system uses two electron guns to provide the initial electrons with electron beam energies of 20–5,000 eV and 3–30 keV, respectively (Figure 3.2).

DOI: 10.1201/9781003562863-3

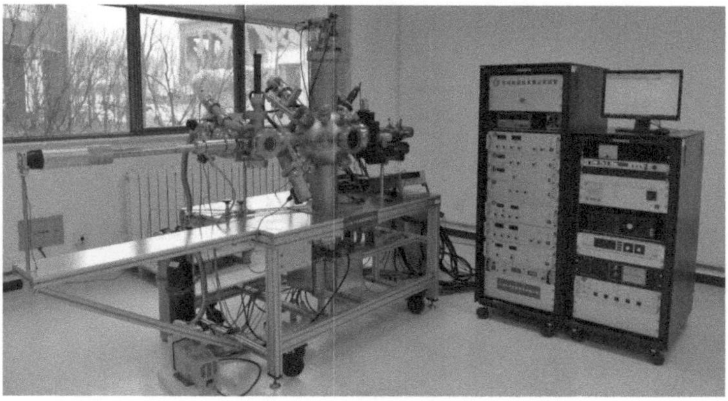

FIGURE 3.1 Research platform for secondary electron emission properties of metals.

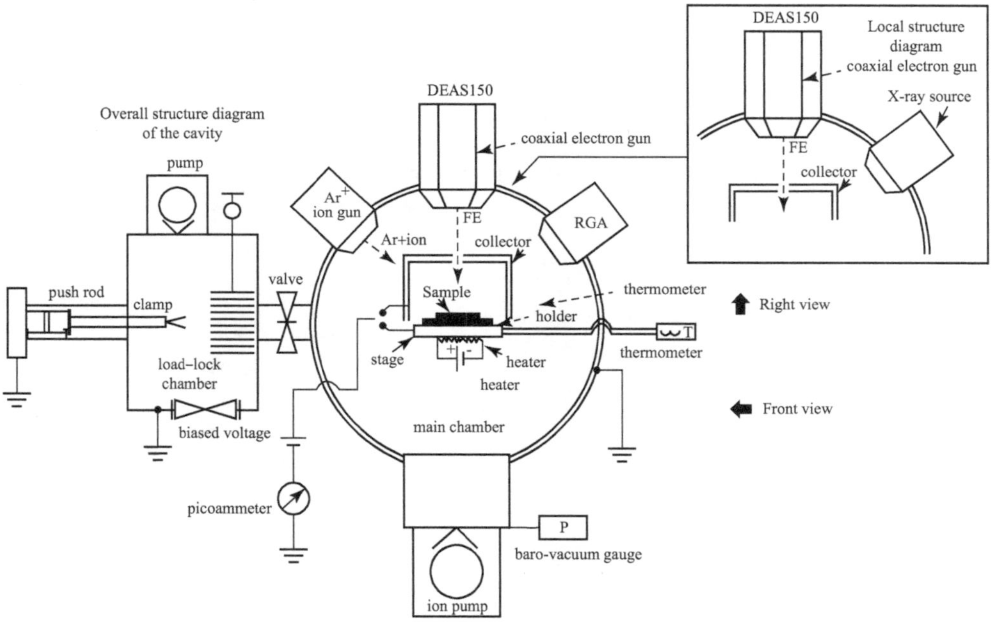

FIGURE 3.2 The structure of the research platform for the secondary electron emission characteristics.

Secondary electron collection uses two methods: collector electrode method and bias current method.

1. **For the collector electrode method, its measurement principle is shown in Figure 3.3**: (1) Connect the sample stage to the collector, and the current measured by the picoammeter is the incident current, referred to as I_p; (2) Under the same incident conditions, disconnect the sample and the collector, and the current measured on the collector is the secondary electron current, referred to as I_c:

$$\text{SEY}=\frac{\text{Secondary electron current}}{\text{Incident electron current}}=I_c/I_p \tag{3.1}$$

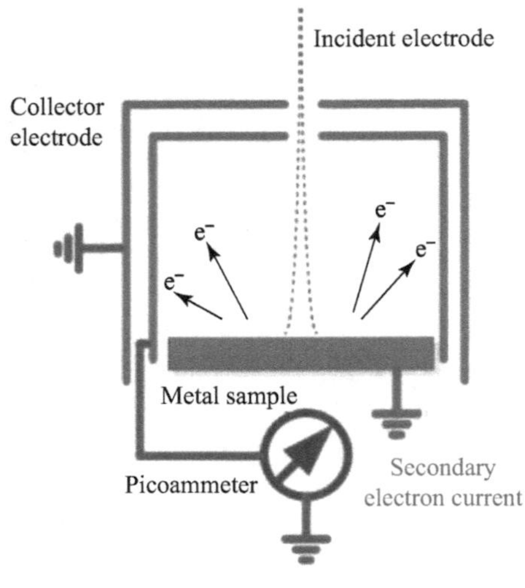

FIGURE 3.3 Test principle of collector electrode method.

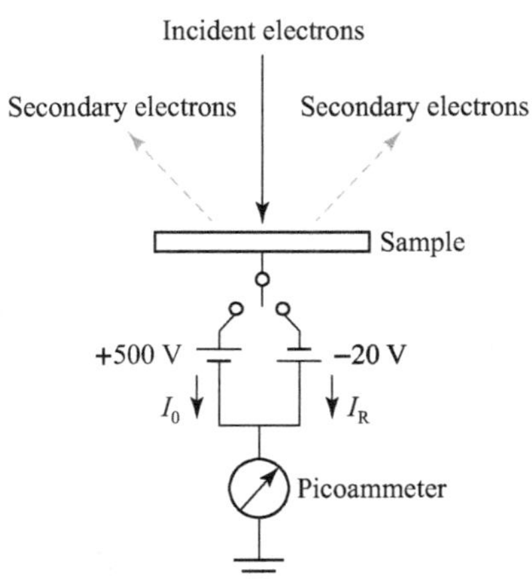

FIGURE 3.4 Testing principle of secondary electron emission coefficient using current method.

2. **Bias current method**: Apply different bias voltages on the sample, you can approximate the incident current and the secondary electron current, as shown in Figure 3.4, and add a large positive bias voltage to the sample; at this time, the current tested by the picometer is approximately considered to be the incident current I_0 of the electron gun; the current applied to the negative bias test on the sample is the difference between the incident current and the secondary electron current. At this point, the secondary electron emission coefficient of the material can be calculated by the following formula, i.e.

$$\delta = \frac{I_0 - I_R}{I_0} = 1 - \frac{I_R}{I_0} \tag{3.2}$$

Both the bias current method and the collector electrode method have advantages and disadvantages. Theoretically, the collector electrode method is more effective in collecting secondary electrons, but it requires the alignment of the electron beam, the small hole of the collector electrode, the sample, and the collector electrode, so the operation is not as convenient as the bias current method. In addition, the size of the collector electrode limits the range of sample movement. Compared with the bias current method, although the collector electrode is simple and fast to operate, the incident electron current of the test may be smaller than the actual value due to the possibility that the emitted high-energy electrons may escape from the sample surface.

3.1.2 Secondary Electron Emission Measurement of Dielectric Materials

3.1.2.1 Composition of Measurement System

The secondary electron emission characterization research platform is mainly composed of three parts, namely, the vacuum system, the XPS photoelectron spectroscopy analysis device, and the secondary electron emission coefficient and energy spectrum test device. The composition of these three major parts is described below.

3.1.2.1.1 Vacuum System The vacuum system consists of three vacuum chambers: an XPS analytical facility, a secondary electron emission coefficient and energy spectrum chamber, and a fast feed/sample storage chamber (the system is shown in Figure 3.5), which are

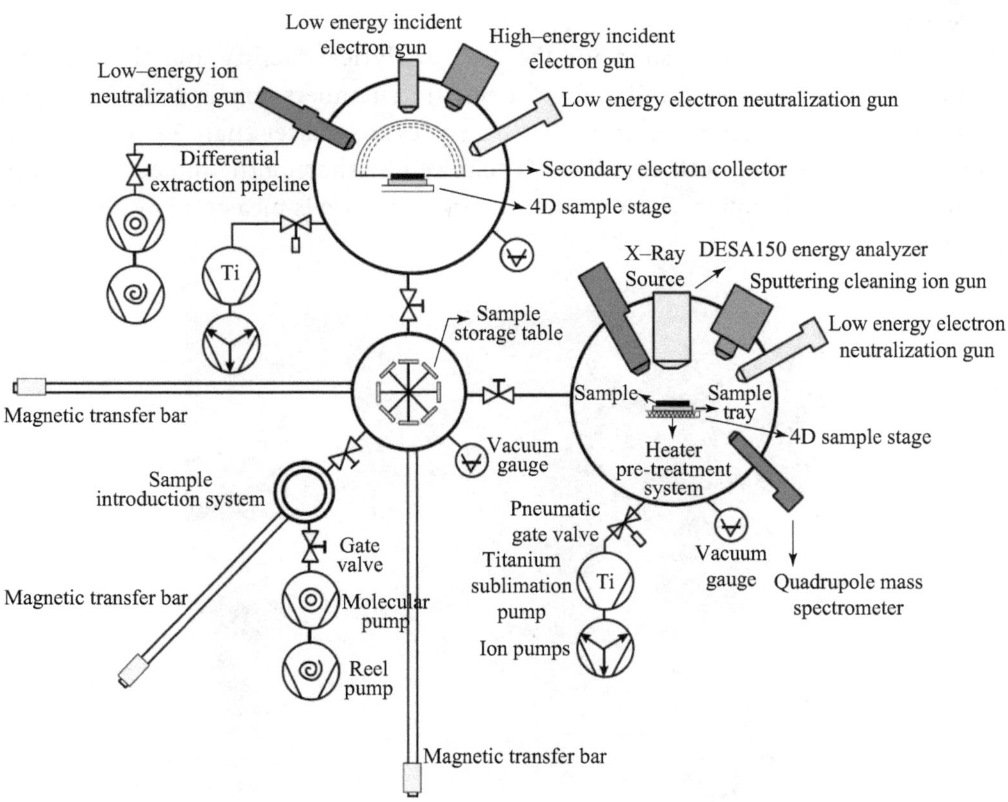

FIGURE 3.5 Schematic diagram of system composition and structure.

separated by a manual gate valve. The entire vacuum system is fixed on a rigid aluminum frame and is equipped with a baking hood for complete baking of the equipment at temperatures up to 200°C.

The rapid sample introduction system is compact in size and has a small quick-opening door for easy sample exchange. After sample exchange, the small size of the sample introduction system allows the molecular pump to quickly pump up to a vacuum level of 10^{-8} Torr, at which point the gate valve between the injection chamber and the sample storage chamber can be opened to transfer the sample from the sample introduction system to the sample storage chamber using a magnetically coupled transfer rod. The sample storage chamber is equipped with a separate molecular pump to maintain the vacuum. Once the sample is in the storage chamber, the other two magnetic rods can be used to transfer the sample to the XPS analytical facility or the secondary electron emission coefficient and spectroscopy analyzer, respectively. Then, the magnetic rod is removed, the gate valve is closed, and the vacuum of the analytical facility can soon reach 10^{-10} Torr, and the sample can be tested. Therefore, the system must be equipped with a rapid sample introduction system in order to maintain an ultra-high vacuum level of 10^{-10} Torr in the analytical facility to facilitate rapid sample exchanges. Without the rapid sample introduction design, the original vacuum must be broken each time the sample analytical facility is changed, and then the vacuum must be reestablished. To reach a vacuum level of 10^{-10} Torr, it must undergo a 2.3-day baking process, which can significantly reduce the testing efficiency (Figure 3.6).

The vacuum extraction system for the XPS analytical facility and the secondary electron emission coefficient and energy spectrum chambers consists of an ion pump and a titanium sublimation pump with a base pressure better than 5×10^{-10} Torr. The vacuum pre-pumping of both chambers is achieved by sharing the molecular pump in the sample storage chamber. The analysis chamber vacuum is measured by an ion gauge vacuum gauge.

FIGURE 3.6 Rapid sample introduction system.

XPS energy spectrum test, secondary electron emission coefficient test, and energy spectrum test must be carried out under a high vacuum degree, because if the vacuum degree is not high enough, there is a high probability of encountering other particles during electron movement, resulting in the annihilation or collision of electrons to generate other electrons, which greatly affects the test accuracy. Therefore, the analysis chamber must have an ultra-high vacuum level. To achieve a vacuum level of 10^{-10}Torr, ion pumps and titanium sublimation pumps are required.

The vacuum extraction system for the rapid sample introduction and sample storage systems consists of a molecular pump and a pre-stage reel pump with a base pressure of better than 5×10^{-10}Torr. The vacuum of the rapid sample introduction and sample storage systems is measured by an ion gauge vacuum gauge.

The molecular pump is a clean pump by which the high vacuum can quickly be achieved. The pumping principle of a turbo-molecular pump is a vacuum pump that uses a high-speed rotor to transfer momentum to the gas molecules, so that they gain directional velocity and are thus compressed, driven toward the exhaust port, and then pumped away for the front stage. Because the absolute cleanliness of the vacuum chamber must be maintained, the pre-stage pump is equipped with an oil-free reel pump.

The sample storage system is equipped with a sample storage table with eight storage positions, which facilitates the replacement of large batches of samples at a time, saving time for sample exchange and improving testing efficiency. The sample storage table is designed with a top and bottom design, i.e., it resembles a drawer with eight upper and lower slots, which is mounted on the flange port with a device that can be rotated to drive the top and bottom lift. By rotating this device, the sample can be selected at different positions for transfer or testing (Figure 3.7).

The XPS analytical facility is equipped with a four-dimensional manual sample holder that can be moved in *X*/*Y*/*Z* directions, and the sample stage can be tilted with an accuracy of 0.1°. The sample holder is equipped with a resistance heater for heating up to 500°C.

FIGURE 3.7 Sample storage system and eight sample trays (equipped with eight sample trays along with the vacuum system).

The temperature is measured by thermocouples mounted near the sample position, and the temperature control is provided by a PID controller.

The equipped sample tray is a device that controls the sample displacement function, which can be displaced in x, y, and z directions and can be rotated around the z-axis. This configuration is designed to allow the adjustment of the position of the X-ray irradiation onto the sample, thus testing the composition and chemical state of the same sample at different positions, which allows us to understand the compositional homogeneity of the sample surface. In addition, the sample is equipped with a heating function that allows the sample to be heated and degassed, while the changes in composition and chemical state of the sample at different temperatures can also be tested (Figure 3.8).

Sample delivery between the sample storage system and the XPS analytical facility is achieved by means of a magnetic rod (Figure 3.9).

FIGURE 3.8 Movable sample holders in x, y, and z directions.

FIGURE 3.9 Magnetic transfer rod.

The system allows the maximum size of the tested sample to be 10×10 mm.

The XPS analytical facility is equipped with a CCD to position the sample for observation. Because the sample transfer operation needs to be performed manually, the operator cannot directly visually adjust the sample position when operating the magnetic transfer rod because it is far from the observation window of the analytical facility, so a CCD is needed to display the real-time picture to the computer, and the operator operates while observing the monitor.

The XPS analytical facility is equipped with a residual gas analyzer (RGA, mass number $l \sim 200$ amu) for residual gas analysis. The core of the residual gas analyzer is the quadrupole mass spectrometer, which is an electric field analyzer consisting of four precision-machined electrode rods and two sets of high-voltage, high-frequency radio frequencies applied in the x and y directions, respectively. The four electrodes can be hyperbolic or cylindrical; the high-voltage and high-frequency signal provides the auxiliary energy for the ions to move through the analyzer, and this energy is selective – only ions that meet certain mathematical conditions can safely pass through the quadrupole analyzer without being accelerated indefinitely.

The RGA is equipped for two purposes: one is to detect leaks in the vacuum chamber when the vacuum level is not being pumped up, that is, to test where leaks are occurring and affecting the vacuum level; the other is to monitor the composition of the gas desorbed from the sample surface when the sample is heated and degassed.

The XPS analytical facility is equipped with an argon ion gun for sputtering cleaning of the sample surface. The working principle of an argon ion gun is to use an electron beam emitted by heating a tungsten wire to bombard an inert gas (such as argon) and ionize it to produce inert gas ions. In this system, the main function of the argon ion gun is to perform sputtering cleaning on the surface of the sample. Because no matter what material is usually produced in the atmosphere, a very thin natural oxide layer or some organic substances will be adsorbed, which will greatly affect its secondary electron emission coefficient characteristics. In order to obtain its intrinsic secondary electron emission characteristics, its surface must first be sputtered and cleaned with an iron gun, so that the natural oxide layer is peeled off to expose a clean surface (Figure 3.10).

The secondary electron emission coefficient and energy spectrum analytical facility is equipped with a tiltable sample stage, which is mainly used to change the angle of incidence of the primary electron beam and to test the variation curve of the secondary electron emission coefficient with the incident angle of the electrons. The sample transfer between the secondary electron emission coefficient, spectroscopy chamber, and the sample storage chamber is achieved by means of a magnetic rod.

The secondary electron emission coefficient and energy spectrum analytical facility are also equipped with CCDs to position the samples for observation. Because the sample transfer needs to be performed manually, the operator cannot directly visually adjust the sample position when operating the magnetic rod because it is far from the observation window of the analytical facility, so a CCD is needed to display the real-time image to the computer, and the operator operates while observing the monitor.

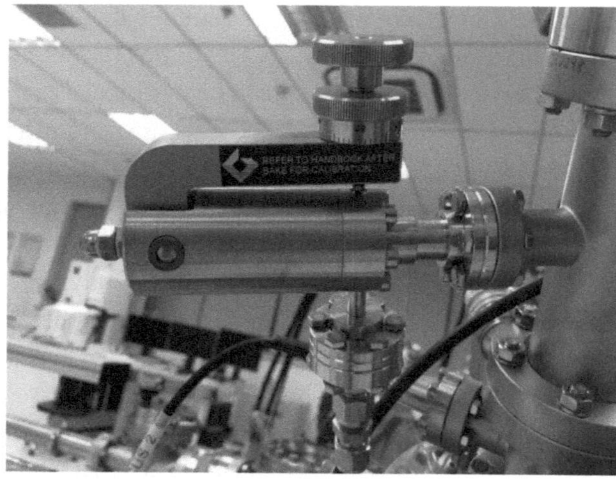

FIGURE 3.10 XPS analytical facility.

3.1.2.1.2 Secondary Electron Emission Coefficient and Energy Spectrum Testing Device The platform is equipped with two low-energy electron guns with different energy ranges as secondary electron excitation sources, with electron energy ranges of 50 eV to 3 keV and 3–30 keV, respectively. These electron guns all use tungsten filaments as electron-emitting materials, which are simple, reliable, stable in beam current, and easy to maintain and replace.

The sample stage is connected to a picoammeter for measuring the electron beam current through the sample. It is equipped with a Keithley Model 6487E picoammeter with a built-in ±500 V DC voltage source. The picoammeter has a wide measurement range from pA to mA scale and has automatic range switching.

A secondary electron collector with three gates is equipped. By varying the gate bias and collector bias, the total secondary electron emission coefficient, backscattered secondary electron emission coefficient, and true secondary electron emission coefficient can be measured. And the secondary electron energy spectrum curve can be measured by voltage scan on G_3.

A wide-beam low-energy electron neutralization gun with electron energy range of 2–100 eV is equipped for charge neutralization when charging the surface of the dielectric sample. When the SEY factor is greater than 1, the sample surface will be positively charged, at which time a low-energy ion gun is used for electrical neutralization.

A scanning low-energy ion neutralization gun with ion energy range of 50–1,000 eV is equipped for charge neutralization when charging the surface of the dielectric sample. When the SEY factor is less than 1, the sample surface will be negatively charged, at which time a low-energy ion gun is used for electrical neutralization.

3.1.2.1.3 XPS Spectroscopy Instruments The working principle of the equipped double-anode X-ray light source is to produce X-rays by bombarding the metal target with an electron beam generated by a heated tungsten filament. The X-ray light source is equipped with dual anodes and the two metal targets are Al target and Mg target with a maximum power of 300 W.

Electron energy spectrum analysis is achieved by a two-stage cylindrical mirror type energy analyzer (DESA150). This energy analyzer has a working distance of up to 55 mm, an energy resolution $\Delta E/E$ of 0.6%–0.05%, and an energy resolution better than 80 meV at 20 eV. The power required for this energy analyzer is provided with the device.

The working principle of the two-stage cylindrical mirror-type energy analyzer: After the slit collection of the energy analyzer, the photoelectrons excited by X-ray irradiation on the sample surface enter the interior of the energy analyzer, and then after a series of internal electron optical lens deflection, focusing, etc., reach the channel plate collector at the bottom. After the photoelectric conversion, such as electron multiplier tube, the information on the number of electrons at different energies is obtained and displayed by software, the curve with the horizontal coordinate of electron energy and the vertical coordinate of the number of electrons (count rate), which together are called electron energy spectrum curves. In summary, the device is used to analyze the XPS energy spectrum curves of different materials to derive information about the material composition and chemical state.

Equipped with a wide-beam low-energy electron neutralizing gun with electron energy range of 2–100 eV. When used to charge the dielectric sample surface for charge neutralization and irradiating the dielectric material with X-ray, the sample surface will be positively charged because photoelectrons are excited out and leave the sample, and the low-energy electron gun is used for electrical neutralization.

3.1.2.2 Working Principle

The platform is equipped with two methods for testing secondary electron emission characteristics: one is the sample bias current method; and the other is the collector electrode method. For metallic materials, both methods are applicable; however, for dielectrics, only the collector electrode method can be used for testing. The working principles of these two methods are described below (Figure 3.11).

3.1.2.2.1 Bias Current Method Firstly, according to the international common principle, the secondary electrons with electron energy lower than 50 eV are considered as true secondary electrons, while the secondary electrons with energy higher than 50 eV are generally Auger electrons and elastic scattering electrons, also collectively referred to as backscattered electrons. By the sample bias current method, different bias pressures are applied to the sample, and the primary incident current, the difference between the primary incident current and the backscattered current, and the difference between the primary incident current and the backscattered current and the true secondary electron current can be obtained, respectively. After a series of addition and subtraction operations, the true secondary current value can be obtained. The working principle is shown in Figure 3.12.

The specific working methods are as follows:

a. First, a voltage of +500 V is applied to the sample. At this point, electrons with secondary electron energy lower than 500 eV generated by the primary incident current bombarding the sample cannot escape. The measured current passing through the sample is the primary incident current (I_p).

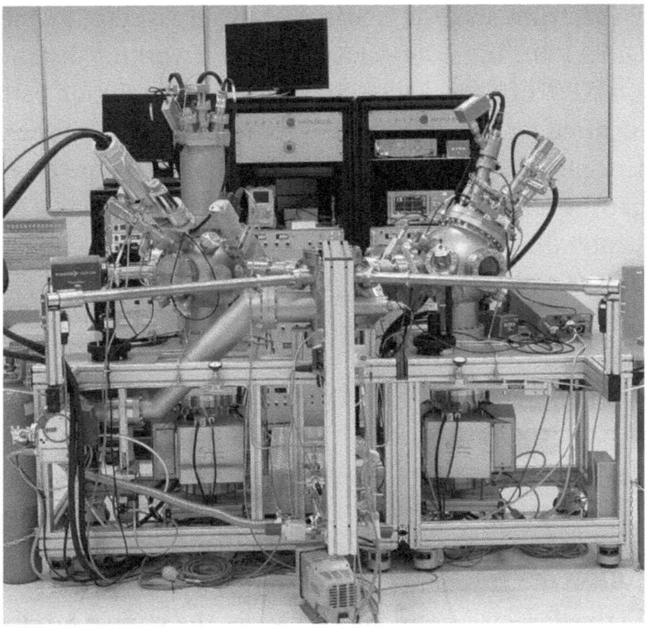

FIGURE 3.11 Physical photo of secondary electron emission test platform.

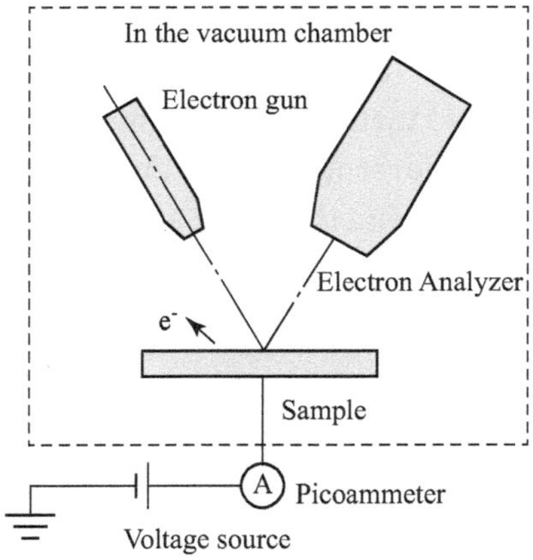

FIGURE 3.12 Schematic diagram of working principle for measuring secondary electron emission coefficient using sample bias method.

b. When a positive bias of +50 V is applied to the sample, the true secondary electrons with electron energy lower than 50 eV cannot escape, but the backscattered electrons with energy higher than 50 eV can, so the current value through the sample measured by the picoammeter at this time is the difference between the primary incident current and the backscattered current ($I_p - I_b$, where I_p is the primary incident current and I_b is the backscattered current).

c. When a negative bias of −20 V is applied to the sample, all secondary electrons (including true secondary electrons and backscattered electrons) can escape from the sample surface, and the current value through the sample measured by the picoammeter is the difference between the primary incident current and the backscattered and true secondary electron currents ($I_p - I_b - I_s$, where I_p is the primary incident current, I_b is the backscattered current, and I_s is the secondary electron current).

d. The true secondary current I_s can be obtained by computing the difference between the two current values measured in step (b) and step (c), and the secondary electron emission coefficient can be calculated using the formula $Y = I_s/I_p$.

3.1.2.2.2 Collector Electrode Method

a. A hemispherical collector is placed over the sample, and this collector has three layers of internal gates, called the first, second, and third gates from inside to outside, abbreviated as G1, G2, and G3. The outermost layer of the collector is called the collector electrode, abbreviated as C. When the sample is excited by the primary incident current, if the sample is grounded and no bias is applied, then the secondary electrons generated escape away from the sample surface and are collected by the collector. By applying different bias voltages to different gates, it is possible to select the electrons of different energies that pass through the gate and are eventually collected by the collector electrode, thus allowing the total secondary electron emission coefficient, the backscattered secondary electron emission coefficient, and the true secondary electron emission coefficient to be derived. The operating principle is shown in Figure 3.13.

b. First, for the primary current incident on the sample, when no bias voltage is applied to the sample, the gate, and the collector, the total current measured at all locations is summed to give the primary incident current value I_p.

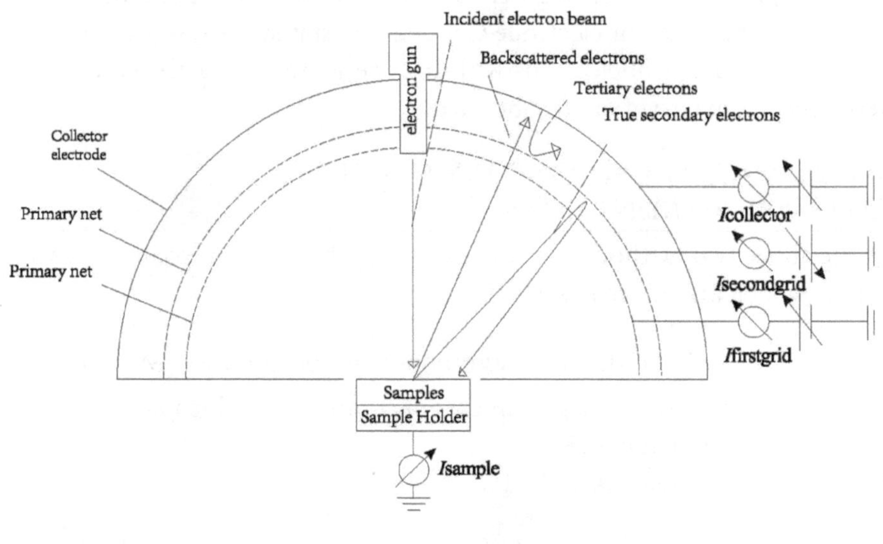

FIGURE 3.13 Schematic diagram of the operating principle of measuring secondary electron emission coefficient by collector electrode method.

c. When a positive bias is applied to the collector electrode C only, all the secondary electrons can pass through the gate to the collector electrode, and the current measured at C is the total secondary electron current, referred to as I_t. However, since several layers of the grid need to be passed, and each layer of the grid has a certain transmittance (known number), the measured current value needs to be multiplied by a correction factor to obtain I_t.

d. When a bias voltage of −50 V is applied to the second gate G2, the secondary electrons with energy lower than 50 eV are repelled by G2 and cannot pass through G2, and the currents collected on the collector electrode C are all secondary electrons with energy greater than 50 eV, i.e. the current value measured at this time is the backscattered current I_b. As above, the transmission rate of the gate needs to be considered and multiplied by the correction factor.

e. The true secondary electron current I_s is obtained by subtracting the backscattered current I_b from the total secondary current I_t. The secondary electron emission coefficient can be calculated using the formula $Y=I_s/I_p$.

f. The sample stage can be rotated. When the sample is rotated, the angle between the incident electron beam and the sample can be changed. After changing the angle, repeat steps (b)–(e) to measure the variation curve of the secondary electron emission coefficient with the angle of the incident electrons.

g. The energy spectrum of secondary electrons can also be tested using the collector electrode method. The principle of operation is to use the deceleration gate method. When the secondary electrons pass through the third gate G3, the negative bias voltage is added to G3, and the negative bias voltage is scanned and changed by, for example, −5 V bias voltage every 2 s, which means that the secondary electrons of different energies can pass through G3 one after another, and the number of electrons collected on the collector electrode C is counted statistically, so that the distribution curve of the number of electrons with the energy range, i.e. the secondary electron energy spectrum curve, can be obtained.

3.2 TYPICAL MATERIAL SECONDARY ELECTRONIC YIELD MEASUREMENT RESULTS

Based on the above two methods, we have measured the SEY of typical metals and dielectric materials, which are summarized below:

3.2.1 SEY and Energy Spectrum Measurement Data of Common Metallic Materials

The SEY and energy spectrum measurements of common metallic materials are shown in Table 3.1 and Figures 3.14 and 3.15.

See Table 3.1 and Figures 3.14 and 3.15.

3.2.2 The Measurement Data of SEY of Common Dielectric Materials

See Table 3.2.

TABLE 3.1 Summary of SEY and Energy Spectrum Test Data of Common Metallic Materials

No.	Metal Material
1	Al
2	Ti
3	Fe
4	Ni
5	Cu
6	Ta
7	W
8	304 stainless steel

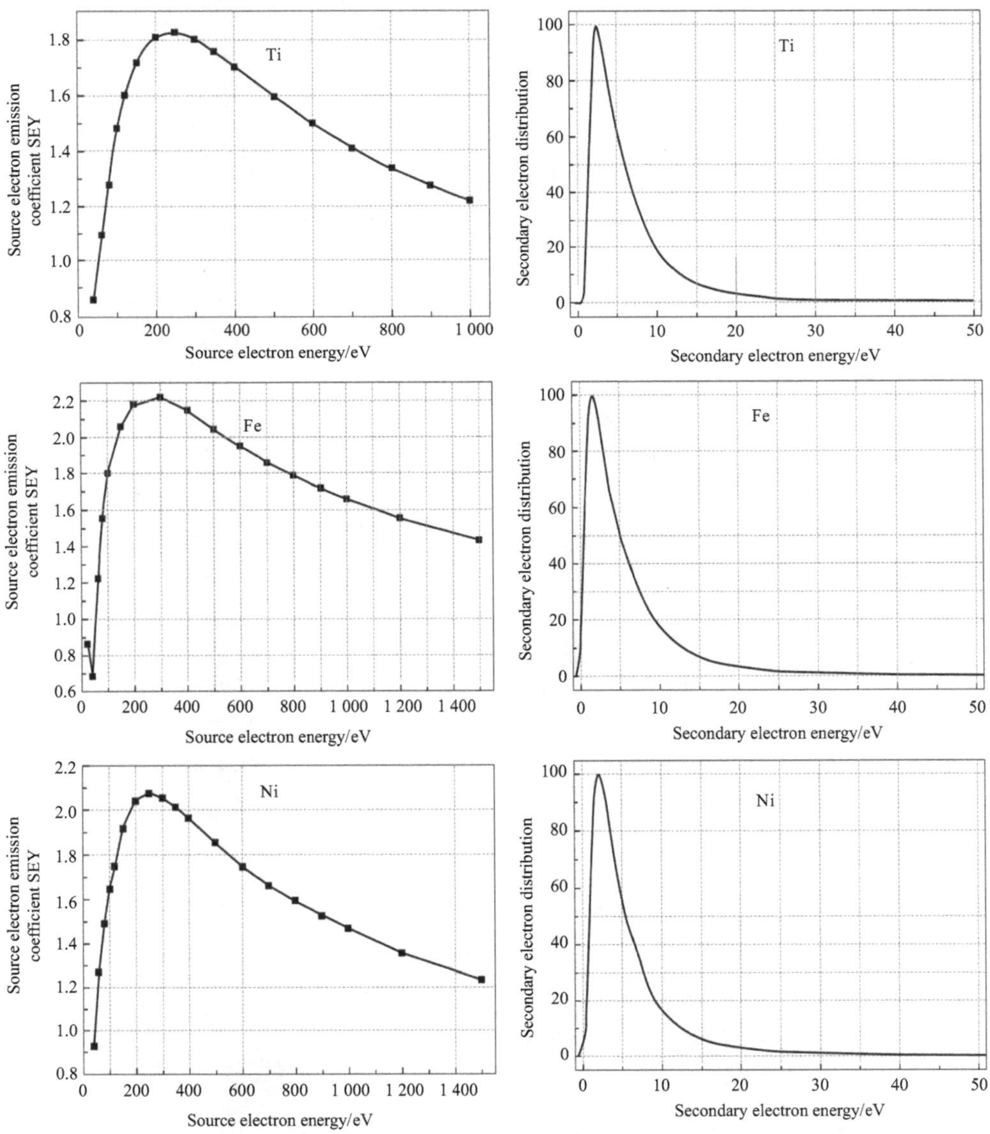

FIGURE 3.14 SEY and energy spectrum data of common metallic materials (I).

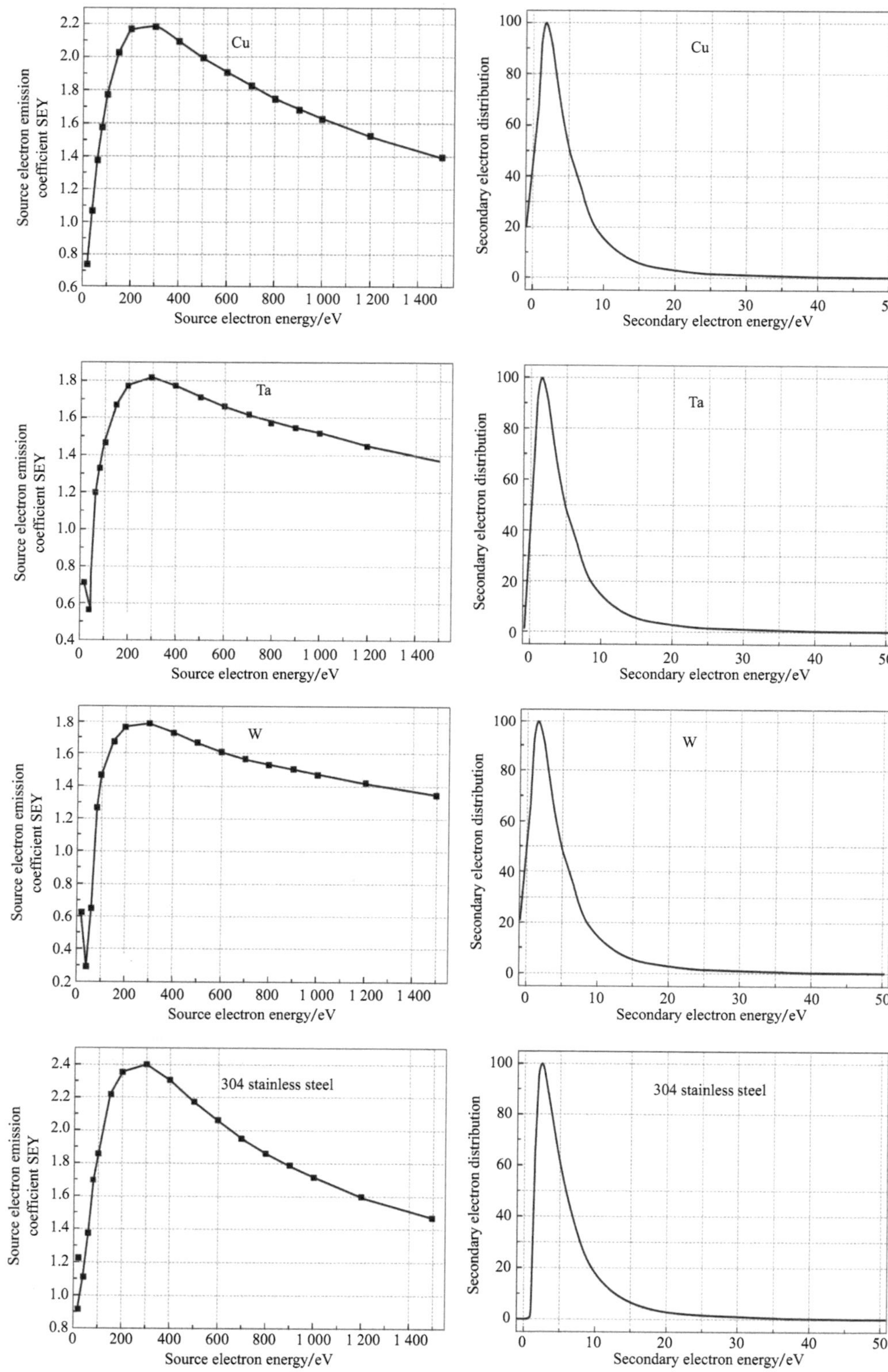

FIGURE 3.15 SEY and energy spectrum data of common metallic materials (II).

TABLE 3.2 Summary of the Measurement Data of SEY of Common Dielectric Materials

No.	Dielectric Material Name
1.	Polyethylene (PE)
2.	Polyimide (PI)
3.	Nylon
4.	Polypropylene (PP)
5.	Polyvinyl chloride (PVC)
6.	Polycarbonate (PC)
7.	Polytetrafluoroethylene (PTFE)
8.	Polyester (PET)
9.	Thermoplastic polyurethane (TPU)
10.	Muscovite
11.	Silicon dioxide (SiO_2)
12.	Polymethyl methacrylate (PMMA)
13.	Alumina (Al_2O_3)
14.	Aluminum nitride (AIN)
15.	Zirconia (ZrO_2)
16.	Silicon carbide (SiC)
17.	Boron nitride (BN)
18.	Silicon nitride (Si_3N_4)
19.	Fe_3YO_6
20.	Ceramic MST-16 (MgO-SiO_2-CaO-TiO_2)
21.	Ceramic MCT-21(MgO-CaO-TiO_2)
22.	Ceramic ZNT-45 (Zr-Nb_2O_5-TiO_2)

The SEY results are shown below:

1. Polyethylene (PE) (see Table 3.3 and Figure 3.16)

TABLE 3.3 SEY of PE Dielectric

Incident Electron Energy (eV)	SEY	Positive Error	Negative Error
50	1.1916	0.0428	−0.0512
70	1.1817	0.0469	−0.0291
100	1.5188	0.1062	−0.1400
150	2.0617	0.0688	−0.1071
200	2.4156	0.0844	−0.0601
250	2.5401	0.0686	−0.0817
300	2.5781	0.1028	−0.1165
350	2.6086	0.0722	−0.0886

(Continued)

TABLE 3.3 (*Continued*) SEY of PE Dielectric

Incident Electron Energy (eV)	SEY	Positive Error	Negative Error
400	2.4983	0.0549	−0.0983
450	2.4111	0.0056	−0.0111
500	2.2606	0.1144	−0.0939
600	1.9960	0.0862	−0.0585
700	1.9335	0.0020	−0.0029
800	1.7830	0.0203	−0.0115
900	1.6773	0.0007	−0.0008
1,000	1.5742	0.0324	−0.0258
1,200	1.3802	0.0044	−0.0052
1,400	1.2316	0.0700	−0.0924
1,600	1.1494	0.0263	−0.0244
1,800	1.0449	0.0085	−0.0136
2,000	0.9359	0.0153	−0.0149
2,200	0.8722	0.0109	−0.0117
2,400	0.8281	0.0156	−0.0142
2,600	0.7974	0.0098	−0.0095
2,800	0.7751	0.0010	−0.0009
3,000	0.7505	0.0196	−0.0191
4,000	0.6823	0.0023	−0.0013
5,000	0.6134	0.0058	−0.0034

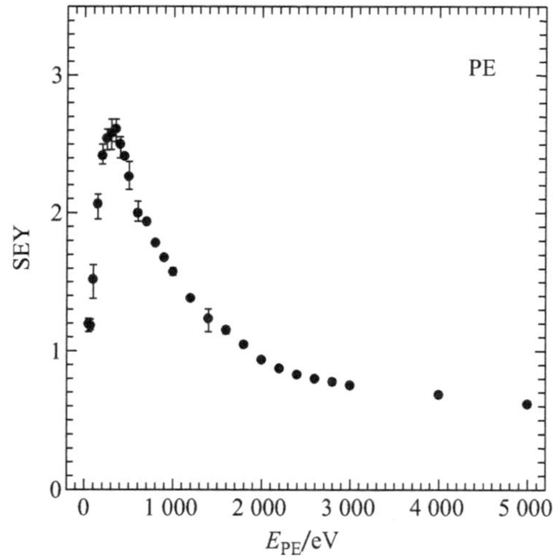

FIGURE 3.16 Variation of SEY of EPE dielectric with incident electron energy.

2. Polyimide (PI) (see Table 3.4 and Figure 3.17)

TABLE 3.4 SEY of PI Dielectric

Incident Electron Energy/eV	SEY	Positive Error	Negative Error
50	1.0575	0.0675	−0.0675
100	1.5875	0.0575	−0.0575
150	1.7375	0.0575	−0.0575
200	1.8400	0.05	−0.05
250	1.8300	0.045	−0.045
300	1.7750	0.045	−0.045
350	1.7250	0.005	−0.005
450	1.5650	0.01	−0.01
550	1.4225	0.0125	−0.0125
650	1.3450	0.01	−0.01
750	1.2475	0.0075	−0.0025
850	1.1525	0.0025	−0.0075
950	1.1150	0.005	−0.005
1,150	1.0400	0.01	−0.01
1,350	0.9875	0.0175	−0.0175
1,550	*0.9550*	0.01	−0.01
1,750	0.9125	0.0075	−0.0075
1,950	0.8675	0.0075	−0.0075
2,150	0.815	0.005	−0.005
2,350	0.7975	0.0125	−0.0125
2,550	0.7475	0.0075	−0.0075
2,750	0.7375	0.0075	−0.0075
2,950	0.7275	0.0075	−0.0075
4,000	0.6625	0.0175	−0.0175
5,000	0.6225	0.0075	−0.0075

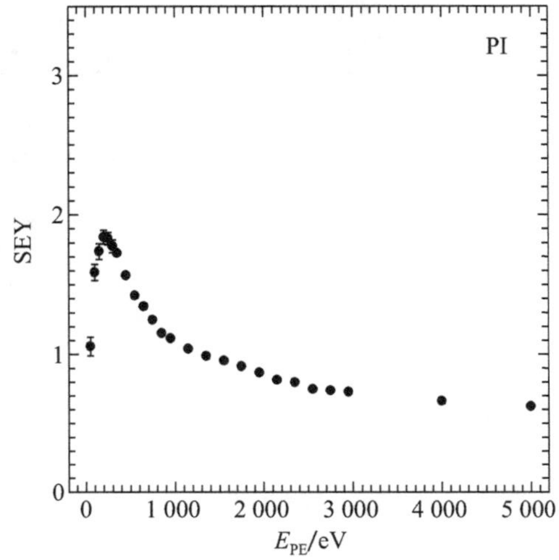

FIGURE 3.17 Variation of SEY of PI dielectric with incident electron energy E_{PE}.

3. Nylon (see Table 3.5 and Figure 3.18)

TABLE 3.5 SEY of Nylon Dielectric

Incident Electron Energy/eV	SEY	Positive Error	Negative Error
50	1.1167	0.0262	−0.0167
70	1.3189	0.1639	−0.1898
100	1.7441	0.0774	−0.0666
150	2.2395	0.1177	−0.0782
200	2.2448	0.0778	−0.0683
250	2.6637	0.1220	−0.1012
300	2.5473	0.0777	−0.0928
350	2.5476	0.0952	−0.0476
400	2.4018	0.1697	−0.1315
450	2.1576	0.1152	−0.2076
500	2.1688	0.1645	−0.1133
600	2.0208	0.0417	−0.0208
700	1.8889	0.0444	−0.0222
800	1.6274	0.0393	−0.0274
900	1.4687	0.0474	−0.0402
1,000	1.4314	0.0086	−0.0147
1,200	1.2075	0.0314	−0.0360
1,400	1.1017	0.0233	−0.0429
1,600	0.9902	0.0098	−0.0196
1,800	0.8716	0.0072	−0.0144
2,000	0.8252	0.0319	−0.0192
2,200	0.7685	0.0128	0.0185
2,400	0.7100	0.0172	0.0131
2,600	0.6837	0.0133	0.0170
2,800	0.6273	0.0178	0.0213
3,000	0.5899	0.0039	0.0078
4,000	0.5067	0.0023	0.0043
5,000	0.4573	0.0015	0.0032

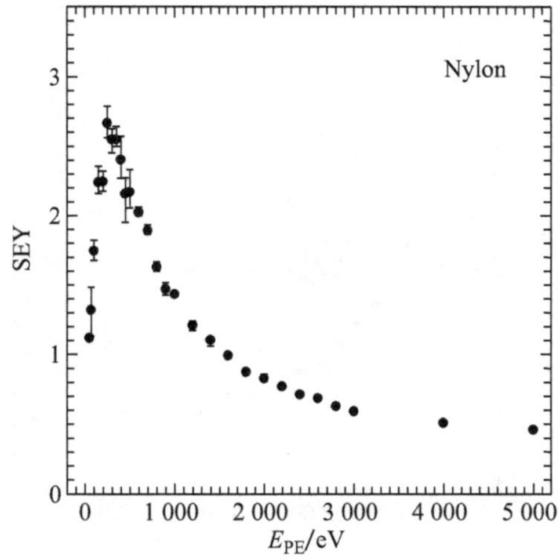

FIGURE 3.18 Variation of SEY of Nylon dielectric with incident electron energy E_{PE}.

4. Polypropylene (PP) (see Table 3.6 and Figure 3.19)

TABLE 3.6 SEY of PP Dielectric

Incident Electron Energy/eV	SEY	Positive Error	Negative Error
50	1.6108	0.3225	−0.1823
100	1.9786	0.1880	−0.1325
150	2.4154	0.1846	−0.1376
200	2.8899	0.0476	−0.0327
250	3.0191	0.2117	−0.1302
300	2.9833	0.2833	−0.2333
350	2.9470	0.1439	−0.1136
400	2.8082	0.0668	−0.1158
450	2.6028	0.1472	−0.1213
500	2.4034	0.0966	−0.0931
600	2.2469	0.0864	−0.0617
700	2.0780	0.0420	−0.0409
800	1.8225	0.1006	−0.0638
900	1.7132	0.0118	−0.0209
1,000	1.6430	0.0427	−0.0484
1,200	1.3907	0.0452	−0.0492
1,400	1.2354	0.0889	−0.0536
1,600	1.0767	0.0122	−0.0069
1,800	0.9487	0.0074	−0.0132
2,000	0.8850	0.0086	−0.0059
2,200	0.8239	0.0170	−0.0154
2,400	0.7919	0.0081	−0.0093
2,600	0.7482	0.0073	−0.0091
2,800	0.7214	0.0020	−0.0040
3,000	0.6878	0.0167	−0.0177
4,000	0.6035	0.0125	−0.0134
5,000	0.5375	0.0173	−0.0174

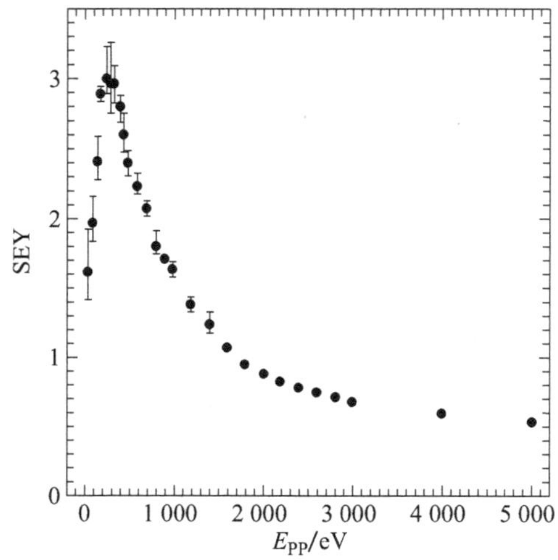

FIGURE 3.19 Variation of SEY of PP dielectric with incident electron energy $E_{PE.}$

5. Polyvinyl chloride (PVC) (see Table 3.7 and Figure 3.20)

TABLE 3.7 SEY of PVC Dielectric

Incident Electron Energy/eV	SEY	Positive Error	Negative Error
50	1.1070	0.1073	−0.0670
100	1.4872	0.0842	−0.0586
150	2.0536	0.0664	−0.0869
200	2.2574	0.0874	−0.0724
250	2.4510	0.1205	−0.1266
300	2.5730	0.1166	−0.0730
350	2.4336	0.0370	−0.0447
400	2.4347	0.0653	−0.1121
500	2.2082	0.1490	−0.1029
600	2.0822	0.1321	−0.0822
700	1.9601	0.0399	−0.0226
800	1.7879	0.0692	−0.0460
900	1.7382	0.1189	−0.1549
1,000	1.6152	0.0515	−0.0363
1,200	1.4379	0.0794	−0.0608
1,400	1.3216	0.0355	−0.0359
1,600	1.1859	0.0175	−0.0252
1,800	1.1202	0.0279	−0.0168
2,000	0.9801	0.0199	−0.0398
2,200	0.9251	0.0241	−0.0128
2,400	0.8809	0.0157	−0.0120
2,600	0.8325	0.0137	−0.0200
2,800	0.8044	0.0191	−0.0187
3,000	0.7653	0.0106	−0.0195
4,000	0.6832	0.0104	−0.0132
5,000	0.6121	0.0100	−0.0126

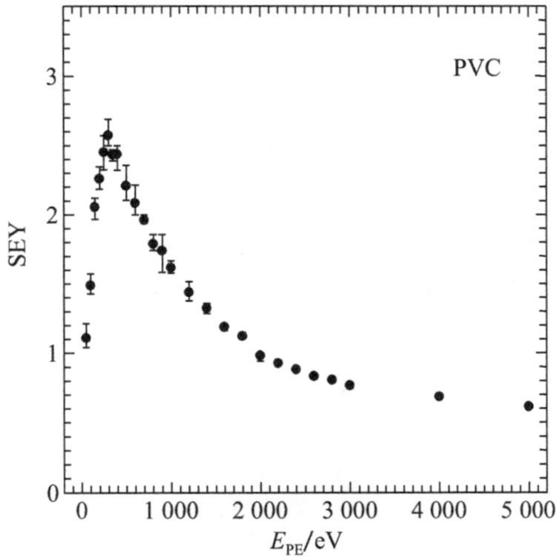

FIGURE 3.20 Variation of SEY of PVC dielectric with incident electron energy E_{PE}.

6. Polycarbonate (PC) (see Table 3.8 and Figure 3.21)

TABLE 3.8 SEY of PC Dielectric

Incident Electron Energy/eV	SEY	Positive Error	Negative Error
50	0.9451	0.0812	−0.0451
100	1.4985	0.0730	−0.0439
150	1.7667	0.0667	−0.1000
200	1.8759	0.1782	−0.2092
250	1.8610	0.0890	−0.1753
300	1.8828	0.0869	−0.0646
350	1.8331	0.1098	−0.1188
400	1.7345	0.0405	−0.0503
500	1.5402	0.0207	−0.0260
600	1.4079	0.0778	−0.0420
700	1.2987	0.0697	−0.0487
800	1.2056	0.0444	−0.0389
900	1.1391	0.0148	−0.0109
1,000	1.0000	0.0000	0.0000
1,200	0.8752	0.0195	−0.0331
1,400	0.7925	0.0233	−0.0147
1,600	0.7041	0.0138	−0.0097
1,800	0.6540	0.0127	−0.0254
2,000	0.5889	0.0111	−0.0175
2,200	0.5566	0.0148	−0.0138
2,400	0.5215	0.0191	−0.0215
2,600	0.5026	0.0052	−0.0026
2,800	0.4741	0.0130	−0.0126
3,000	0.4529	0.0042	−0.0085
4,000	0.3934	0.0034	−0.0053
5,000	0.3474	0.0025	−0.0023

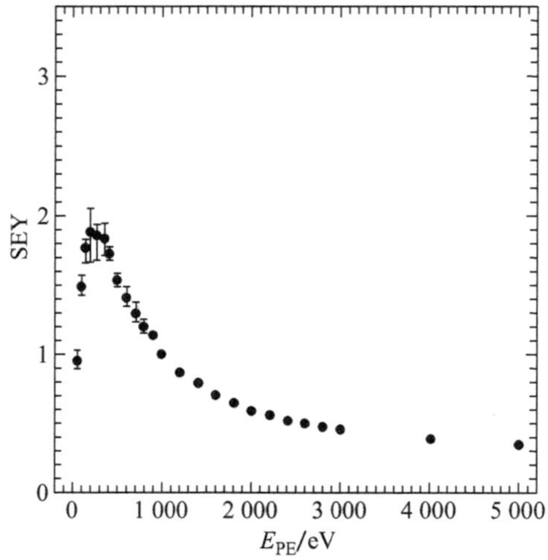

FIGURE 3.21 Variation of SEY of PC dielectric with incident electron energy E_{PE}.

7. Polytetrafluoroethylene (PTFE) (see Table 3.9 and Figure 3.22)

TABLE 3.9 SEY of PTFE Dielectric

Incident Electron Energy/eV	SEY	Positive Error	Negative Error
50	0.9308	0.0573	−0.0191
100	1.1576	0.1168	−0.0460
150	1.6665	0.2788	−0.0908
200	1.9512	0.1820	−0.0603
250	2.0881	0.1636	−0.0556
300	2.2552	0.0982	−0.0781
350	2.2606	0.0406	−0.0505
400	2.2585	0.0256	−0.0458
450	2.2435	0.0418	−0.0676
500	2.2219	0.0666	−0.0962
600	2.0947	0.0612	−0.0871
700	1.8816	0.0597	−0.0545
800	1.6962	0.2551	−0.0991
900	1.6586	0.2586	−0.0915
1,000	1.5281	0.1296	−0.1493
1,200	1.4289	0.0723	−0.1005
1,400	1.2803	0.0874	−0.0808
1,600	1.1936	0.0315	−0.0455
1,800	1.1118	0.0065	−0.0029
2,000	1.036	0.0138	−0.0073
2,200	0.9386	0.0063	−0.0073
2,400	0.9125	0.0599	−0.0187
2,600	0.9255	0.0729	−0.0745
2,800	0.9032	0.0797	−0.0697
3,000	0.8573	0.0688	−0.0616
4,000	0.7657	0.0090	−0.0089
5,000	0.7298	0.0105	−0.0104

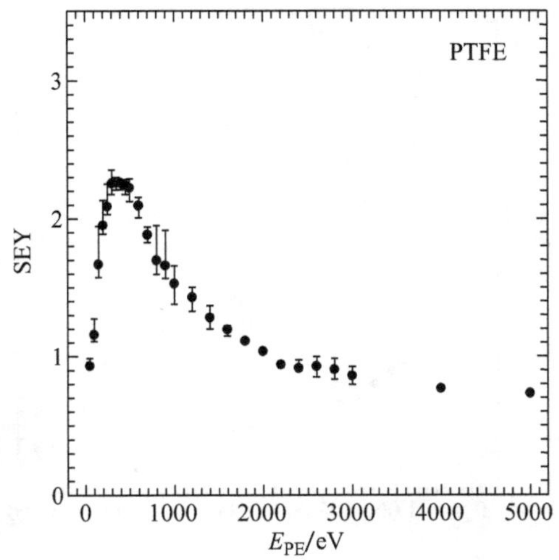

FIGURE 3.22 Variation of SEY of PTFE dielectric with incident electron energy E_{PE}.

8. Polyester (PET) (see Table 3.10 and Figure 3.23)

TABLE 3.10 SEY of PET Dielectric

Incident Electron Energy/eV	SEY	Positive Error	Negative Error
50	0.9189	0.0203	−0.0300
100	1.3809	0.0515	−0.0296
150	1.6203	0.0309	−0.0203
200	1.7744	0.0718	−0.0535
250	1.8479	0.0732	−0.0674
300	1.7406	0.0286	−0.0197
350	1.6800	0.0105	−0.0134
400	1.5820	0.0370	−0.0265
500	1.3967	0.0107	−0.0174
600	1.2595	0.0097	−0.0095
700	1.1478	0.0597	−0.0601
800	1.0346	0.0071	−0.0142
900	0.9385	0.0053	−0.0093
1,000	0.8729	0.0108	−0.0129
1,200	0.7867	0.0049	−0.0055
1,400	0.7054	0.0120	−0.0107
1,600	0.6433	0.0025	−0.0050
1,800	0.6057	0.0113	−0.0139
2,000	0.5502	0.0030	−0.0038
2,200	0.5276	0.0058	−0.0076
2,400	0.5000	0.0000	0.0000
2,600	0.4753	0.0082	−0.0072
2,800	0.4530	0.0035	−0.0041
3,000	0.4353	0.0047	−0.0051
4,000	0.3724	0.0032	−0.0043
5,000	0.3354	0.0034	−0.0036

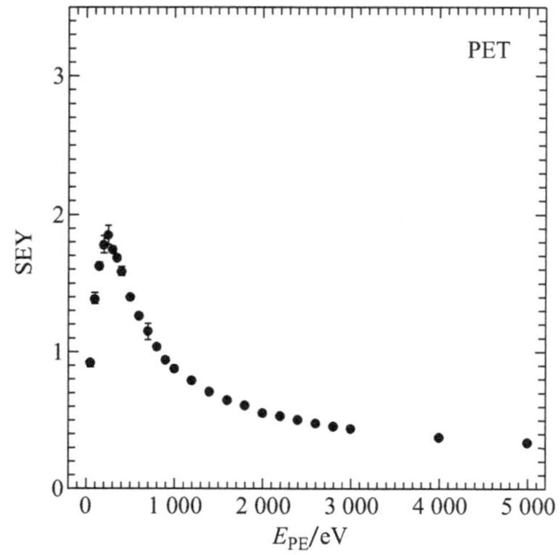

FIGURE 3.23 Variation of SEY of PET dielectric with incident electron energy E_{PE}.

9. Thermoplastic polyurethane (TPU) (see Table 3.11 and Figure 3.24)

TABLE 3.11 SEY of TPU Dielectric

Incident Electron Energy/eV	SEY	Positive Error	Negative Error
50	1.4580	0.1373	−0.0884
100	1.7728	0.0453	−0.0293
150	2.3853	0.0050	−0.0043
200	2.5192	0.1308	−0.1238
250	2.6355	0.0311	−0.0541
300	2.5834	0.0213	−0.0379
350	2.5197	0.1303	−0.0652
400	2.3363	0.0590	−0.0636
500	2.1563	0.0297	−0.0401
600	1.9630	0.0846	−0.0519
700	1.7987	0.0550	−0.0368
800	1.6366	0.0134	−0.0268
900	1.4899	0.0450	−0.0551
1,000	1.4453	0.0181	−0.0362
1,200	1.2538	0.0189	−0.0114
1,400	1.1177	0.0035	−0.0034
1,600	0.9956	0.0044	−0.0089
1,800	0.9129	0.0037	−0.0063
2,000	0.8396	0.0054	−0.0087
2,200	0.7852	0.0148	−0.0074
2,400	0.7340	0.0029	−0.0042
2,600	0.7008	0.0115	−0.0064
2,800	0.6635	0.0032	−0.0063
3,000	0.6310	0.0079	−0.0093
4,000	0.5560	0.0054	−0.0043
5,000	0.5120	0.0043	−0.0032

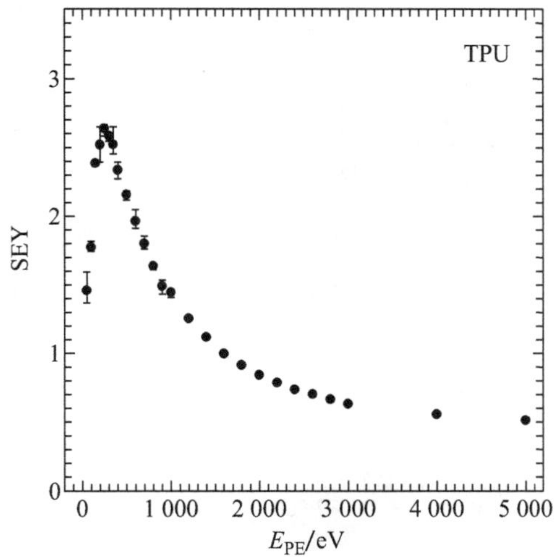

FIGURE 3.24 Variation of SEY of TPU dielectric with incident electron energy E_{PE}.

10. Muscovite (see Table 3.12 and Figure 3.25)

TABLE 3.12 SEY of Muscovite Dielectric

Incident Electron Energy/eV	SEY	Positive Error	Negative Error
50	1.8428	0.3794	−0.2639
100	2.6129	0.0213	−0.0275
150	3.4842	0.1408	−0.1900
200	3.7376	0.1958	−0.2082
250	3.9288	0.1823	−0.0955
300	4.1054	0.1887	−0.3054
350	4.2281	0.0351	−0.0702
400	4.1719	0.1439	−0.1719
450	4.0667	0.1333	−0.0667
500	4.1832	0.1501	−0.0780
600	3.9889	0.1778	−0.1889
700	3.9259	0.0741	−0.0370
800	3.5611	0.1056	−0.2111
900	3.6127	0.4540	−0.3746
1,000	3.2876	0.0458	−0.0523
1,200	2.8357	0.0532	−0.0357
1,400	2.5876	0.0869	−0.0876
1,600	2.3064	0.0572	−0.1064
1,800	2.2487	0.0370	−0.0265
2,000	2.1913	0.1871	−0.1458
2,200	2.0425	0.1075	−0.1702
2,400	1.8460	0.0826	−0.0588
2,600	1.6821	0.0389	−0.0376
2,800	1.5534	0.0553	−0.0436
3,000	1.5224	0.0231	−0.0224
4,000	1.3236	0.0134	−0.0135
5,000	1.2332	0.0100	−0.0100

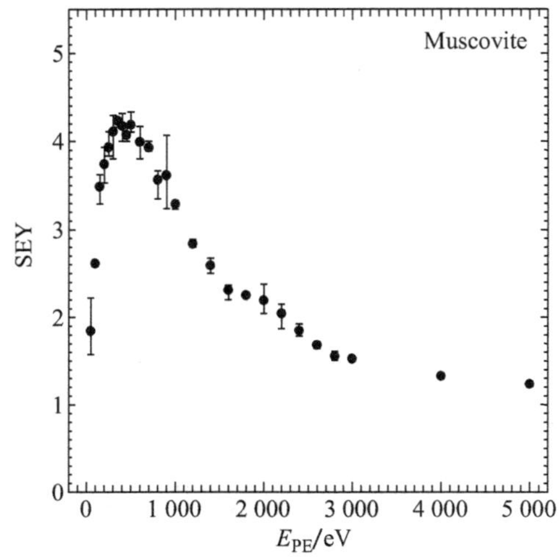

FIGURE 3.25 Variation of SEY of muscovite dielectric with incident electron energy E_{PE}.

11. Silicon dioxide (SiO_2) (see Table 3.13 and Figure 3.26)

TABLE 3.13 SEY of SiO_2 Dielectric

Incident Electron Energy/eV	SEY	Positive Error	Negative Error
50	1.6389	0.2611	−0.2222
70	2.1960	0.0768	−0.0626
100	2.6310	0.1786	−0.1310
150	3.3664	0.4757	−0.5093
200	3.8120	0.6255	−0.5812
250	3.8635	0.5981	−0.5186
300	4.0867	0.3800	−0.3944
350	4.0367	0.5018	−0.5194
400	4.2331	0.6492	−0.4254
500	3.8452	0.6163	−0.4606
600	3.5959	0.4041	−0.3736
700	3.6319	0.2181	−0.2473
800	3.4306	0.2536	−0.2567
900	3.1044	0.3242	−0.3771
1,000	2.8457	0.3058	−0.2457
1,200	2.7169	0.0331	−0.0199
1,400	2.4041	0.1765	−0.0938
1,600	2.3751	0.2613	−0.1529
1,800	2.2130	0.0370	−0.0592
2,000	2.0037	0.1502	−0.1465
2,200	1.9048	0.1786	−0.1905
2,400	1.6929	0.1220	−0.1214
2,600	1.6559	0.0298	−0.0405
2,800	1.5813	0.0473	−0.0813
3,000	1.5840	0.1160	−0.0705
4,000	1.4284	0.1130	−0.0428
5,000	1.3543	0.1040	−0.0367

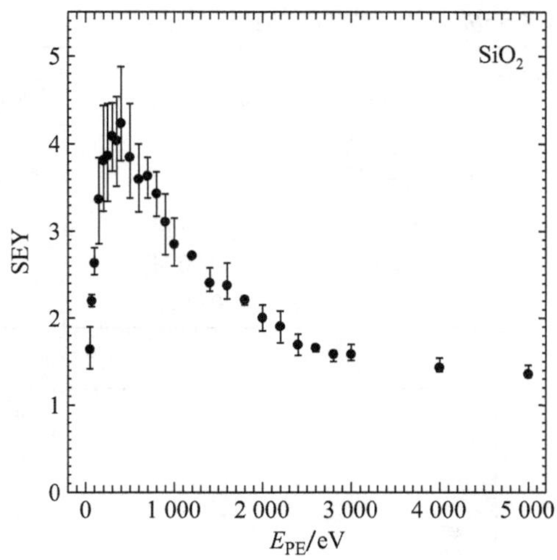

FIGURE 3.26 Variation of SEY of SiO_2 dielectric with incident electron energy E_{PE}.

12. Polymethyl methacrylate (PMMA) (see Table 3.14 and Figure 3.27)

TABLE 3.14 SEY of PMMA Dielectric

Incident Electron Energy/eV	SEY	Positive Error	Negative Error
50	1.2328	0.1006	−0.0629
70	1.3423	0.2992	−0.1791
100	1.5471	0.1863	−0.1137
150	2.0918	0.0661	−0.0441
200	2.2957	0.0452	−0.0829
250	2.4388	0.0385	−0.0345
300	2.4528	0.0361	−0.0615
350	2.4705	0.0179	−0.0236
400	2.3248	0.0448	−0.0391
450	2.2575	0.0536	−0.0268
500	2.0999	0.0557	−0.0285
600	1.9626	0.0563	−0.0503
700	1.8068	0.0422	−0.0211
800	1.6711	0.0836	−0.0440
900	1.5369	0.0822	−0.0453
1,000	1.4491	0.0421	−0.0491
1,200	1.3331	0.0188	−0.0254
1,400	1.1589	0.0197	−0.0160
1,600	1.0461	0.0254	−0.0138
1,800	0.9111	0.0175	−0.0111
2,000	0.8572	0.0049	−0.0097
2,200	0.8152	0.0124	−0.0186
2,400	0.7877	0.0359	−0.0240

(Continued)

TABLE 3.14 (*Continued*) SEY of PMMA Dielectric

Incident Electron Energy/eV	SEY	Positive Error	Negative Error
2,600	0.7287	0.0120	−0.0074
2,800	0.7095	0.0048	−0.0095
3,000	0.6943	0.0148	−0.0101
4,000	0.6132	0.0034	−0.0230
5,000	0.5712	0.0023	−0.0340

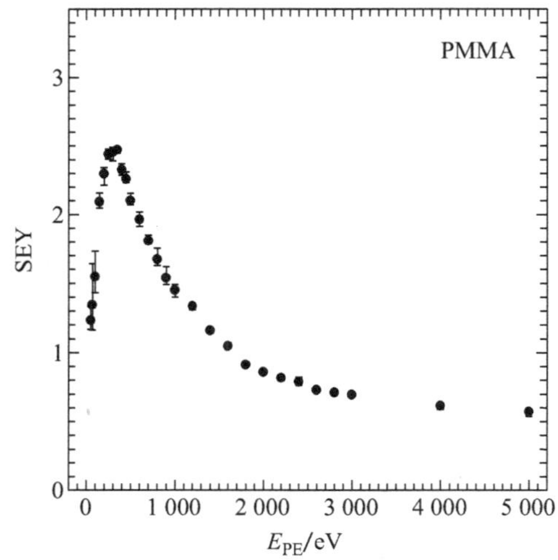

FIGURE 3.27 Variation of SEY of PMMA dielectric with incident electron energy E_{PE}.

13. Alumina (Al_2O_3) (see Table 3.15 and Figure 3.28)

TABLE 3.15 SEY of Al_2O_3 Dielectric

Incident Electron Energy/eV	SEY	Positive Error	Negative Error
50	1.2459	0.0247	−0.0742
100	1.8244	0.0904	−0.0904
200	2.6192	0.1046	−0.1046
300	3.3341	0.3199	−0.1066
400	3.6723	0.0565	−0.0565
500	4.1758	0.1469	−0.1469
600	4.5309	0.5551	−0.1851
700	4.3023	0.1372	−0.1373
800	3.9343	0.1517	−0.0506
900	4.0462	0.0462	−0.0464
1,000	3.9125	0.0625	−0.0625
1,200	3.7538	0.0170	−0.0057
1,400	3.4765	0.1394	−0.4182

(*Continued*)

TABLE 3.15 (*Continued*) SEY of Al_2O_3 Dielectric

Incident Electron Energy/eV	SEY	Positive Error	Negative Error
1,600	3.4457	0.1063	−0.3190
1,800	3.0491	0.1844	−0.1843
2,000	2.9041	0.3101	−0.3102
2,200	2.5559	0.1294	−0.3885
2,400	2.5575	0.3064	−0.3065
2,600	2.2508	0.1516	−0.4551
2,800	1.9035	0.2323	−0.2323
3,000	1.8452	0.1890	−0.1891
4,000	1.5846	0.0642	−0.0215
5,000	1.3744	0.0389	−0.0389

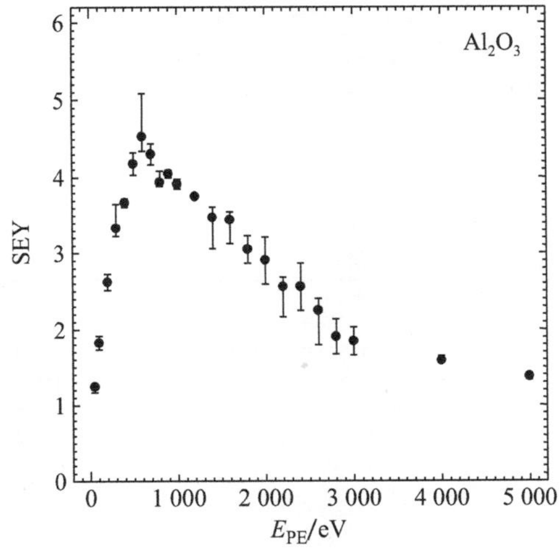

FIGURE 3.28 Variation of SEY of Al_2O_3 dielectric with incident electron energy E_{PE}.

14. Aluminum nitride (AlN) (see Table 3.16 and Figure 3.29)

TABLE 3.16 SEY of AlN Dielectric

Incident Electron Energy/eV	SEY	Positive Error	Negative Error
50	1.096	0.0522	−0.0521
70	1.476	0.0759	−0.0758
150	2.1636	0.0406	−0.1217
200	2.6123	0.0202	−0.0203
300	2.9353	0.0539	−0.0540
400	3.0219	0.0877	−0.0876

(*Continued*)

TABLE 3.16 (*Continued*) SEY of AlN Dielectric

Incident Electron Energy/eV	SEY	Positive Error	Negative Error
500	2.8491	0.0284	−0.0285
600	2.682	0.0410	−0.0136
700	2.5848	0.0434	−0.0434
800	2.5153	0.0291	−0.0292
900	2.2874	0.0311	−0.0311
1,000	2.2892	0.0577	−0.0576
1,200	2.0605	0.0368	−0.0369
1,400	1.8413	0.0094	−0.0094
1,600	1.7236	0.0271	−0.0271
1,800	1.6474	0.0420	−0.0421
2,000	1.5553	0.0257	−0.0258
2,200	1.4850	0.0023	−0.0022
2,400	1.4066	0.0195	−0.0195
2,600	1.3325	0.0104	−0.0311
2,800	1.2831	0.0036	−0.0037
3,000	1.2231	0.0094	−0.0094
4,000	1.0336	0.0030	−0.0031
5,000	0.9999	0.0175	−0.0058

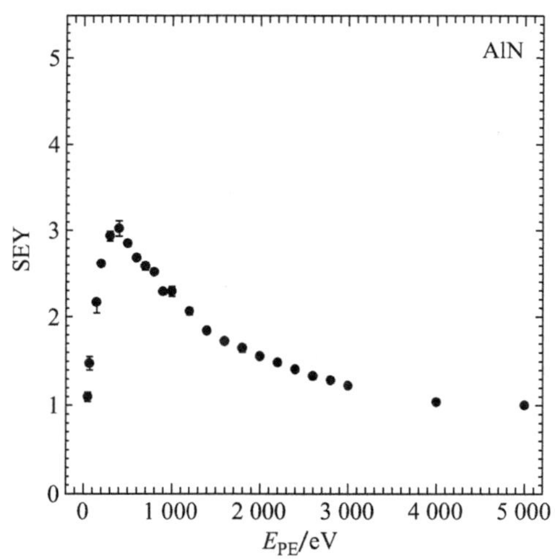

FIGURE 3.29 Variation of SEY of AlN dielectric with incident electron energy E_{PE}.

15. Zirconia (ZrO_2) (see Table 3.17 and Figure 3.30)

TABLE 3.17 SEY of ZrO_2 Dielectric

Incident Electron Energy/eV	SEY	Positive Error	Negative Error
50	1.4900	0.1700	−0.1600
100	2.0300	0.1200	−0.1200
150	2.5000	0.0500	−0.0400
200	2.8700	0.0300	−0.0300
250	3.1800	0.1800	−0.1800
300	3.2900	0.0600	−0.0600
350	3.3900	0.0400	−0.0300
400	3.2900	0.0400	−0.0400
500	3.2000	0.0900	−0.0900
600	3.1300	0.0100	−0.0100
700	2.9800	0.0600	−0.0500
800	2.7700	0.0800	−0.0700
900	2.6400	0.0600	−0.0500
1,000	2.5200	0.0300	−0.0200
1,200	2.3700	0.0200	−0.0200
1,400	2.1300	0.0700	−0.0700
1,600	2.0300	0.0300	−0.0300
1,800	1.9700	0.0300	−0.0300
2,000	1.7900	0.0300	−0.0200
2,200	1.7400	0.0300	−0.0300
2,400	1.6000	0.0000	0.0100
2,600	1.5700	0.0100	−0.0100
2,800	1.5100	0.0000	0.0000
3,000	1.4500	0.0200	−0.0100
4,000	1.3000	0.0100	0.0000
5,000	1.2100	0.0100	0.0000

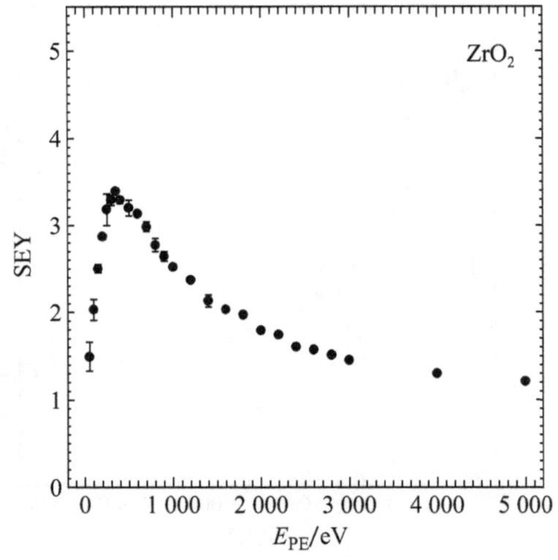

FIGURE 3.30 Variation of SEY of ZrO_2 dielectric with incident electron energy E_{PE}.

16. Silicon carbide (SiC) (see Table 3.18 and Figure 3.31)

TABLE 3.18 SEY of SiC Dielectric

Incident Electron Energy/eV	SEY	Positive Error	Negative Error
50	1.5800	0.0800	–0.0800
100	2.5300	0.0800	–0.0700
150	2.8800	0.0600	–0.0600
200	2.8000	0.0300	–0.0200
250	2.9800	0.0200	–0.0200
300	2.9800	0.0200	–0.0100
350	2.9200	0.0200	–0.0200
400	2.8800	0.0700	–0.0700
500	2.5900	0.0400	–0.0300
600	2.5300	0.0700	–0.0700
700	2.3100	0.0000	0.0000
800	2.1900	0.0200	–0.0200
900	1.9800	0.0000	0.0000
1,000	1.8900	0.0300	–0.0200
1,200	1.7300	0.0200	–0.0100
1,400	1.5800	0.0300	–0.0200
1,600	1.4500	0.0000	0.0000
1,800	1.3800	0.0200	–0.0200
2,000	1.2800	0.0200	–0.0100
2,200	1.2000	0.0000	0.0100
2,400	1.1200	0.0100	0.0000
2,600	1.1100	0.0100	0.0000
2,800	1.0500	0.0100	0.0000
3,000	1.0200	0.0000	0.0000
4,000	0.9000	0.0000	0.0000
5,000	0.8200	0.0000	0.0000

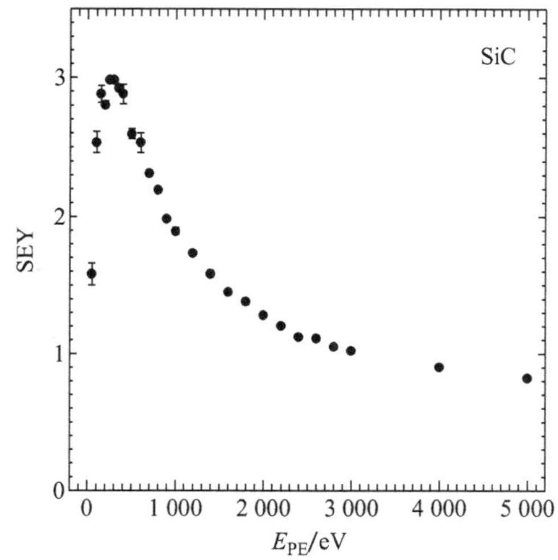

FIGURE 3.31 Variation of SEY of SiC dielectric with incident electron energy E_{PE}.

17. Boron nitride (BN) (see Table 3.19 and Figure 3.32)

TABLE 3.19 SEY of BN Dielectric

Incident Electron Energy/eV	SEY	Positive Error	Negative Error
50	0.6800	0.0300	−0.0200
100	1.2000	0.0100	−0.0100
150	1.7200	0.0300	−0.0200
200	2.1000	0.0600	−0.0600
250	2.3300	0.0300	−0.0200
300	2.5200	0.1400	−0.1300
350	2.5900	0.0500	−0.0500
400	2.5900	0.0200	−0.0200
500	2.6300	0.0100	−0.0100
600	2.4700	0.0000	0.0000
700	2.4400	0.1100	−0.1000
800	2.3900	0.0700	−0.0700
900	2.2900	0.0500	−0.0400
1,000	2.1800	0.0200	−0.0100
1,200	2.0200	0.0200	−0.0200
1,400	1.9100	0.0100	−0.0100
1,600	1.8500	0.0300	−0.0200
1,800	1.7700	0.0500	−0.0500
2,000	1.7300	0.0200	−0.0100
2,200	1.6800	0.0200	−0.0200
2,400	1.6000	0.0200	−0.0200
2,600	1.4900	0.0200	−0.0100
2,800	1.4400	0.0100	0.0000
3,000	1.4300	0.0200	−0.0100
4,000	1.3200	0.0200	−0.0100
5,000	1.2500	0.0100	−0.0100

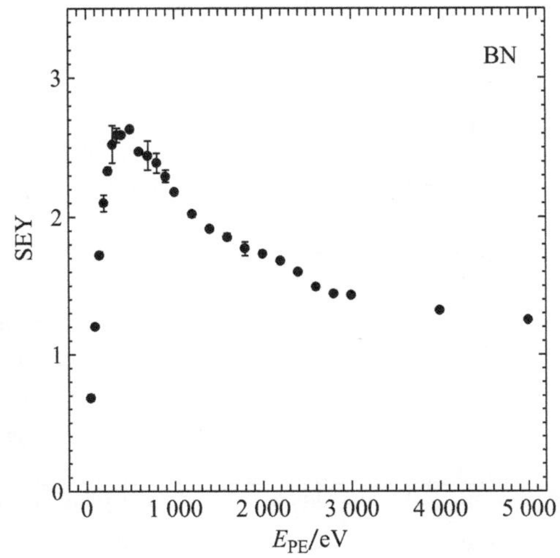

FIGURE 3.32 Variation of SEY of BN dielectric with incident electron energy E_{PE}.

18. Silicon nitride (Si_3N_4) (see Table 3.20 and Figure 3.33)

TABLE 3.20 SEY of Si_3N_4 Dielectric

Incident Electron Energy/eV	SEY	Positive Error	Negative Error
50	0.8500	0.1000	–0.0900
100	1.1100	0.0100	–0.0100
150	1.5600	0.0100	–0.0100
200	1.8800	0.0300	–0.0300
250	2.1000	0.0100	–0.0100
300	2.2500	0.0000	–0.0100
350	2.3100	0.0200	–0.0100
400	2.3700	0.0300	–0.0200
500	2.3900	0.0500	–0.0400
600	2.4000	0.0200	–0.0200
700	2.4100	0.0500	–0.0400
800	2.3300	0.0200	–0.0200
900	2.3100	0.0600	–0.0600
1,000	2.2200	0.0400	–0.0300
1,200	2.1700	0.0200	–0.0100
1,400	2.0100	0.0200	–0.0100
1,600	1.9300	0.0000	0.0000
1,800	1.8200	0.0200	–0.0100
2,000	1.7600	0.0500	–0.0500
2,200	1.7000	0.0000	0.0100
2,400	1.6400	0.0000	0.0100
2,600	1.5600	0.0300	–0.0200
2,800	1.5500	0.0500	–0.0500
3,000	0.7200	0.0300	–0.0200
4,000	1.4000	0.0400	–0.0300
5,000	1.3200	0.0300	0.0200

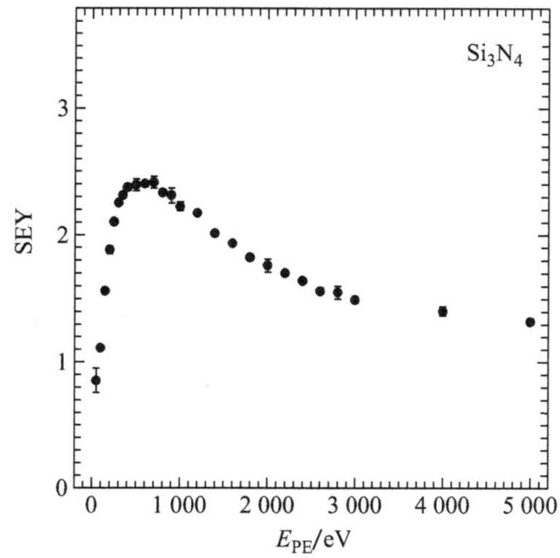

FIGURE 3.33 Variation of SEY of Si_3N_4 dielectric with incident electron energy E_{PE}.

19. Fe_3YO_6 (see Table 3.21 and Figure 3.34)

TABLE 3.21 SEY of Fe_3YO_6 Dielectric

Incident Electron Energy/eV	SEY	Positive Error	Negative Error
50	1.1509	0.0235	−0.0708
100	1.7358	0.0585	−0.1103
200	2.1843	0.0332	−0.0884
300	2.3791	0.0466	−0.0653
400	2.3620	0.1552	−0.1464
500	2.2452	0.0452	−0.0382
600	2.2105	0.0424	−0.0395
700	2.2130	0.0930	−0.0744
800	2.0583	0.0689	−0.0617
900	1.9111	0.0462	−0.0504
1,000	1.7834	0.0467	−0.0627
1,200	1.6923	0.0278	−0.0349
1,400	1.6092	0.0485	−0.0574
1,600	1.4589	0.0284	−0.0411
1,800	1.4277	0.0124	−0.0141
2,000	1.3957	0.0043	−0.0090
2,200	1.3351	0.0255	−0.0285
2,400	1.2937	0.0206	−0.0244
2,600	1.2512	0.0159	−0.0238
2,800	1.1994	0.0124	−0.0179
3,000	1.1699	0.0165	−0.0301
4,000	1.0182	0.0117	−0.0117
5,000	0.9632	0.0030	−0.0089

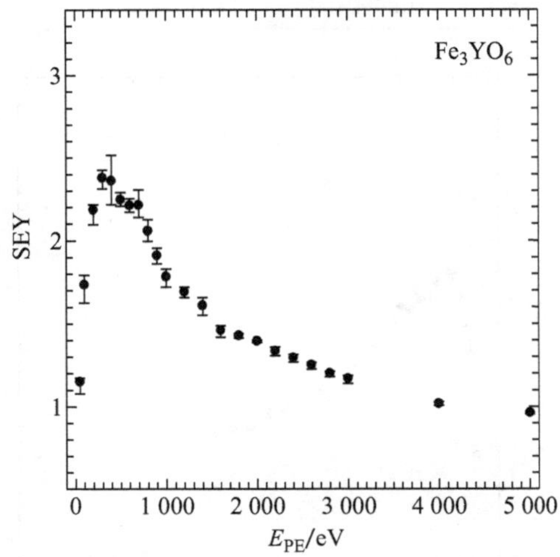

FIGURE 3.34 Variation of SEY of Fe_3YO_6 dielectric with incident electron energy E_{PE}.

20. Ceramic MST-16 (MgO-SiO_2-CaO-TiO_2) (see Table 3.22 and Figure 3.35)

TABLE 3.22 SEY of Ceramic MST-16 Dielectric

Incident Electron Energy/eV	SEY	Positive Error	Negative Error
50	1.2400	0.0400	−0.0400
100	1.8600	0.0200	−0.0300
150	2.4400	0.0900	−0.1000
200	2.7000	0.1000	−0.1000
250	3.0000	0.0500	−0.0500
300	3.2600	0.0600	−0.0700
350	3.2400	0.1000	−0.1100
400	3.2900	0.0500	−0.0600
500	3.2000	0.0000	−0.0100
600	3.0000	0.0000	0.0000
700	3.0100	0.0500	−0.0500
800	2.8900	0.0400	−0.0400
900	2.9200	0.0200	−0.0300
1,000	2.7400	0.0900	−0.1000
1,200	2.4400	0.0700	−0.0800
1,400	2.2600	0.0100	−0.0200
1,600	2.2300	0.0500	−0.0500
1,800	1.9700	0.0200	−0.0300
2,000	1.9200	0.0200	−0.0200
2,200	1.7400	0.0600	−0.0600
2,400	1.7100	0.0300	−0.0400
2,600	1.5700	0.0200	−0.0300
2,800	1.5400	0.0100	−0.0200
3,000	1.4800	0.0100	−0.0200
4,000	1.3200	0.0100	−0.0100
5,000	1.2400	0.0100	0.0000

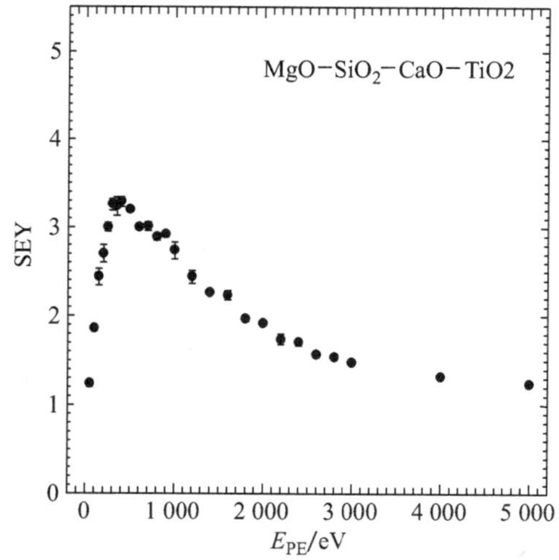

FIGURE 3.35 Variation of SEY of ceramic MST-16 dielectric with incident electron energy E_{PE}.

21. Ceramic MCT-21 (MgO-CaO-TiO_2) (see Table 3.23 and Figure 3.36)

TABLE 3.23 SEY of Ceramic MST-21 Dielectric

Incident Electron Energy/eV	SEY	Positive Error	Negative Error
50	1.3275	0.0525	−0.0525
100	1.9600	0.0300	−0.0300
150	2.5400	0.0300	−0.0150
200	2.8275	0.0075	−0.0075
250	2.9775	0.0275	−0.0275
300	3.0550	0.0300	−0.0300
350	3.1075	0.0525	0.0525
400	3.0225	0.0125	−0.0125
500	2.8650	0.0200	−0.0200
600	2.8025	0.0058	−0.0175
700	2.6250	0.0200	−0.0200
800	2.4225	0.0325	−0.0975
900	2.4000	0.0067	−0.0200
1,000	2.2400	0.0250	−0.0250
1,200	2.0175	0.0125	−0.0125
1,400	1.8475	0.0125	−0.0125
1,600	1.7300	0.0100	−0.0100
1,800	1.6100	0.0100	−0.0100
2,000	1.5450	0.0150	−0.0150
2,200	1.4925	0.0058	−0.0175
2,400	1.4175	0.0175	−0.0175
2,600	1.3500	0.0100	−0.0100
2,800	1.3000	0.0100	−0.0100
3,000	1.2525	0.0075	−0.0075
4,000	1.0700	0.0100	−0.0100
5,000	0.9825	0.0525	−0.0175

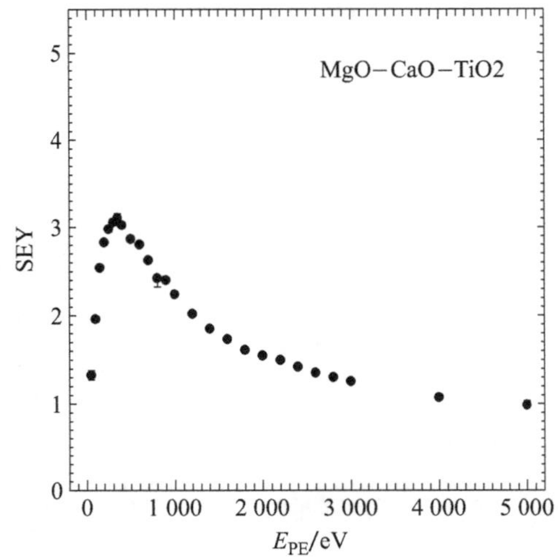

FIGURE 3.36 Variation of SEY of ceramic MST-21 dielectric with incident electron energy E_{PE}.

22. Ceramic ZNT-45 (Zr-Nb_2O_5-TiO_2) (see Table 3.24 and Figure 3.37)

TABLE 3.24 SEY of Ceramic ZNT-45 Dielectric

Incident Electron Energy/eV	SEY	Positive Error	Negative Error
50	1.4200	0.0000	0.0000
100	2.0400	0.2000	−0.2000
150	2.6200	0.0900	−0.1000
200	2.9600	0.0300	−0.0400
250	3.0600	0.1100	−0.1200
300	3.0100	0.0100	−0.0200
350	3.1200	0.0500	−0.0600
400	2.9800	0.1300	−0.1300
500	2.9700	0.0200	−0.0300
600	2.7800	0.0400	−0.0500
700	2.7600	0.0200	−0.0300
800	2.5700	0.0200	−0.0300
900	2.3700	0.0900	−0.1000
1,000	2.4100	0.0300	−0.0300
1,200	2.1500	0.0400	−0.0400
1,400	1.9900	0.0400	−0.0500
1,600	1.8700	0.0200	−0.0200
1,800	1.7100	0.0000	−0.0100
2,000	1.6500	0.0100	−0.0200
2,200	1.5700	0.0100	−0.0100
2,400	1.5100	0.0000	0.0000
2,600	1.4400	0.0000	0.0000
2,800	1.3600	0.0000	0.0000
3,000	1.3000	0.0000	0.0000
4,000	1.1800	0.0000	0.0000
5,000	1.1000	0.0000	0.0000

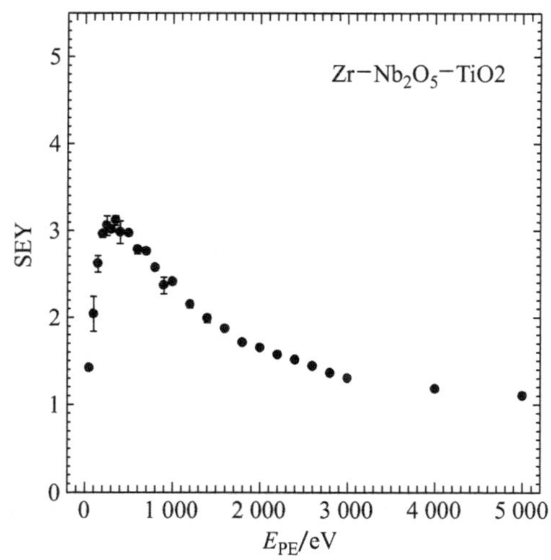

FIGURE 3.37 Variation of SEY of ceramic ZNT-45 dielectric with incident electron energy E_{PE}.

BIBLIOGRAPHY

1. N. Zhang, M. Cao, W. Z. Cui, et al. Lab-built test-stand for secondary electron emission of metals in ultra-high vacuum. *Chinese Journal of Vacuum Science and Technology*, 2014, 34(5):554–558.
2. F. Wang, D. J. Li, M. Weng, G. B. Feng, et al. Database of secondary electron yield and energy spectra for dielectric samples by electron irradiation. *Chinese Journal of Vacuum Science and Technology*, 2021, 41(12):1142–1156.
3. J. Yang, W. Z. Cui, T. C. Hu, et al. Research of secondary electron energy spectrum on pure silver material. *Journal of Terahertz Science and Electronic Information Technology*, 2014(1):141–144.
4. W. Z. Cui, J. Yang, N. Zhang. Testing method of the secondary electron emission yield of space metal materials. *Space Electronic Technology*, 2013, 10(2):75–78.
5. Y. He, J. Yang, G. Miao, et al. High-performance multifunctional apparatus for studying secondary electron emission characteristics of dielectric. *High Power Laser and Particle Beams*, 2020, 32: 033003.
6. D. Wang, Y. He, W. Cui. Secondary electron emission characteristics of TiN coatings produced by RF magnetron sputtering. *Journal of Applied Physics*, 2018, 124(5): 053301.
7. W. Cui, N. Zhang, G. Feng, et al. Research progress of secondary electron emission research and its space application. *Space Electronic Technology*, 2022, 19(4): 1–9.
8. Y. Zhang, J. Yang, Z. Yang, G. Miao, N. Zhang, W. Cui. The secondary electron emission characteristics of molybdenum. *Rare Metal Materials and Engineering*, 2024, 53(2): 465–473.
9. G. Miao, W. Cui, J. Yang, et al. Research and development of the measurement equipment of secondary electron emission. *Space Electronic Technology*, 2018, 15(1): 25–32, 41.
10. C.-J. Bai, G.-B. Feng, W.-Z. Cui, Y.-N. He, W. Zhang, S.-G. Hu, M. Ye, T.-C. Hu, G.-S. Huang, Q. Wang. Suppressing second electron yield based on porous anodic alumina. *Acta Physica Sinica*, 2018, 67(3): 037902.

CHAPTER 4

Surface Modification for Secondary Electron Emission

Wanzhao Cui, Guobao Feng, Yongning He, and Ming Ye

4.1 SUPPRESSING SECONDARY ELECTRON EMISSION BY SURFACE MICROSTRUCTURE

Multipactor is essentially caused by the multiplication of secondary electrons in the resonant field of the microwave cavity. As depicted in Figure 4.1, when the seed electrons in a cavity are accelerated to move to one side of the electrode plate under the action of an applied radio frequency electric field, the resulting secondary electrons will be accelerated to move to the other side of the electrode plate under the action of the next radio frequency inversion electric field, thereby generating more electrons, and so on repeatedly, and eventually produce a large number of electron clouds, the micro discharge phenomenon. Therefore, the essence of microdischarge research is the study of secondary electron emission characteristics, and the core of suppressing microdischarge effects is to suppress the multiplication process of secondary electrons in the microwave cavity. In aerospace engineering, the main method currently is to improve the threshold value of microdischarge by changing the structure of the microwave cavity and designing special sizes and operating frequencies for sensitive areas of micro discharge effects. However, on the one hand, this approach requires a large

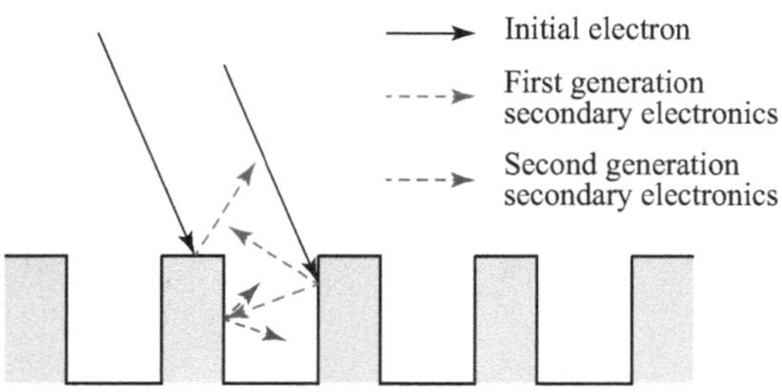

FIGURE 4.1 Mechanism of secondary electron emission suppression by microporous structure.

DOI: 10.1201/9781003562863-4

amount of time and cost to conduct a large number of experiments on each microwave component to obtain specific sensitive areas; on the other hand, changes in component size and operating frequency also restrict its operating range and performance.

In this chapter, several ways to suppress secondary electron emission from material surfaces by surface micro-trap structures are proposed to further suppress multipactor phenomena. The suppression properties of the surface are obtained by generating regular arrays of surface micro-traps on the simple configuration surface. In addition, the suppression law of secondary electron emission by trap structures and the influence characteristics on the multipactor quality factor are obtained by modeling the probability of secondary electron emission under surface micro-trap structures.

4.1.1 Basic Principle of Surface Configuration Suppressing Secondary Electrons

The basic principle of the surface configuration technology for suppressing secondary electrons mainly depends on the following two physical effects, namely, the oblique incidence effect, which means that the SEY increases with the increase of the initial electron incidence angle, and the shielding effect, which means that the emitted secondary electrons are shielded by materials near the emission location, resulting in a decrease of the SEY due to absorption. Figure 4.1 shows the basic principle of suppressing SEY with a microporous array structure: the secondary electrons excited by the initial electrons injected into the hole are suppressed due to the shielding effect of the hole sidewall, resulting in a decrease in SEY. It should be noted that when the scale of the surface configuration is in the nanoscale, the wave or quantum properties of the electrons may dominate, which is beyond the scope of this section.

4.1.2 Regular Slot Structure on the Sample Surface

In order to obtain metal surface micro-traps with a highly regular slot structure, a sample preparation method of silicon substrate photolithography and metal sputtering is used in this section. The preparation process is shown in Figure 4.2. First, the silicon substrate is subjected to surface polishing, cleaning, and other pretreatments. Then, the pre-treated sample is rotationally glued and baked. Next, the coated sample is exposed to ultraviolet radiation under the cover of a mask plate (the size of the mask plate is in the order of tens of microns). After baking again, the sample is developed using a positive adhesive developer, the developed sample is baked, and etched with an etchant. After removal of the photoresist, the etched sample was ultrasonically cleaned and dried. Finally, the silicon substrate with the surface microstructure was subjected to metal silver magnetron sputtering, with a sputtering thickness of about 300 nm. On the one hand, because the sputtering thickness is much smaller than the size structure, the impact on the regularity of the surface microstructure is very small; on the other hand, for surface secondary electron emission studies, inner secondary electrons below a depth of 300 nm will not emit again, which does not affect the secondary electron emission characteristics of metals. Figure 4.3 shows a laser confocal image of a sample with a cylindrical hole and rectangular slot surface microarray structure.

The etched and sputtered samples with surface micro-trap arrays are tested using a secondary electron emission platform. Figure 4.4 shows the structure diagram of the

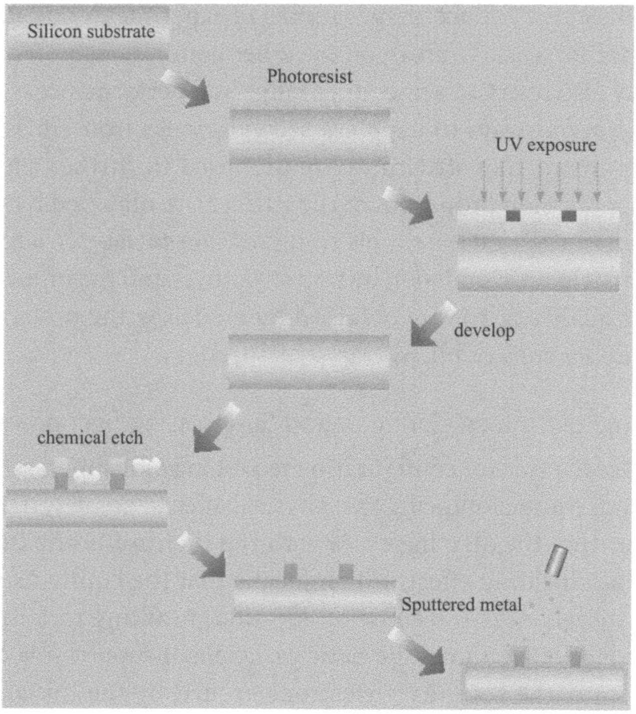

FIGURE 4.2 Schematic diagram of the surface configuration process.

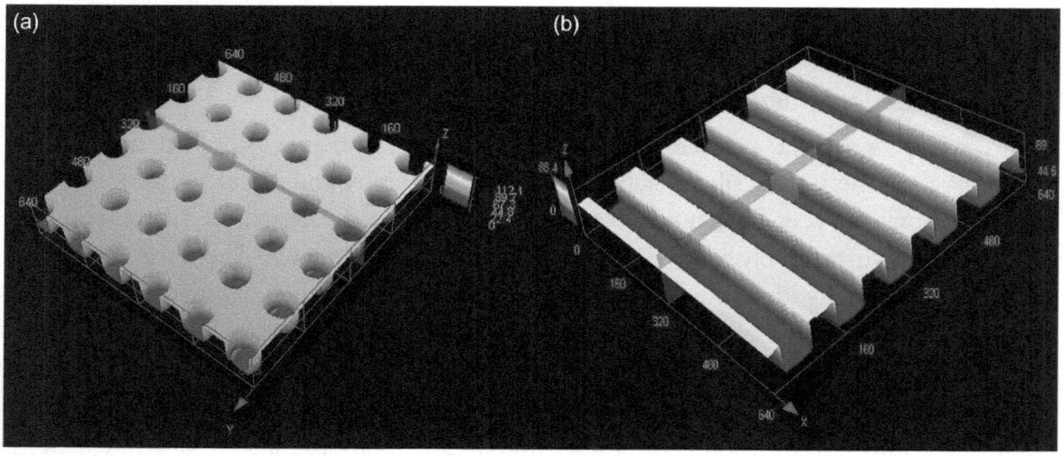

FIGURE 4.3 Laser confocal imaging for different surface configurations (a) round hole structure and (b) rectangular slot structure.

secondary electron emission platform used for testing and the physical object of the test sample location. The entire sample is placed in a high vacuum environment with a vacuum degree of 10^{-8} Pa, and simultaneously pumped and maintained by Gamma Vacuum's ion pump in conjunction with a titanium sublimation pump. The main electron gun of the platform is a coaxial grid-controlled electron gun with a model of EK5IK produced

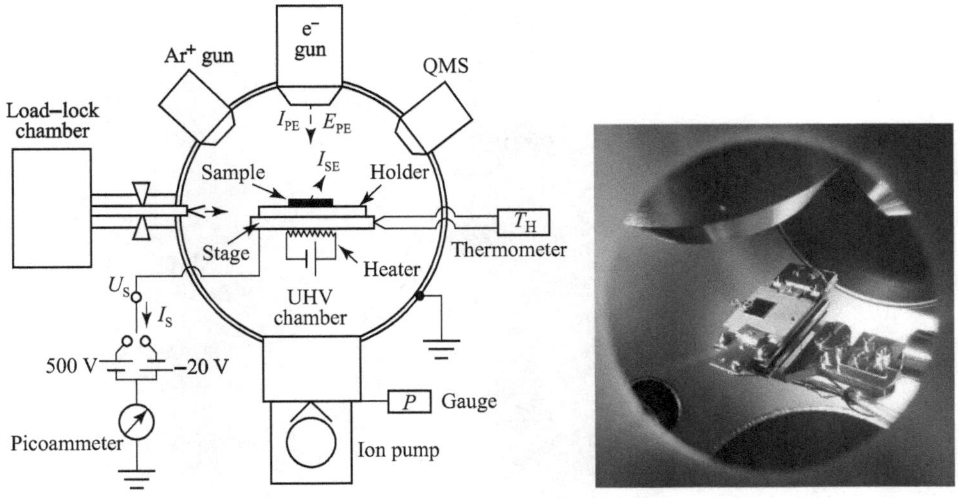

FIGURE 4.4 Schematic diagram of the secondary electronic test platform and the actual test part.

by Staib Instruments, Germany, with an energy range of 20–5,000 keV and a beam current range of 50 Pa to 2 nA. The main ammeter tested is a picoammeter, which is a 6487E ammeter produced by Keithley Instruments Inc. of the United States, with a current range of 20 fA to 20 mA and the measurement accuracy of 10 fA. The measurement of the secondary electron yield mainly adopts the current method.

First, a bias voltage of +500 V is applied to the sample to ensure that most of the incident electrons are irradiated on the sample surface, and the current measured by a picoammeter is the incident electron current, denoted as I_1. Then, a bias voltage of −20 V is applied to the sample to repel the outgoing secondary electrons from the sample surface, and the measured current is the sample current with the secondary electron current removed, denoted by I_2. Thus, the secondary electron current $I_{SE} = I_1 - I_2$ and the SEY can be expressed as $\delta = I_{SE}/I_1$.

Secondary electron emission is mainly a phenomenon that occurs when incident electrons interact with the material and exit from the surface of the sample. The schematic diagram of its principle is shown in Figure 4.5. The incident initial electrons will undergo elastic and inelastic collisions with the material's atomic nuclei and extranuclear electrons, in which the outer nuclear electrons are excited to generate inner secondary electrons during the inelastic collision process. The electrons emitted from the surface after a series of collisions should include primary electrons that bounce back directly from the surface (elastic backscattered electrons, primary electrons that have undergone inelastic collisions), inelastic backscattered electrons, and internally generated secondary electrons (intrinsic secondary electrons).

Therefore, these three types of electrons are mainly processed in numerical simulation. In this paper, a probabilistic model is used to simulate the probability of each of the three types of electrons at varying incident electron energies and angles, thereby enabling the estimation of the total SEY and energy spectrum.

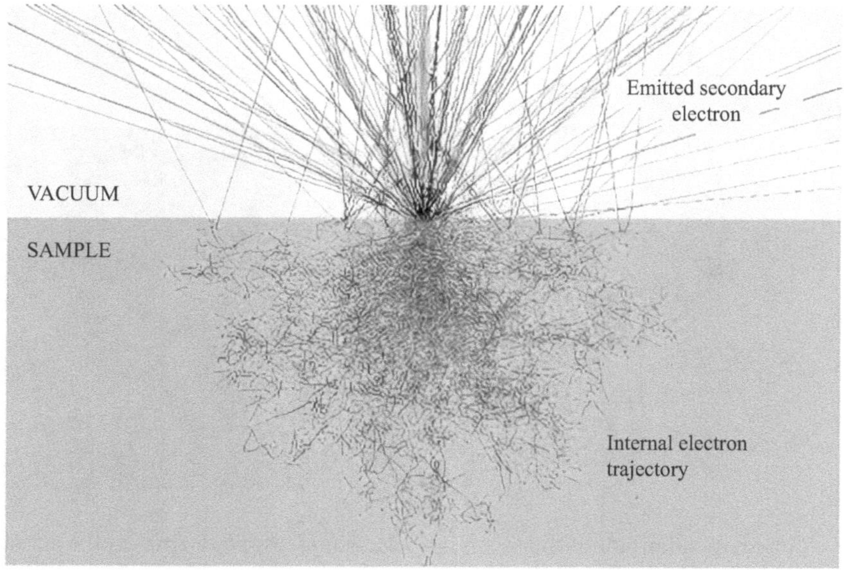

FIGURE 4.5 Schematic diagram of secondary electron emission principle.

For elastic backscattering, the probability can be expressed as

$$P_e(E_p,\theta_p)=\varepsilon(E_p,\theta_p)=\varepsilon(E_p)^{\cos\theta_p}C_2^{1-\cos\theta_p} \tag{4.1}$$

$$C_2=\chi\varepsilon(E_p)/(\varepsilon(E_p)+\eta(E_p)) \tag{4.2}$$

$$\varepsilon(E_p)=\varepsilon_1/(1+E_p/E_{e1})+\varepsilon_2/(1+E_p/E_{e2}), \tag{4.3}$$

where C_2 is the elastic backscattering correlation coefficient; $\varepsilon(E_p)$ is the homogenized elastic backscattering coefficient; $\varepsilon_1=\varepsilon_0-\varepsilon_2$, ε_0 and ε_2 are the elastic backscattering coefficient related to the initial electron incident energy; E_p is the initial electron incident energy; θ_p is the electron incidence angle.

The probability of inelastic backscattering is

$$P_i(E_p,\theta_p)=\eta(E_p,\theta_p)=\eta(E_p)^{\cos\theta_p}C_1^{1-\cos\theta_p} \tag{4.4}$$

$$C_1=\chi\eta(E_p)/(\varepsilon(E_p)+\eta(E_p)) \tag{4.5}$$

$$\eta(E_p)=a_1(1-b_1E_p)E_p^{\gamma}\exp\left(-(E_p/E_b)^{\mu}\right), \tag{4.6}$$

where C_1 is the inelastic backscattering correlation coefficient; $\eta(E_p)$ is the homogenized inelastic backscattering coefficient; $E_{b=}c_1+d_1Z$, Z is the atomic number, a_1 depends on the material properties.

For a single collision of an incident electron with a material atom, only one of elastic backscattering, inelastic backscattering or intrinsic secondary electron emission is emitted. Therefore, the probability of the remaining intrinsic secondary electron emission can be expressed as

$$P_s(E_p,\theta_p)=1-\varepsilon(E_p,\theta_p)-\eta(E_p,\theta_p) \tag{4.7}$$

After obtaining the probability of emission for each collision type, it is necessary to further determine the yield, energy spectrum, and angular distribution of the emitted secondary electrons. SEY can be divided into three types according to the type of emitted electrons: the intrinsic secondary electron yield δ, the backscattered secondary electron yield η and the total secondary electron output σ, and the calculation formulas are as follows:

$$\begin{aligned}
&\delta=\delta_m\frac{sE_p/E_m}{s-1+(E_p/E_m)^s}\frac{k+1}{k+\cos\theta_p};\\
&\eta=\left[\left(1-P_s(E_p,\theta_p)\right)/P_s(E_p,\theta_p)\right]\delta;\\
&\sigma=\delta/P_s(E_p,\theta_p),
\end{aligned} \tag{4.8}$$

where δ_m and E_m are the maximum yield and corresponding initial electron energy, respectively; s is a constant related to the material; θ_p is the incident angle; $k=pZ+r$; and r is a constant related to the surface state.

Next, consider the energy spectrum of the emitted secondary electrons, which are also classified into three types of electrons. Elastic backscattering electrons. The energy of the emitted elastic backscattering electrons can be directly approximated to the incident energy. Inelastic backscattering electrons. The energy of the emitted electrons can be expressed as

$$E_b=E_pG_b(u)=E_p\alpha^{-1/n_b}\left(\arccos(1-\beta u)\right)^{1/n_b}, \tag{4.9}$$

where ß=1−cosα; is distribution within the (0,1) interval.

Intrinsic secondary electrons. The energy of the emitted electrons can be expressed as

$$E_{se}=E_{re}G_s(u), \tag{4.10}$$

where $E_{re,i}$ is the elastic backscattering energy E_{re} when the ith intrinsic secondary electron is emitted, and considering that energy conservation is satisfied, the $i-1$th intrinsic secondary electron energy $E_{se,i-1}$ can be obtained by analogy to $E_{se,i-1}=E_{re,i-1}-E_{re,i}$. Here, the energy probability density function $G_s(u)$ in the above Equation can be expressed as

$$G_s(u)=\left(\frac{2}{\pi}\tan^{-1}\left(\left(\tan\left(\frac{\pi}{2}X_{cs}\right)\tan\left(\frac{\pi}{2}u\right)\right)^{1/2}\right)\right)^{1/n_s}, \tag{4.11}$$

TABLE 4.1 Secondary Electron Emission Model Parameter Settings

Parameter	n_b	X_{cb}	n_s	A	B	δ_m	E_{m0}	r_1	S
Value	1.5	0.9	0.51	4.0	5.8	1.4	300	0.5	1.46
Parameter	r	χ	ε_2	g	h	γ	μ	Z	P
Value	2	0.89	0.07	50	0.25	0.56	0.70	47	0.0027

where

$$X_{cs} = X_c/(0.9+1.1X_c), \tag{4.12}$$

$$X_c = 4\left(B-e^{-E_{tr}/A}\right)/E_{tr}, \tag{4.13}$$

$$E_{tr} = \begin{cases} E_{re}, & E_{re} > 1\,\text{eV} \\ 1, & E_{re} \le 1\,\text{eV} \end{cases}. \tag{4.14}$$

For the distribution of the electron exit angle, the angle between the emitted electron and the surface normal direction in this model follows a cosine distribution.

The material mainly targeted in this section is metal Ag, and the corresponding secondary electron emission model parameter values are also shown in Table 4.1.

Based on the model of secondary electron emission, the total SEY ultimately emitted from the sample surface is obtained by simulating different surface structures under certain incident electron conditions (Figure 4.6).

The simulation process is as follows:

1. Initialize the parameters related to the initial incident electrons, including the incident electron energy, angle, and position. In this simulation, the incident electron energy ranges from 20 to 4,000 eV, the incident angle is vertical, and the incident electrons are uniformly irradiated on the sample surface.

2. Track the trajectory of the incident electrons. The energy of the initial electron reaching the surface of the sample, the angle of incidence, and the spatial position coordinates are obtained.

3. Simulate the secondary electron emission process. The number, energy, and angle of secondary electrons emitted from the surface are obtained by calculating the interaction between the incident electrons and the material using the probabilistic model described above.

4. Determine whether electrons can escape from the surface without re-incidence based on the secondary electron parameters emitted from the surface. If so, record the emission data and carry out the next electron incidence; otherwise, continue to track the electron trajectory for re-incidence simulation.

5. After a single electronic tracking is completed, the next electronic tracking is performed until all electronic tracking is completed, and finally, data is stored and output.

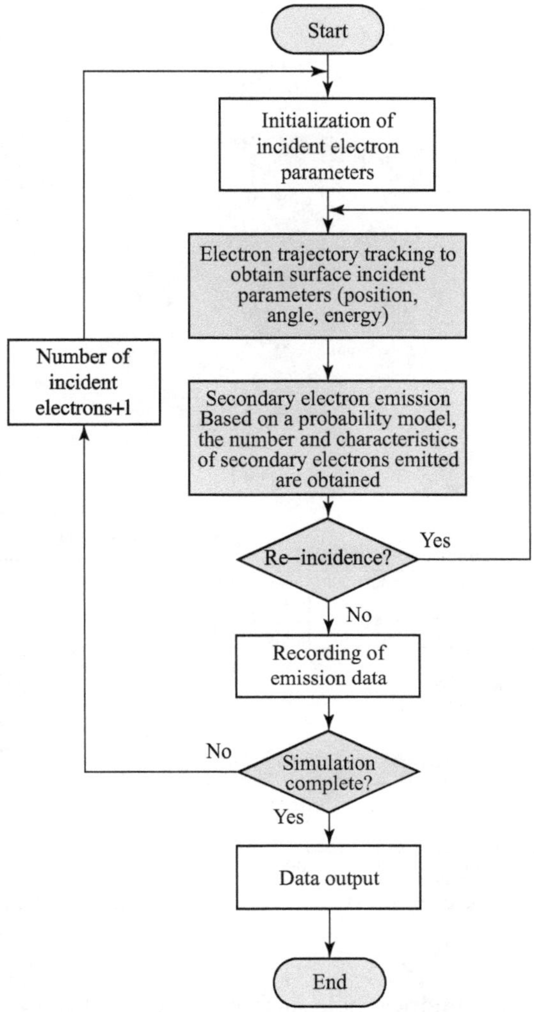

FIGURE 4.6 Block diagram of the numerical simulation flow of secondary electron emission.

By measuring the SEY of different surface micro-trap structures (here mainly cylindrical hole and rectangular slot structures), and combining them with numerical simulations, the influence of the surface structure on SEY was analyzed. The metal Ag material used on the surface here is a commonly used surface material in aerospace microwave components.

Figure 4.7 shows the experimental measurement and numerical simulation results of the SEY curve of the cylindrical hole surface structure. Figure 4.7a shows a schematic diagram of a series of collision processes in which incident electrons are driven into the interior of a cylindrical hole to be emitted. Some of the secondary electrons emitted from the bottom of the sample exit directly, while the other part will hit the sidewall for re-incidence and re-emission, and eventually emit from the trap or be absorbed by the sample.

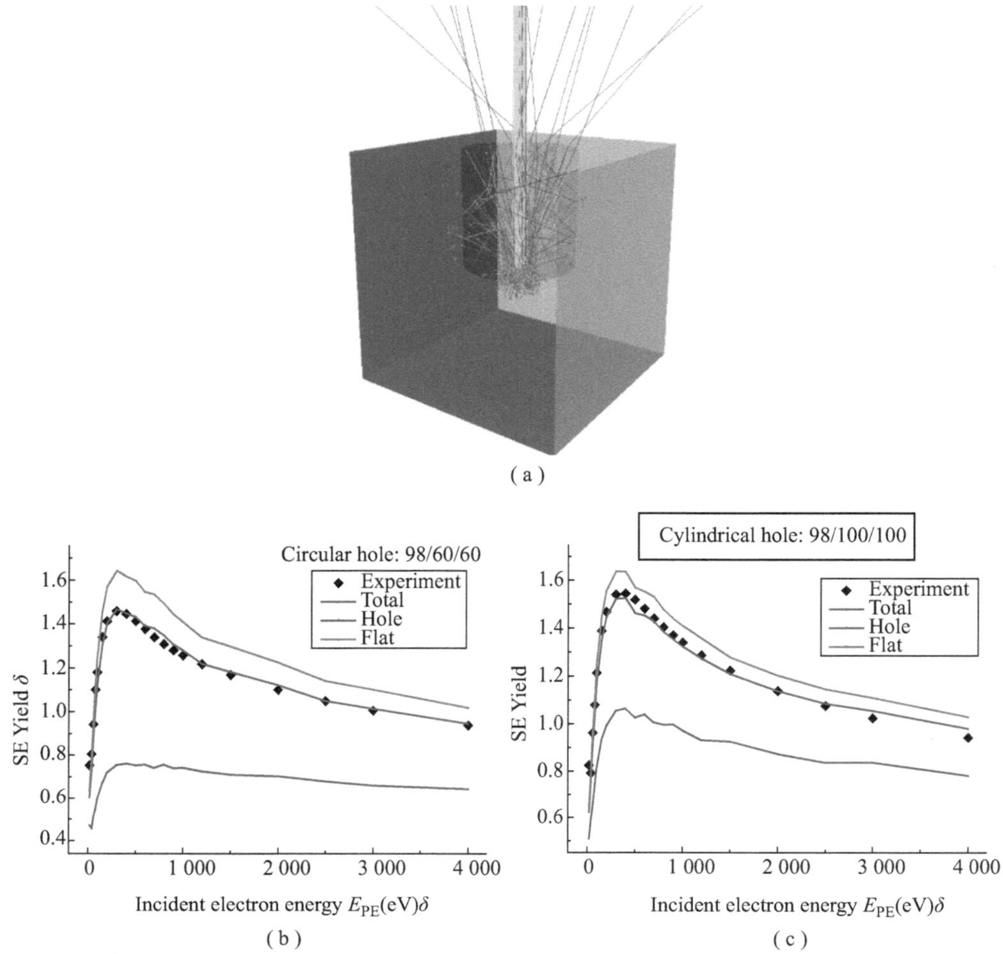

FIGURE 4.7 SEY curve for cylindrical hole surface structure (a) shows a schematic diagram of a series of collision processes emitted when incident electrons enter a cylindrical hole. (b) displays a cylindrical trap with depth, hole diameter, and hole edge spacing of 98µm, 60m, and 60pm respectively. The solid line represents the numerical simulation results of the secondary electron yield curve, while black dots indicate experimental measurements. (c) illustrates a cylindrical trap with depth, hole diameter, and hole edge spacing of 98m, 100wm, and 100wm respectively. The solid line shows the numerically simulated secondary electron yield curve, and black dots represent experimental measurement results.

The solid lines in Figure 4.7b and c show the cylindrical trap depth, hole diameter, and hole edge spacing of 98/60/60(µm), and 98/100/100(µm), respectively, and the solid lines in Figure 4.7b and c show SEY curves from numerical simulations, while the black dots indicate the results from experimental measurements. The green and blue solid lines in Figure 4.7b and c indicate the SEYs emitted from the non-trapped and trapped structure regions, respectively, while the red solid line indicates the overall total SEY. It can be seen from the figure that the simulated value of the total SEY curve is in good agreement with the actual measured value, which can also check the numerical simulation and experiment against each other. The overall trend shows that the SEY in the trap region is significantly reduced and effectively suppressed. Due to the limitation of the trap structure

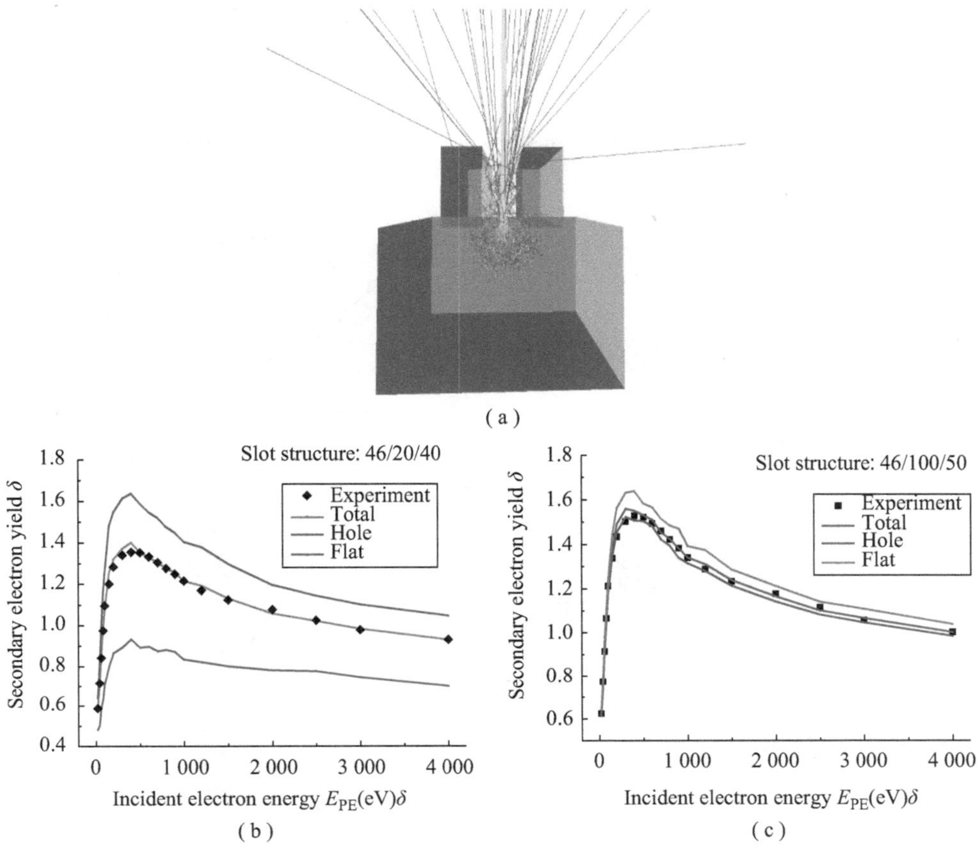

FIGURE 4.8 SEY curve for rectangular slot surface structure (a) presents a schematic diagram of electron trajectories within a groove structure. (b) shows experimental measurements and numerical simulation results of the secondary electron yield curve for a trap structure with depth, width, and spacing of 46wm, 20pm, and 40mm respectively. (c) demonstrates experimental measurements and numerical simulation results of the secondary electron yield curve for a trap structure with depth, width, and spacing of 46wm, 100mm, and 50μm respectively.

area, the total SEY is reduced to a lesser extent. In addition, the trap structure with larger depth-to-width ratio suppresses the electron emission inside the trap more strongly.

Figure 4.8 shows the experimental test and numerical simulation results of secondary electron suppression with a rectangular slot trap structure. Figure 4.8a shows a schematic diagram of the electron trajectories in a slot structure. Secondary electrons generated in the trap structure partially escape from the surface after multiple re-incidence processes with the sidewall. Experimental measurements and numerical simulations of the secondary electron yield curves for the cases of 46/20/40 and 46/100/50 for the depth, width and spacing of the trap structure are shown in Figure 4.8b and c, respectively, from which it can be found that similar to the cylindrical hole trap structure, the rectangular slot structure can also suppress the secondary electron emissions, but not as strong as the cylindrical hole, and when the aspect ratio decreases, such as the depth-to-width ratio of 46/100 shown in Figure 4.8c, the suppression of the inner secondary electrons decreases significantly, and the suppression effect on the overall secondary electrons is more limited.

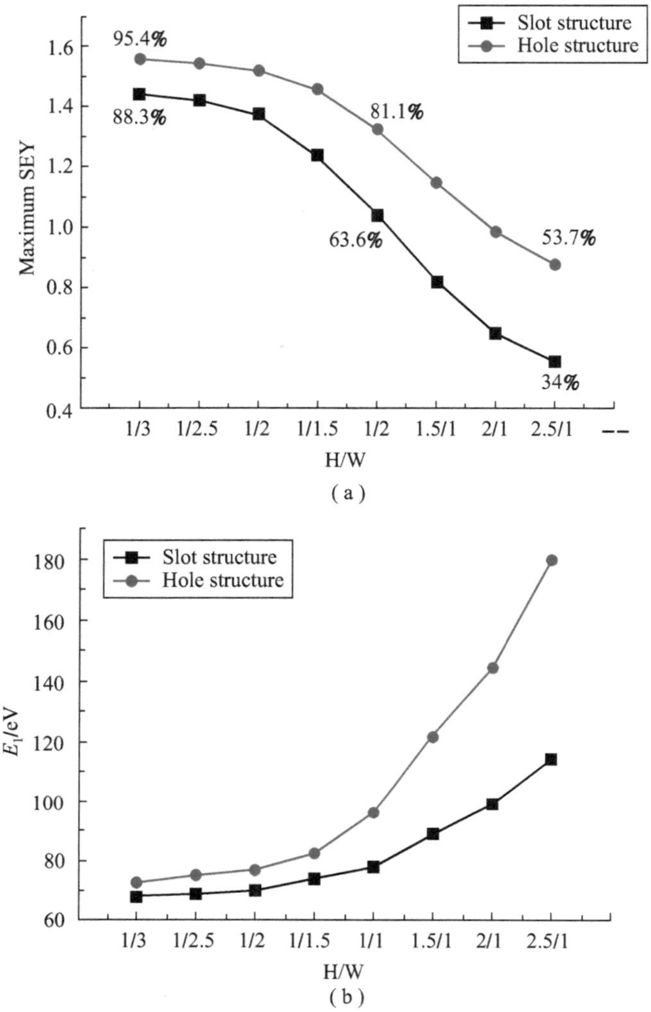

FIGURE 4.9 Secondary electron emission for different surface structure depth-to-width ratio H/W (a) maximum yield and (b) first cross-energy E_1.

Considering the complexity of sample preparation for continuous-size structures, the following numerical simulations are used to study the suppression characteristics of secondary electrons by different depth-to-width ratio trap structures.

Figure 4.9 shows the relationship between the maximum SEY and the first crossover energy E_1 of the SEY curve in the cylindrical hole trap and rectangular slot trap structures for different depth-to-width ratios, where E_1 is the incident electron energy corresponding to SEY=1 in the SEY curve. From the figure, it can be found that the SEY in the region of both cylindrical hole and rectangular slot trap structures tends to decrease with the increase of the depth-to-width ratio H/W. As the depth-to-width ratio increases from 1/3 to 2.5/1, the maximum SEY in the cylindrical hole decreases from 95.4% to 53.7% of the maximum SEY in the planar structure, while that in the surface of the rectangular slot structure decreases from 88% to 34%. This indicates that the cylindrical hole structure can suppress the secondary electron emission more effectively and corresponds to a larger first crossover energy.

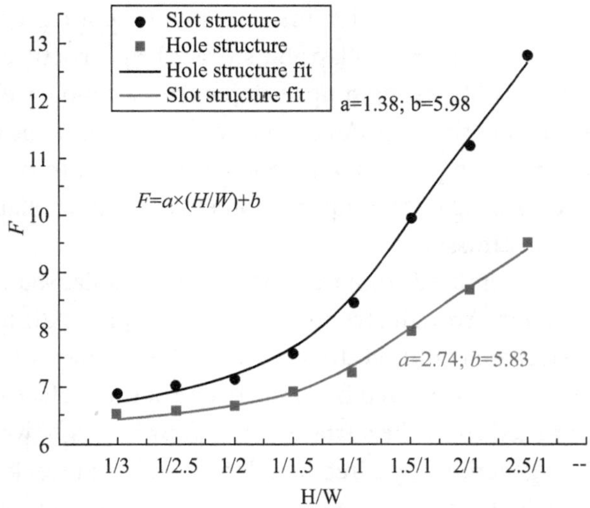

FIGURE 4.10 Multipactor quality factor *F* for different surface structure depth-to-width ratio *H*/*W*.

For aerospace microwave components, the maximum yield δ_{max} of the SEY curve and the first crossover energy E_1 are directly related to the ease of the multipactor phenomenon generation. The multipactor quality factor *F* can be used to measure the ease of multipactor of the material:

$$F = \sqrt{\frac{E_1}{\delta_{max}}} \tag{4.15}$$

Figure 4.10 shows the multipactor quality factors *F* for two trap types with different trap structure depth-to-width ratios *H*/*W* of the metal Ag. To fit the actual situation and to highlight the effect of the aspect ratio, the two trap structures are set to have the same trap area occupancy ratio. From the figure, it can be found that the increase of the trap structure depth-to-width ratio will improve the multipactor quality factor *F*. And by fitting, it is found that the multipactor quality factor *F* and the trap structure depth-to-width ratio H/W coincide well with the linear relationship in a certain range for a certain trap area occupancy ratio:

$$F = a \times (H/W) + b, \tag{4.16}$$

where a, b are two parameters related to the slot structure, $a = 1.38$ and $b = 5.98$ for the cylindrical hole structure and $a = 2.74$ and $b = 5.83$ for the rectangular slot structure. Comparing the two trap structures, it is found that the cylindrical hole trap structure has better suppression effect on multipactor and the quality factor improves more significantly as the depth-to-width ratio increases.

4.1.3 Collapsible Surface Trap Structure

The prevailing surface treatment methods include the use of triangular slot, rectangular slot or random surface and other open rough structure to a certain extent to suppress the secondary electron emission, but because this kind of suppression is mainly the use of rough structure sidewalls on the secondary electron emission shielding effect, its biggest problem comes from this shielding structure may also cause the secondary electron re-incidence cascade multiplier, so the suppression of secondary electron emission effect are limited.

In this section, based on the suppression mechanism of the surface structure on the secondary electron emission trajectory, a collapsible surface trap structure is further proposed, which can perform a collapsible strong suppression of the secondary electron re-incidence cascade multiplication inside the trap. A composite phenomenological probability model is used to describe the emission process of secondary electrons. In this section, numerical simulations are used to investigate the suppression effect of this collapsible trap structure on the secondary electron emission.

The secondary electrons emitted from the surface with a collapsed micro-trap structure (this section takes the gourd trough structure as an example) mainly contain two parts: one is the secondary electrons emitted directly from the flat part of the surface, and the other is the secondary electrons emitted from the inside of the surface trap. To simulate the secondary electrons emitted from the surface of the gourd trough, we need to first obtain the secondary electrons generated by direct incidence, and then track the trajectory of the secondary electrons inside the gourd trough, and after several times of secondary electron emission, the final secondary electrons that can be emitted from the surface are those obtained. In the simulation process of this section, the secondary electron generation is described by the phenomenological probability model.

For the convenience of calculation, the directions of the secondary electron emission are set as follows: the directions of both elastic and inelastic backscattering are determined according to the law of specular reflection; the emission angle of the intrinsic secondary electron follows the cosine distribution, and the azimuth angle follows the uniform distribution. Figure 4.11 shows the results of the SEY variation law with energy on the silver-plated aluminum alloy surface fitted by the above-mentioned phenomenological probability calculation method, and the calculated results are in good agreement with the measured results, which proves the feasibility of applying the phenomenological probability calculation method to the simulation of secondary electrons on the surface structure of the gourd trough.

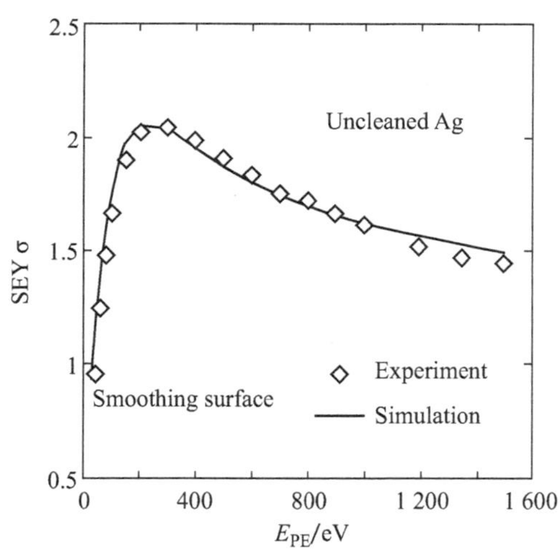

FIGURE 4.11 Comparison of experimental and simulated values of Ag secondary electron emission coefficient vs. incident energy curve.

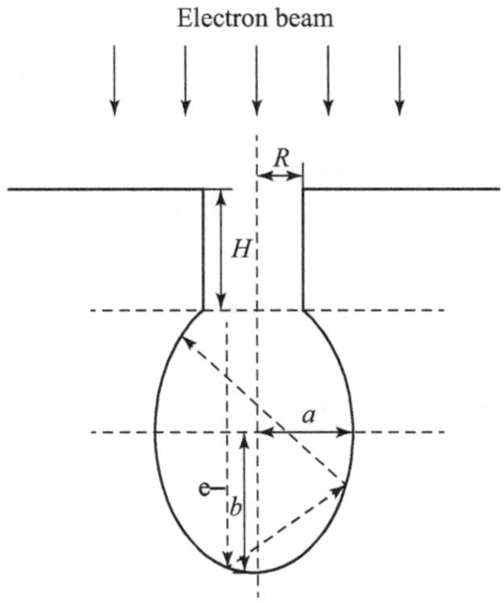

FIGURE 4.12 Schematic diagram of the surface section structure of the gourd trough.

Considering that the incident electrons generate multiple generations of secondary electrons inside the micro-surface structure, such multiple generations of secondary electrons increase the secondary electron yield to a certain extent. Therefore, in this section, a gourd trough structure, which contributes to the "collapsible" effect of the inner secondary electron, is used to suppress the secondary electron production in the slot structure. The collapsible gourd trough structure in Figure 4.12 consists of two parts: the first is the upper rectangular structure with a rectangular half-width of R and a height of H; the other is the lower elliptical structure with a horizontal semi-axis of a (Here, the horizontal semi-axis is required to be longer than the rectangular half-width, i.e., $a \geq R$) and a vertical semi-axis of b. The surface micro-slot structures studied in this section are on the micron scale and do not produce penetration of the surface structures by secondary electrons in the range of incident electron energies studied in this book.

Since the surface is a micro gourd trough array, the gourd trough structures are all tightly aligned on the surface to reduce the emission of secondary electrons from the flat surface. Therefore, the total secondary electron yield δ_T emitted from the surface can be found by using the weighting of the porosity $P = R/a$, i.e.

$$\delta_T = P\delta_{calabash} + (1-P)\delta_{flat}, \tag{4.17}$$

where δ_{flat} is the SEY emitted from the flat surface, which can be obtained directly from the above secondary electron phenomenological probability model; $\delta_{calabash}$ is the SEY emitted from the surface of the gourd trough structure, which needs to be obtained by simulating the multi-generational secondary electron generation process inside the slot structure.

4.1.3.1 The Results and Analysis

The surface structure simulated in this section is constructed on the surface of the unwashed metallic Ag sample in the same state as that of Ag in Figure 4.11, where the work function of Ag is 4.26 eV. In order to suppress the SEY to the maximum extent, the gourd trough structure of the surface in this book's model is all closely aligned. The secondary electron emission characteristics under this surface structure and the effects of the surface structure parameters on the secondary electron and multipactor characteristic parameters are simulated and analyzed.

4.1.3.1.1 Secondary Electron Emission Characteristics Figure 4.13 shows the yield curves of secondary electrons incident vertically on the surface of the structure with gourd troughs at different places, where the rectangular half-width $K=30\,\mu m$, the rectangular height $H=4\,\mu m$, the elliptical horizontal half-axis length $a=40\,\mu m$, the vertical half-axis length $b=60\,\mu m$ for the gourd trough structure, and the porosity $P=0.75$ under this surface structure.

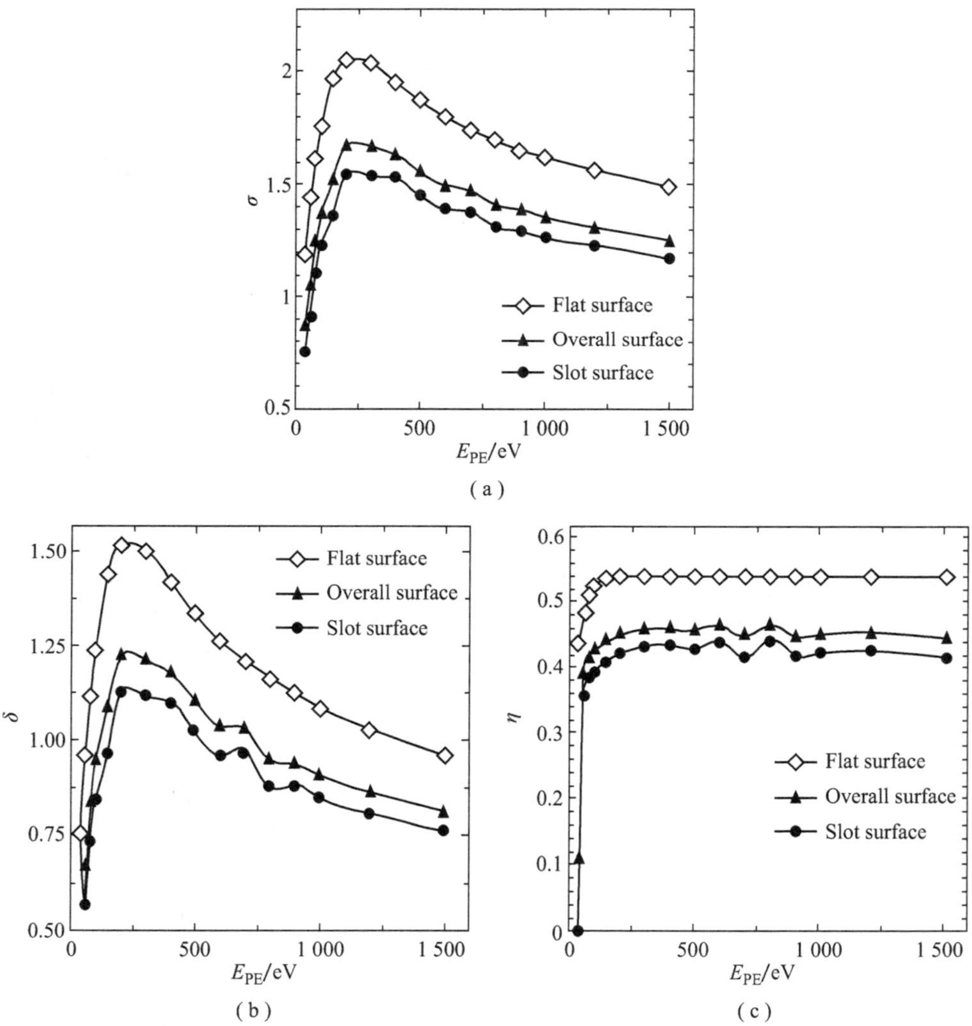

FIGURE 4.13 Effect of surface gourd trough structure on secondary electron emission coefficient (a) total secondary electron coefficient; (b) true secondary electron coefficient; and (c) backscattered secondary electron coefficient.

From Figure 4.13b and c, it can be found that the SEY curves of both true secondary electrons and backscattering secondary electrons show an overall decrease at the surface of the gourd trough structure: the corresponding TSEY (True Secondary Electron Yield) and BSEY (Back Secondary Electron Yield) peaks decreased from about 1.5 and 0.54 to about 1.1 and 0.42, respectively. This results in a corresponding overall decrease in the total SEY curve (Figure 4.13a), with a 25.5% and 18.6% decrease in the peak yields on the gourd trough surface and the whole surface, respectively.

4.1.3.1.2 Effect on Secondary Electron Energy Spectrum and Angle The secondary electrons are emitted from the surface of the gourd trough structure, and the energy spectrum of the secondary electrons finally emitted from the surface may also change to some extent due to the possibility of multiple re-incidence and re-emission. Figure 4.14 shows the comparison of the secondary electron energy spectrum from the surface of the gourd trough structure with that of the flat surface, where the rectangular half-width of the gourd trough structure $R = 15\,\mu m$, the rectangular height $H = 2\,\mu m$, the horizontal half-axis of the ellipse $a = 20\,\mu m$, and the vertical half-axis length $b = 50\,\mu m$. It can be found that the surface of the gourd trough structure causes a certain amount of the secondary electron energy spectrum to be concentrated in the peak region, and the overall energy spectrum shows a narrowing trend. This is mainly because the multiple incidences in the trap reduce the average incidence energy, resulting in the narrowing of the generated multi-generational secondary electron spectrum.

Figure 4.15 shows the SEY emitted at different slot locations (a) and the SEY curve emitted at the slot structure at different incident angles (b). The position-normalized curves are shown in Figure 4.15a, from which it can be found that the yield curves are basically symmetric along the center, and the "collapse" suppression effect of the secondary electrons is best at both sides of the center. From Figure 4.15b, it can be found that the SEY curve increases to a certain extent with the increase of the incident angle, and the incident energy corresponding to the peak of the yield shows a slight tendency to move to the right.

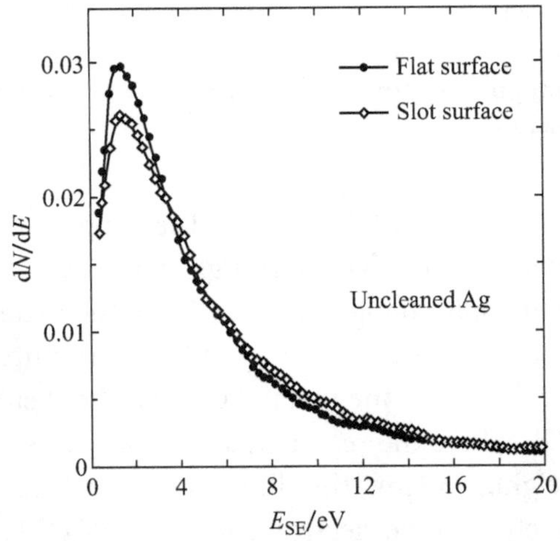

FIGURE 4.14 Effect of surface structure on the true secondary electron energy spectrum.

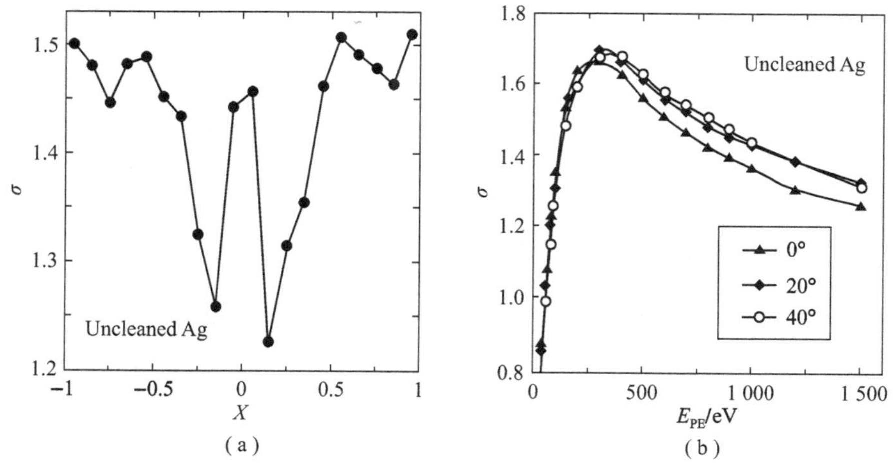

FIGURE 4.15 Effect of incident position and angle on total secondary electron emission (a) at different slot positions and (b) at different incident electron angles.

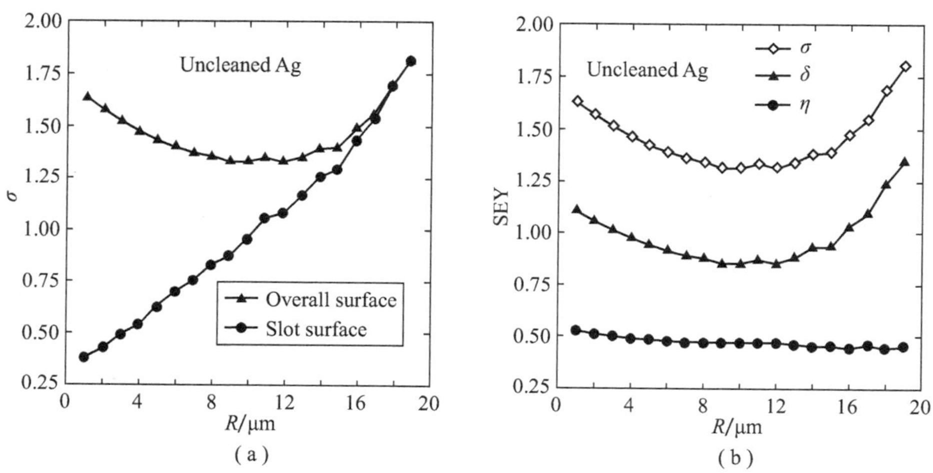

FIGURE 4.16 Effect of rectangular width on secondary electron emission (a) total secondary electrons and (b) comparison of outgoing SEYs.

4.1.3.1.3 Influence of Surface Structure Parameters The suppression of secondary electron emission by the surface with different gourd trough structures may also vary because different structural parameters may change the porosity of the material surface or affect the "collapse" suppression of secondary electrons by the slot structure.

Figure 4.16 shows the SEY out of the metal Ag (unwashed) surface for different gourd trough rectangle widths R, where the rectangular half-width is taken to range from 1 to 19 μm, the rectangular height $H = 2$ μm, the ellipse horizontal and vertical half-axis length of 20 and 30 μm, respectively, and the electron energy of 800 eV for vertical incidence.

From Figure 4.16a, it can be found that although the SEY emitted from the surface of the slot structure gradually increases with the increase of the rectangular half-width R, the

SEY emitted from the overall surface first decreases and then increases and finally coincides with the yield from the surface of the slot structure. This is because, on the one hand, the increase of the rectangular half-width directly weakens the suppression effect of the slot structure on the secondary electrons, as shown by the triangular line in Figure 4.16a, and the yield increases significantly with the increase of R; on the other hand, the increase of the rectangular half-width R also increases the porosity P of the surface, which reduces the proportion of the incident electrons into the trap structure, so that the overall SEY first decreases and then increases, and finally approaches the yield emitted from the slot surface. Figure 4.16b shows the comparison of the three overall yield curves (including total SEY σ, TSEY δ, and BSEY η) at different rectangular half-widths. It can be found that the increase of the rectangular half-width causes a slight decrease in the BSEY, which is mainly reflected in the effect on the TSEY.

In the surface structure of the gourd trough, the lower elliptical structure mainly plays a role in the collapse effect of the secondary electrons of multiple generations, so the elliptical structure parameters affect the secondary electron suppression effect to some extent. In Figure 4.17a, the material is Ag, the vertical half-axis length $b = 2 \sim 70\,\mu m$, the elliptical horizontal half-axis length $a = 20\,\mu m$, the rectangular half-width $R = 10\,\mu m$, the rectangular height $H = 2\,\mu m$, and the vertical incident electron energy $E_{PE} = 300\,eV$.

From Figure 4.17a, it can be found that the total SEY decreases with the increase of the vertical semi-axis length, and can be roughly divided into two phases. In the first phase, the total SEY shows a rapid decrease in the process of gradually increasing the vertical semi-axis length b to the horizontal semi-axis length $a = 20\,\mu m$; in the second phase, when the vertical semi-axis length b is larger than the horizontal semi-axis length and continues to grow, the decrease in the total SEY is no longer significant, which indicates that when $b < a$ in the elliptical structure, the effect on the secondary electron confinement is significant; and when $b > a$, the effect of increasing depth on the secondary electrons in the trap is weakened, so that the suppression effect on the secondary electron emission

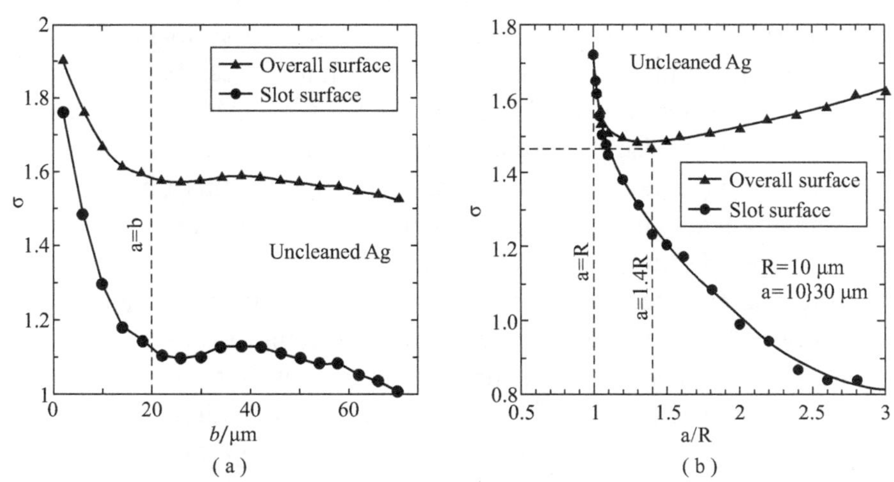

FIGURE 4.17 Effect of ellipse size on total SEY (a) the length of the vertical semi-axis of the ellipse and (b) the length of the horizontal semi-axis of the ellipse.

becomes stronger and slower, and the "collapse" suppression effect on the secondary electrons becomes less obvious when the ellipse is further stretched in the vertical direction.

The material in Figure 4.17b is Ag, with the ellipse horizontal semi-axis length $a = 10–30\,\mu m$, the vertical semi-axis length $b = 30\,\mu m$, the rectangular half-width $R = 10\,\mu m$, the rectangular height $H = 10\,\mu m$, and the incident electron energy $E_{PE} = 300\,eV$ for vertical incidence. Although the SEY emitted from the surface of the slot structure (black line in the figure) decreases continuously as the horizontal semi-axis length a of the ellipse increases, the overall SEY emitted shows a trend of first decreasing and then increasing. This is because, on the one hand, the increase of the horizontal semi-axis length strengthens the 'collapse' suppression effect of the secondary electrons inside the slot structure; on the other hand, the increase of the horizontal semi-axis length also increases the proportion of the flat surface and reduces the porosity, which in turn weakens the suppression effect of the overall surface. It is worth noting that when the horizontal semi-axis length is close to the rectangular half-width ($a \approx R$), this surface structure changes similarly to the rectangular surface structure (considering that the bottom of the actual rectangular slot structure cannot be flat), and the suppression effect on secondary electrons is not the best at this time, but the best when $a/R \approx 1.4$. Therefore, it can also be further shown that the surface of the gourd trough structure has a better suppression effect on secondary electron emission than the surface of the rectangular slot structure, and here, its suppression effect is also 14.5% higher than that of the ideal rectangular slot structure with the porosity tending to 1. As shown in Figure 4.17, compared with the rectangular slot structure with the same depth-to-width ratio and porosity, the gourd trough structure can improve the suppression effect of SEY by 21.2%.

The elliptical part of the gourd trough structure has a 'collapse' effect on the electrons inside the trap, while the height of the rectangular part affects the difficulty of escaping from the trap.

The structural parameters of the metal Ag slot in Figure 4.18 are of ellipse horizontal half-axis length $a = 14\,\mu m$, vertical half-axis length $b = 30\,\mu m$, rectangular half-width $R = 10\,\mu m$, and the incident electrons are still vertically incident.

The first critical energy E_1 in Figure 4.19a is the first incident energy value $E_1 = E|_{\sigma=1,\,\text{first}}$ corresponding to a SEY of 1. From the figure, it can be found that the first critical energy E_1 increases correspondingly as the rectangular height H increases, which means that the SEY curve is shifted to the right side accordingly. The maximum SEY value SEY_{max} decreases with the increase of the rectangular height H. This is because the increase of the rectangular height lengthens the escape path of the secondary electrons in the slot structure and suppresses the emission of secondary electrons.

Considering that the secondary electron emission characteristics directly affect the multipactor threshold, for example, the larger the value of the secondary electron emission coefficient σ, the lower the multipactor threshold of microwave devices is usually, and the larger the first critical energy E_1 of the secondary electron emission coefficient curve, the higher the threshold is usually. The quality factor F of the multipactor threshold defined by the secondary electron emission characteristics of the material is

$$F = \left(E_1 / \sigma_{max}\right)^{1/2} \tag{4.18}$$

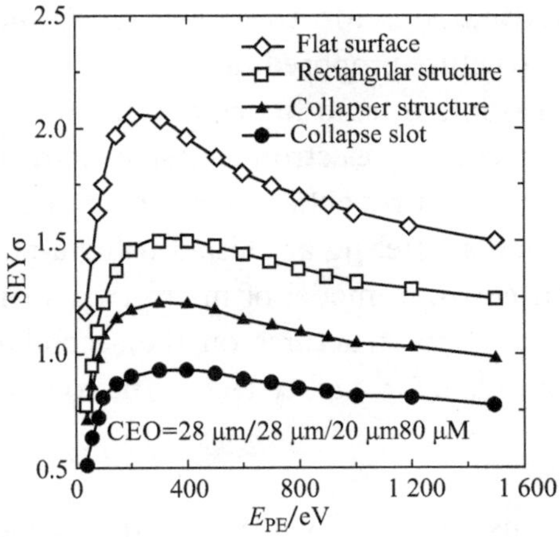

FIGURE 4.18 Comparison of suppression on secondary electron emission between collapse structure and rectangular slot structure (with same depth-to-width ratio and porosity).

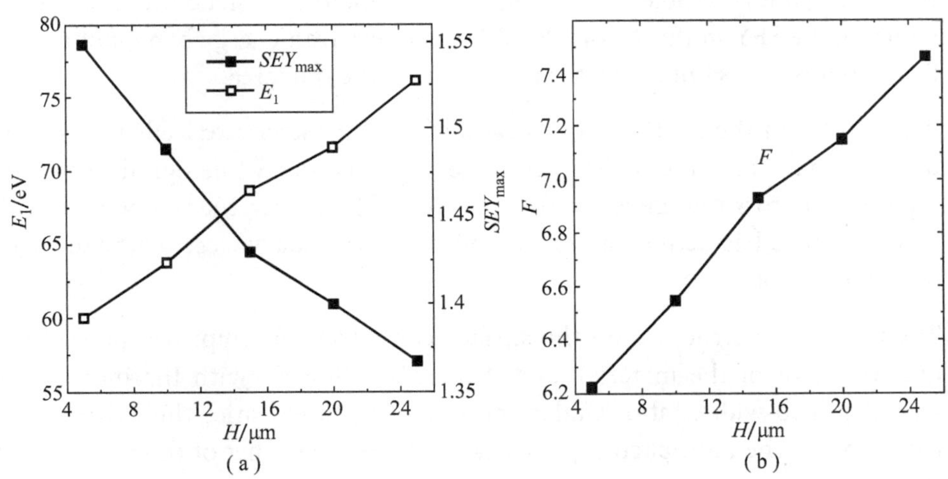

FIGURE 4.19 Effect of surface structure on the parameter multipactor quality factor F (a) E_1 and SEY_{max} and (b) multipactor quality factor F.

From Figure 4.19b, it can be found that an increase in the rectangular height H significantly increases the quality factor F of the multipactor, with a roughly nearly linear relationship, i.e.

$$F \propto H, \tag{4.19}$$

where the increase in the quality factor F implies an increase in the microwave component multipactor threshold. Moreover, by comparing the rectangular slot structure with the same depth-to-width ratio and porosity, the multipactor quality factor F is increased by 24.97% with the gourd trough surface structure. Therefore, to obtain better multipactor suppression, it is necessary to design material surface structures with higher quality factor

F values under the allowed conditions. Moreover, with the miniaturization of surface processing, the gourd trough structure proposed in this section can be processed by surface micro-nano processes such as micro-nano-imprinting stitching and 3D printing.

In this section, after the secondary electron emission characteristics of the metal samples with surface micro-trap structures obtained by etching trap structures on the surface of silicon substrate and metal sputtering are tested using a platform, by combining the surface secondary electron emission model of micro-trap structures and analyzing the suppression law of the micro-trap structures on the secondary electron emission and the influence law on the multipactor quality factor, the following main conclusions are obtained:

1. The secondary electrons generated within the trap structure are re-incident and re-emitted with the sidewall emission cascade, which to some extent produces a shielding effect on the secondary electrons and ultimately reduces the secondary electrons escaping from the trap surface.
2. Both the cylindrical hole and rectangular slot micro-trap structures are effective in reducing the SEY in the 20 ~ 4,000 eV incident electron energy band, but the overall SEY suppression is limited by the percentage of the trap area.
3. An increase in the depth-to-width ratio of the trap structure leads to a significant decrease in the maximum SEY within the trap structure, while significantly enhancing the first crossover energy of the yield curve. Moreover, the trap with cylindrical hole structure has better suppression effect on secondary electrons than that with rectangular slot.
4. The micro-trap structure on the surface can effectively improve the multipactor quality factor of the material, and it increases linearly with the increase of the trap depth-to-width ratio. Under the same trap area ratio, the cylindrical hole can improve the multipactor quality factor better than that of the rectangular slot structure.
5. In this section, we propose a collapsed gourd trough micro-trap surface to suppress secondary electron emission by "collapsing" the multi-generational secondary electrons generated in the slot structure through an internal elliptical structure. The results of this study show that the surface gourd trough structure leads to an overall reduction of secondary electron production in each incident energy band and a better suppression of TSE. The energy spectrum of the secondary electrons emitted from the surface of the slot structure shows a narrowing trend. The increase of the rectangular half-width in the parameters of the gourd trough structure causes the overall surface SEY to decrease first and then increase. When the vertical half-axis of the elliptical part of the structure is longer than the horizontal half-axis, its growth does not significantly reduce the SEY. The SEY at the optimum suppression value of the horizontal half-axis length is 21.2% lower than that of the rectangular slot surface with the same aspect ratio and porosity, while the corresponding multipactor quality

factor is increased by 24.97%. In addition, the increase in the height of the rectangular part of the gourd trough structure hinders the escape of secondary electrons and increases the quality factor of multipactor nearly linearly.

4.1.4 Surface Configuration Technology Process for Suppressing Secondary Electrons

Compared with surface configuration technology to enhance SEY, more significant progress has been made in the research field of surface configuration technology to suppress SEY in recent years, especially the actual demand in the field of micro discharge effect of space-borne high-power microwave components and particle accelerator electron cloud effect has promoted the development of this field. In engineering practice, the surface configuration and its realization process are subject to certain constraints because different applications have different constraints. Depending on whether the material is removed or added during the specific process, the surface configuration techniques can be divided into two categories: subtractive and additive processes.

4.1.4.1 Subtractive Processes

4.1.4.1.1 Wet Chemical Corrosion Wet chemical etching means that a suitable etching solution is prepared to etch the material to achieve a certain surface configuration. This type of technology is usually applied to the preparation of micro- and nanoscale surface configuration on the surface of metallic materials. Specifically, it can be subdivided into two categories: one is chemical etching without a mask; the other is chemical etching with a mask. A typical example of the latter is the photolithography process commonly used in the semiconductor field, where a patterned mask is used to protect areas that are not to be etched, thereby achieving corrosion in specific areas. Currently, wet etching can achieve corrosion to various metallic materials and some inorganic media materials.

The technology of wet etching silver-plated aluminum alloy surfaces can be applied to suppress SEY. Research has shown that random corrosion can reduce the SEY of the original silver-plated surface from 2.17 to 1.21. After roughening the silver-plated surface by wet etching and then plating a gold film, SEY can be suppressed. This process can be applied to Ku-band impedance converters, reducing the surface SEY of components from 2.20 to 1.40 and increasing the micro discharge power from 700 W to over 6,500 W.

The advantage of wet etching is that the process is relatively easy to achieve. Samples with different surface morphologies can be obtained by controlling parameters such as etchant composition, temperature, agitation, and etching time. From a microscopic perspective, chemical corrosion reactions take place preferentially at the grain boundaries, so the corrosion morphology has a significant relationship with the crystal structure of the coating itself. Compared to the dry etching described below, a disadvantage of wet etching is that the chemical reactions are generally isotropic, making it difficult to obtain a high aspect ratio structure using this method (Figure 4.20).

4.1.4.1.2 Dry Etching The basic principle of dry etching is to remove material by bombarding the material to be etched with particles with certain energy formed by plasma. According to the physical and chemical processes involved in material removal, dry etching

FIGURE 4.20 Micropore surface structure obtained by wet etching process (left: top view; right: side view).

can be divided into three categories, namely, physical dry etching (material removal by particle kinetic energy bombardment), chemical dry etching (material removal by generating gaseous reactants through chemical reactions), and Physicochemical dry etching (material removal by both physical and chemical effects). Compared to wet etching, the significant advantage of dry etching is that the lateral etching effect is much smaller than that of wet etching, making it easy to form high aspect ratio structures; however, dry etching equipment is relatively complex and costly, and the applicable material types are relatively limited.

The materials used for the lithographic mask layer include photoresist and aluminum film. SF_6 is used as the etching gas, and C_4F_8 as the passivation gas. After obtaining a silicon-based electron trap structure, a metal electron trap structure surface is obtained by sputtering metal. The results show that after sputtering Ag with a silicon-based electron trap structure, the maximum SEY value can be reduced from about 2.1 to about 1.3; after sputtering silver on a silicon substrate, SiO_2 with a thickness of 10–150 nm was sputtered and tested for SEY. The results showed that the electron trap structure with a surface of SiO_2 also had an SEY suppression effect; the SEY suppression effect of the nested structure shows that the maximum SEY value decreases from about 1.7 to about 1.4 after the introduction of the nested structure, indicating that the suppression efficiency has been further improved.

4.1.4.1.3 Laser Etching Compared with other low SEY surface configuration technologies, laser etching technology is a relatively simple surface treatment method to prepare micro-nano structures on the surface of materials. Moreover, laser etching has the advantages of low environmental requirements and stable microstructures. Currently, laser etching technology has been successfully applied to the study of surface configurations of metal and dielectric, and has shown its potential in the field of SEY suppression.

After laser treatment, the SEY of copper, aluminum, and stainless steel were significantly decreased from 1.90, 2.55, and 2.25 of the original surfaces to 1.25, 1.34, and 1.22, respectively, demonstrating that laser etching technology can be used to prepare low SEY surfaces. After laser etching, a micro/nano composite structure, nanostructure on a micron

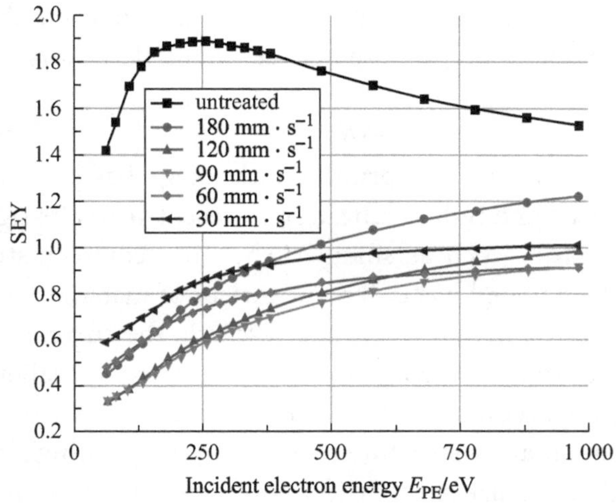

FIGURE 4.21 Effect of laser etching process on SEY.

scale slot and slot wall, is formed on the surface, in which the diameter of copper spherical particles is about 5 μm. After optimizing the process parameters, the minimum SEY is about 0.9. Although laser etching technology has been shown to be applicable for SEY suppression, whether other performance parameters of laser-processed surfaces meet the requirements of system applications remains to be further studied (Figure 4.21).

4.1.4.1.4 Machining When the surface configuration scale is in the millimeter order, SEY suppression can be achieved using traditional machining techniques. This scale of surface configuration has been applied to the suppression of electron cloud effects in particle accelerators. For example, in the field of linear accelerators, the effects of millimeter scale triangular, serrated, and rectangular slot structures on SEY have been studied. Using electrical discharge machining technology, the opening angle is 40° on the surface of aluminum alloy. Compared with the non-slotted case, the maximum SEY value decreases from 3.25 to 2.35; After opening a rectangular slot on the copper surface (with a depth of 5 mm, a width of 1.8 mm, and a gap of 0.254 mm), the SEY can be reduced from 1.65 to 0.65, and it is found that the SEY suppression effect of the inclined rectangular slot structure is not as good as that of the vertical rectangular slot. A hole array structure was achieved using a drilling process. When the drilling depth is 0.5 mm, the diameter is 1mm, and the porosity is 75%, the SEY on the copper surface is about 1.2.

4.1.4.2 Additive Process

The additive process refers to the technology of forming a surface configuration by transferring foreign materials to the surface to be treated. The very popular 3D printing technology in recent years is a typical representative of the additive process. In addition, traditional processes such as electroplating, physical vapor deposition, chemical vapor deposition, atomic layer deposition, and pulsed laser deposition are also typical additive processes. In the additive process, different coating materials can be deposited on the substrate surface, fully exploiting the performance advantages of the substrate material and the coating material to achieve the best compromise between multiple factors such as performance and cost.

4.1.4.2.1 3D Printing Technology 3D printing technology has become a hot topic in the field of scientific research and industrialization in recent years. Its significant advantages lie in application scenarios such as small batches and complex structures. Currently, mature 3D printing technology has covered a variety of materials such as resin plastics, ceramics, and metals. Although in laboratory research, the current 3D printing technology can achieve the accuracy of the micrometer scale or even smaller, in the widely used printing technology, the accuracy usually reaches the millimeter or submillimeter level. The metal silver columnar array structure prepared by 3D printing also exhibits SEY suppression characteristics. The size of the achieved surface configuration is in the millimeter level, which shows good SEY suppression characteristics. This has a good reference value for numerical simulation methods to verify the influence of surface configuration on SEY, but the practical application of this configuration method still needs to be studied.

4.1.4.2.2 Physical Vapor Deposition Physical vapor deposition processes typically include vacuum evaporation (thermal evaporation, electron beam evaporation, etc.), sputtering (DC sputtering, RF sputtering, magnetron sputtering, etc.), and ion plating. Such process technology is widely used in the field of thin film preparation, usually carried out in a vacuum system, for preparing continuous and dense thin film materials. If the process parameters are properly controlled, such coating methods can also form special surface configurations for SEY suppression.

As early as 1930, Pfund proposed to prepare blackened bismuth nanostructures on the substrate surface by thermal evaporation in an inert atmosphere at low pressure. In 1954, Bruining first proposed that carbon black like metal nanostructures could be prepared on metal surfaces to reduce SEY Preparation of nanostructures on copper, aluminum, and 316L stainless steel surfaces can reduce SEY. The experimental results show that after the growth of homogeneous nanostructures on the surface, the SEYs of copper, aluminum, and 316L stainless steel are reduced from 1.90, 2.55, and 2.25 on the original surface to 1.12, 1.45, and 1.12, respectively. Silver nanostructures and gold nanostructures prepared by low pressure evaporation process can successfully suppress SEY below 1. In addition, Glancing Angle Deposition (GLAD) by sputtering can also achieve special surface configurations, as shown in the Figure. Although there is no application of this process technology in the field of SEY suppression, the author believes that it is a worthwhile research direction to try.

4.1.4.2.3 Microelectroforming Technology In order to form a micron scale surface configuration, the microelectroforming process can be viewed as a combination of semiconductor patterning and electroforming processes. The main idea is to use patterned photoresist as a mask to achieve local plating of the surface, and then remove the photoresist to form a surface microstructure of metallic materials. The surface of the cylindrical hole array structure achieved has an ideal geometric structure, demonstrating good SEY suppression function. More importantly, the numerical calculations of SEY suppression have been experimentally verified.

A surface with a copper electron trap structure was formed through a precision electroforming process (after dry etching of the silicon substrate, copper was electrochemically deposited), and it was found that a lower current density was conducive to achieving a smaller roughness. Once the patterning was achieved, 50 nm chromium was first sputtered, followed by 100 nm copper, and then copper electroplating to increase the copper thickness. Finally, the silicon substrate was etched to obtain electron trap structured copper. The measured results showed that the maximum SEY value was about 1.4.

4.2 SURFACE COATING TO MODULATE SECONDARY ELECTRON EMISSION

The development and maturity of vacuum coating technology provides the technical basis for the use of surface coating to modulate the secondary electron emission properties of materials. The development and application of micro/nano manufacturing and smart manufacturing technologies provide more possibilities to modulate the secondary electron emission properties of materials by means of surface configuration. In recent years, the research trend of SEE modulation is to combine surface coating technology with surface configuration technology and to study the modulation technology from the perspective of theory, simulation, and experiment to meet the system application requirements for complex and specific application scenarios and needs.

As described in the subsequent chapters of this book on the different applications of the secondary electron, depending on the operating scenario of the component or system, sometimes a larger secondary electron emission coefficient is desired, and sometimes a smaller secondary electron emission coefficient is desired. Therefore, the modulation of the secondary electron emission coefficient involves the suppression and enhancement of two very different directions. The following mainly describes the basic principles and methods of surface coating technology to moderate the secondary electron emission coefficient.

4.2.1 The Principle of Surface Coating Technology Modulating SEY

As described earlier in this book, the basic process of secondary electron emission is a continuous process of elastic and inelastic scattering after the initial electrons are emitted into the interior of the material. This can cause electrons inside the material to be excited and emitted as secondary electrons (called intrinsic secondary electrons), or the incident electrons themselves can undergo elastic/inelastic scattering and emit (called backscattered electrons). Since the scattering processes vary within different materials, or even within the same material, differences in microstructure can lead to differences in SEY due to scattering processes (a typical case in this regard are carbon materials: the SEY of carbon in diamond form can reach 6–18, while the SEY of carbon in graphite form is only 0.9–1.6 [1]). Therefore, for practical engineering surfaces, the SEY can be modulated by surface coating techniques (the modulation mechanism is mainly based on the following two scenarios: the film layer itself has a different SEY than the substrate, or the introduction of the film layer affects the secondary electron emission of the substrate). Note that the term coating is used in a broad sense, and in addition to the common introduction of a thin layer of additional material on the surface by physical or chemical vapor deposition, we also classify SEY modulation by changing the chemical composition or material structure of the material by high temperature, irradiation, etc.).

4.2.2 SEY Suppression by Surface Coating Technology

SEY suppression by surface coating technology needs to address two main issues: first, finding a coating material with low SEY; second, developing a suitable coating technology in combination with the system performance requirements. In engineering practice, both two aspects may need to be considered in the context of other technical specifications for the sake of system performance. For example, in addition to low SEY characteristics, the selected coating material may also be required to have a low gas release rate, high irradiation resistance, etc. In addition to the conventional technical specifications such as uniformity and adhesion of the coating layer, the developed coating technology may also require consideration of the shape and size of the specific workpiece/surface to be treated and the process compatibility. Especially when technical indicators are conflicting and difficult to unify, it is necessary to carefully make compromises and even develop new processes, new materials, new equipment, etc.

Due to various factors such as the coating process and the change in substrate performance caused by the film layer, the thickness of the coated layer usually needs to be carefully considered. Generally speaking, the escape depth of secondary electrons of metallic materials is on the order of several nanometers (when the initial electron energy is less than 1 keV), so when the thickness of the film layer is greater than the escape depth, the SEY externally exhibited is the SEY of the film layer itself; on the contrary, if the thickness of the film layer is less than the escape depth, the SEY externally exhibited may be the combined effect of the SEY properties of the surface film layer and the substrate. For the latter, it can be theoretically modeled as a bilayer material structure, and there is a lack of research on the relationship between the SEY of this bilayer material structure and the SEY properties of the material layers themselves that make up the structure. The following is a brief description of the classification of common low SEY plating materials according to their classification.

4.2.2.1 Titanium Nitride Film

The TiN film has been widely used in the electronic cloud effect of inhibiting the accelerator. Research institutions that apply Tin film to reduce the risk of electronic cloud effects in the world mainly include SLAC, CERN, KEKB, and the Institute of High Energy Physics of the Chinese Academy of Sciences [1].

Researchers at SLAC investigated the TiN thin film aspect [2] and prepared TiN thin films by the reactive magnetron sputtering process with Ar/N_2 atmosphere using 6063 aluminum alloy and 304 stainless steel as substrates, respectively, with Ti targets. The test results showed that the SEY distribution of TiN films with aluminum alloy as substrate was in the range of 1.5–2.5 (perpendicular incidence); the SEY distribution of TiN films with stainless steel as substrate was in the range of 1.7–2.0 (incidence angle of 23°); the SEY distribution of TiN films prepared by direct magnetron irradiation with TiN targets was in the range of 1.8–1.9. Our Beijing Electron-Positron Collider (BEPC) had also used titanium nitride to suppress the electron cloud effect [3]: the SEY decreased from 1.8 to 1.06 after titanium nitride plating on stainless steel [4]. Magnetron sputtering of titanium nitride at different nitrogen concentrations changes the intrinsic SEY, and there is a correlation between the resistivity of TiN films.

4.2.2.2 Graphite/Graphene

Carbon in the form of graphite is a typical low SEY material with excellent thermal conductivity and chemical stability. Its secondary electronic emission properties have been widely studied. The graphene, which was invented in 2004, attracted widespread attention from the world and won the Nobel Prize in 2010. The study of secondary electronic emission of graphene originally originated from the actual needs of observing graphene with SEM. Later, systematic in-depth research was gradually carried out on its electronic transport and SEY suppression mechanism.

A comparative study of the SEY of amorphous carbon films [5] (prepared by DC magnetron sputtering using high-purity graphite targets with thicknesses of 20 ~ 100 nm) and graphene showed that graphene has a lower SEY than amorphous carbon films, with SEY maxima as low as 0.5 for 1- to 2-layer graphene and about 0.8 for 6- to 7-layer graphene. The SEY of graphene prepared by CVD method on the surface of copper substrates The results of SEY studies with different thicknesses of graphene showed that the increase of graphene thickness led to a monotonic decrease of SEY [6], and the SEY of 6 ~ 8-layer graphene could reach 1.25. In addition, the coating of monolayer graphene on the copper surface had the effect of suppressing the SEY (from 2.1 to 1.5 suppression) [7]. From the existing studies, like many other materials, the SEY measurements of graphene are still somewhat scattered, and the related experimental studies and their practical applications still need to be carried out in depth. Moreover, there are different explanations for the mechanism of SEY suppression by graphene; some believe that the modulation of the work function by the surface graphene coating is the main factor, while others believe that the influence of the graphene layer itself on the electron transport cannot be neglected. This theoretical disagreement is also the direction of further research.

4.2.2.3 Titanium-Zirconium-Vanadium (TiZrV) Thin Film

TiZrV films have the potential to improve chamber vacuum performance in addition to the potential suppression of the electron cloud effect. The SEY of TiZrV films before heat treatment was 2.03, while it decreased to 1.55 after heat treatment at 200°C for 2 hours [3]. For sputtered 2-μm TiZrV films on stainless steel substrates [2], the SEY can decrease from about 2.0 before activation to about 1.3 after activation. For sputtered 1-μm TiZrV films on aluminum alloy substrates, the SEY can decrease to about 1.2 after activation. Similarly, for 1-μm TiZrV films deposited on copper surfaces, the SEY decreases to about 1.1 after activation at 200°C [8].

4.2.2.4 Other Films

The secondary electron emission characteristics of a metal/dielectric composite coating showed good SEY suppression [9]. The suppression effect of magnetron-sputtered platinum on the secondary electron emission coefficient of the silver-plated surface showed that SEY suppression could be achieved after sputtering 100 nm of platinum on the silver-plated surface of aluminum alloy [10]. Significant SEY suppression was also achieved after sputtering PTFE films on aluminum alloy surfaces, and the SEY was correlated with the thickness of the dielectric film [11]. In addition, TiN prepared by arc ion plating, Au, Pt, and Ni prepared by electroplating, and Ir (iridium) films prepared by magnetron sputtering all exhibited SEY suppression [12], which was consistent with the trend predicted by work function theory.

4.2.3 Surface Coating Technology to Enhance SEY

Materials with high secondary electron emission coefficients are required for applications such as plasma displays and electron multipliers. Surface coating technology is the current mainstream technology for SEY enhancement, and significant SEY enhancement can be achieved by coating high emissivity materials such as magnesia and diamond, and by using doping modification methods for these films. The following is a brief description of typical SEY-enhanced thin film materials.

4.2.3.1 Magnesium Oxide Film

Magnesium oxide has a band gap of about 7 eV and good secondary electron emission properties, and it is one of the popular research materials in the field of electron multipliers and plasma displays. With the spherical injection process of gold-doped MgO, the SEY maximum of 300 nm thick gold-doped MgO films can reach 8 [13]. By studying the secondary electron emission properties of MgO single crystals with different crystal orientations under neon ion bombardment [14], it was found that MgO single crystals with <111> orientation have the highest electron emission efficiency. In addition, the doping of MgO films with zinc increased the SEY by more than a factor of 1 due to the change in the work function and the introduction of impurity energy levels [15]. In contrast, reactive sputtering in an oxygen atmosphere prepared MgO/Al_2O_3 composite films with higher SEY than MgO films: 11.6 for the former and 4.9 for the latter [16]. The effect of gold doping on the secondary electron emission of MgO was studied, and it was found that for the gold-doped MgO films prepared by reactive sputtering [17], the maximum SEY reached 11.5 and the optimum doping concentration was 3%. The thickness of MgO films has an important effect on the SEY [18] and needs to be controlled in a suitable thickness range to obtain good electron emission properties, which can reach 20 for single-crystal MgO.

4.2.3.2 Aluminum Oxide Film

Like magnesium oxide, alumina is also one of the more studied SEY-enhanced thin film materials. In fact, for aluminum alloy engineering surfaces, the SEY is much higher than that of pure aluminum due to the natural oxide layer of 2–3 nm on the surface, thus showing the SEY enhancement effect of aluminum oxide thin films. In addition to aluminum oxide thin films, the SEY properties of alumina ceramic blocks have also attracted the interest of many researchers due to their wide range of applications [19].

For the SEY of aluminum oxide thin films using RF sputtering (with a glass deposited with SiO_2 films as a substrate [20], a water-cooled parallel plate sputtering system, a sputtering rate of 0.2 nm/min, and aluminum oxide thin film thickness of 7–10 mn), the maximum SEY reaches 4.3. The SEY of aluminum oxide thin films is related to the crystallographic properties and the stoichiometric ratio [18], so the SEY properties significantly depend on the preparation process. The available methods for the preparation of aluminum oxide thin films are evaporation of pure aluminum followed by oxidation, RF sputtering of alumina, and CVD preparation of alumina, and the SEY of alumina can reach 8.

4.2.3.3 Diamond Film

The main methods for the preparation of diamond films are hot wire CVD method, flame deposition method, plasma-enhanced CVD method, and DC arc plasma jet CVD method [21]. Compared with the aforementioned magnesium oxide and alumina [21], the SEY of diamond films can reach a higher level. We prepared diamond films on the surface of single-crystal silicon substrates using microwave plasma chemical vapor deposition process and investigated the effects of boron doping, orientation, and grain size on the SEY, which can reach up to 98. In addition, it is found that due to the enhanced electrical conductivity, diamond films can be doped to avoid charging and thus have greater SEY. Furthermore, the boron doping concentration affected the secondary electron emission of diamond films characteristics with a maximum SEY of about 90 [22]. For the SEY of MPCVD-grown B-doped diamond films, the SEY of the prepared films was measured to be 18.3 after several weeks in air at an initial electron energy of 1 keV [23].

4.2.3.4 Other Technologies

4.2.3.4.1 Electron Beam Cleaning In the early studies of secondary electrons, it was not realized that the electron beam current used for SEY measurements could also affect the SEY measurements. Later, with the help of advanced analytical tools in a vacuum environment, the mechanism of the effect of electron beam irradiation on SEY was essentially elucidated [8,24,25]. Moreover, this effect has also been applied to gas pedal vacuum systems – the cleaning function of the electron cloud leads to a gradual decrease of the cavity wall SEY, allowing the system performance to be improved. The results of existing studies show that the main reason for the effect of electron beam cleaning on SEY is the change in surface composition (e.g., removal of surface staining by the electron beam-induced desorption process) and that the irradiation dose is the main parameter affecting the amount of SEY change. TiN films prepared by an electron beam irradiation sputtering process at an energy of 130 eV [2] showed a decrease in SEY from ~1.7 to ~1.0 when the irradiation dose reached 6,529 $\mu C/mm^2$. Copper can decrease from ~2.0 to ~1.0 after electron beam cleaning [8] due to the removal of surface oxides and the graphitization of the surface. Currently, this method is mainly used in the field of particle gas pedals.

4.2.3.4.2 Ionic Beam Cleaning Ion beam cleaning uses high-energy ions to bombard the material surface for the purpose of surface microroughening or removal of surface adsorbates. The effect pattern of nitrogen ion bombardment on the adsorbed elements and SEY on the surface of titanium nitride showed that after nitrogen ion bombardment, the ratio of oxygen and carbon adsorbed on the surface of the sample decreased significantly [26], and the SEY maximum value decreased from the original 1.50–1.10. The study of cleaning etching on the surface of gold samples using argon ions [27] showed that after etching the sample surface using argon ions, the surface roughness of the sample became larger, and the oxygen and carbon stains on the surface were effectively removed, and the SEY maximum value was reduced from 2.0 to 1.5 on the original surface. After ion beam cleaning (1,720 μC N_2^+), the SEY of TiN thin film deposited on aluminum alloy substrate decreased from about 1.5 to about 1.1 [26], the SEY of TiCN thin film decreased to 1.29, and the SEY of TiZrV thin film decreased to 1.15. After argon ion beam cleaning of copper, the SEY decreased to approximately 1.4 [8] due to the removal of the natural oxide layer, hydrocarbons, and contamination layer on the surface. Different ion cleaning conditions change the secondary electron emission properties of the metal surface, and by analyzing the mechanism of the effect, it was found that ion cleaning has two effects: removing surface staining to decrease SEY or increasing surface roughness to increase SEY [28].

4.2.3.4.3 Heating Treatment Heating and baking usually lead to the desorption of adsorbents on the surface of the sample, thereby reducing SEY. For example, heating causes the desorption of water molecules, resulting in a decrease in SEY [28]. After heating the copper at 300°C, the SEY decreased from approximately 2.4 to approximately 1.8. Another typical case of reduced SEY caused by heat treatment is NEG thin films. However, some studies have shown that heat treatment can increase the SEY of insulation materials. Another perspective on studying the influence of heating and baking on SEY is the variation of material SEY with temperature, which has practical significance for certain high-temperature application scenarios.

4.2.3.4.4 Alloy For fields such as magnetrons, a main goal in cathode development is to increase the secondary electron emission coefficient of materials. In this field, controlling the material SEY through alloying technology is a commonly used method. The experiment on the influence of different doping elements on the SEY of W cathode shows that Re doping has the best effect, and different concentrations of Re doping indicate a maximum SEY of 1.8 [29] when the mass fraction is 5%. (It is worth noting that the SEM images in the book show that the sample surface is a porous structure with a pore size of 1–2 μm. This is like the surface configuration described below to suppress the porous surface in SEY, indicating that the SEY of the W cathode may also have significant room for improvement).

4.2.4 The Combination of Surface Coating Technology and Surface Configuration Technology

It has been shown that selecting a surface coating material with low SEY and coating it on a surface configuration with SEY inhibition effect results in a more significant SEY inhibition effect. A surface treatment process using a wet chemical etching process to randomly

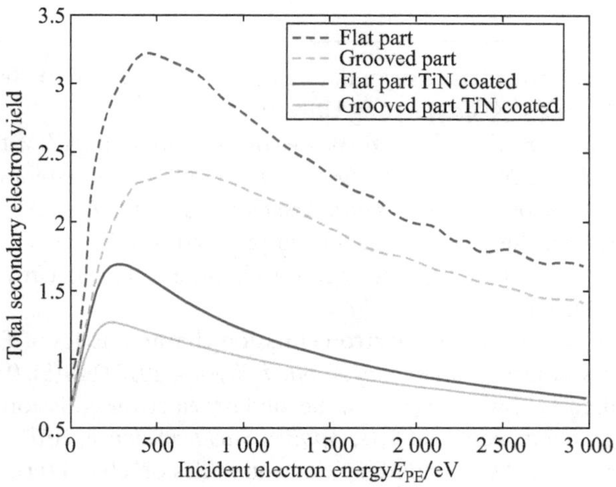

FIGURE 4.22 A combination of surface coating and surface conformation techniques is used to suppress SEY.

etch the silver-plated surface, and then preparing silver nanostructures on the etched silver-plated surface. Similarly, a periodic triangular groove structure with a depth of 1 mm and an angle of 40° was formed on the aluminum alloy substrate, and then a 130 nm thick TiN film was sputtered and coated, and the SEY measurement results are as below (four scenarios including smooth aluminum alloy surface, smooth aluminum alloy surface coated with TiN, triangular groove aluminum alloy surface, and triangular groove aluminum alloy surface coated with TiN), it can be seen that the SEY is reduced from about 3.2 to about 1.3 by using a combination of surface coating and surface configuration (Figure 4.22).

In recent years, the research on SEY modulation by surface treatment has gained great progress driven by practical needs. This chapter discusses the principles and methods of SEY modulation by surface coating technology and surface configuration technology, respectively. Driven by the demand for suppression of multipactor effects in satellite-borne high-power microwave components, promising progress has been made in the theoretical and technological aspects of SEY suppression by surface configuration in China. In the future, the possible directions of SEY regulation research include the following two aspects:

- **Theoretical aspects**: The numerical simulation of the SEY surface configuration control effect of dielectric materials is much more complicated than that of metallic materials due to the consideration of the charging effect and leakage effect, etc. The calculation process is still in the initial stage at home and abroad, and in-depth research is needed in the future;

- **Process technology:** Combined with the system application requirements, the optimization of other aspects of performance indicators while considering the SEY characteristic requirements is a key research issue for the further development of SEY regulation technology for practical applications in the future.

REFERENCES

1. Wang D. Research on the surface secondary electron emission characteristics and regulation of space materials. Xi'an: Xi'an Jiaotong University, 2019.
2. Le Pimpec F, Kirby R E, King F, et al. Properties of TiN and TiZrV thin film as a remedy against electron cloud. *Nuclear Instruments and Methods in Physics Research Section A: Accelerators, Spectrometers, Detectors and Associated Equipment*, 2005, 551(2–3): 187–199.
3. Zhang B. Research on the process and related properties of TiZrV vacuum absorbing agent thin films deposited on the walls of vacuum chambers. Hefei: University of Science and Technology of China, 2011.
4. Wang D, He Y, Cui W. Secondary electron emission characteristics of TiN coatings produced by RF magnetron sputtering. *Journal of Applied Physics*, 2018, 124(5): 053301.
5. Montero I, Aguilera L, Dávila M E, et al. Secondary electron emission under electron bombardment from graphene nanoplatelets. *Applied Surface Science*, 2014, 291: 74–77.
6. Wang J, Wang Y, Xu Y, et al. Secondary electron emission characteristics of graphene films with copper substrate. *Chinese Physics C*, 2016, 40(11): 117003.
7. Cao M, Zhang X S, Liu W H, et al. Secondary electron emission of graphene-coated copper. *Diamond and Related Materials*, 2017, 73: 199–203.
8. Baglin V, Malyshev O B. Beam-induced electron multipacting, electron cloud, and vacuum design. *Vacuum in Particle Accelerators: Modelling, Design and Operation of Beam Vacuum Systems*. Weinheim, Germany: Wiley-VCH Verlag GmbH & Co. KGaA, 2019: 349–420.
9. Olano L, Montero I, Dávila M E, Jacas A, Rodriguez M A, Dennison J R, Dynamic secondary electron emission in dielectric/conductor mixed. *Proceedings of the 9th International Workshop on Multipactor, Corona and Passive Intermodulation (MULCOPIM)*. Noordwijk, The Netherlands: ESA/ESTEC, April 7, 2017.
10. He Y, Yu B, Wang Q, Bai C J, Yang J, Hu T C, Xie G B, Cui W Z. Magnetron sputtering platinum suppresses secondary electron emission on silver plated surfaces. *Acta Physica Sinica, Journal of Physics*, 2018, 67(8): 087901.
11. Wang F, Zhao Q, Li J, et al. Significantly reduced secondary-electron-yield of aluminum sheet with fluorocarbon coating. *Coatings*, 2018, 8(7): 249.
12. Wang J M. Study on the preparation of high power function thin films and the mechanism of suppressing secondary electron emission from collecting poles. Chengdu: University of Electronic Science and Technology of China, 2012.
13. Henrich V E, Fan J C. High-efficiency secondary-electron emission from sputtered MgO–Au cermets. *Applied Physics Letters*, 1973, 23(1): 7–8.
14. Choi E H, Lim J Y, Kim Y G, et al. Secondary electron emission coefficient of a MgO single crystal. *Journal of Applied Physics*, 1999, 86(11): 6525–6527.
15. Yu H K. Secondary electron emission properties of Zn-doped MgO thin films grown via electron-beam evaporation. *Thin Solid Films*, 2018, 653: 57–61.
16. Wang F, Zhou F, Wang J, et al. Characterization of MgO/Al2O3 composite film prepared by DC magnetron sputtering and its secondary electron emission properties. *Journal of Electronic Materials*, 2018, 47: 4116–4123.
17. Li J, Hu W, Wang K, et al. Au doping effect on the secondary electron emission performance of MgO films. *Materials*, 2018, 11(11): 2104.
18. Dai X Y. Study on the secondary electron generation rate of magnesium oxide and nano diamond microcrystalline composite films on iron nickel alloys. Taiwan: National Tsing Hua University, 2006.
19. Michizono S, Saito Y, Yamano Y, et al. Influence of Mechanical Finishing on Secondary Electron Emission of Alumina Ceramics. *2006 International Symposium on Discharges and Electrical Insulation in Vacuum*. IEEE, 2006, 1: 97–100.

20. Rajopadhye N R, Bhoraskar S V. Secondary electron emission of sputtered alumina films. *Pramana*, 1985, 25: 327–334.
21. Wang Y Q. Preparation and performance study of diamond thin films for secondary electron emission cathode materials. Mianyang, Sichuan: Southwest University of Science and Technology, 2009.
22. Ye Q Y, Wang B, Gan K Y, Li K, Zhou L, Wang D. The effect of boron doping concentration on the secondary electron emission characteristics of diamond thin films. *Material Guide B: Research Section*. 2012, 26 (3): 38–44.
23. Ding M Q, Li L, Feng J. Secondary electron emission from Boron doped diamond films grown by MPCVD. *Chinsese Journal of Liquid Crystals and Displays*, 2013, 28(5): 688–692.
24. Belhaj M, Tondu T, Inguimbert V, et al. The effects of incident electron current density and temperature on the total electron emission yield of polycrystalline CVD diamond. *Journal of Physics D: Applied Physics*, 2010, 43(13): 135303.
25. Cimino R, Commisso M, Grosso D R, et al. Nature of the decrease of the secondary-electron yield by electron bombardment and its energy dependence. *Physical Review Letters*, 2012, 109(6): 064801.
26. Le Pimpec F, Kirby R E, King F K, et al. The effect of gas ion bombardment on the secondary electron yield of TiN, TiCN and TiZrV coatings for suppressing collective electron effects in storage rings. *Nuclear Instruments and Methods in Physics Research Section A: Accelerators, Spectrometers, Detectors and Associated Equipment*, 2006, 564(1): 44–50.
27. Yang J, Cui W, Li Y, et al. Investigation of argon ion sputtering on the secondary electron emission from gold samples. *Applied Surface Science*, 2016, 382: 88–92.
28. Hu X. Research on the complex influencing factors of secondary electron emission characteristics on metal surfaces. Xi'an: Xi'an Jiaotong University, 2014.
29. Qi S K, Wang X, Luo J R, Zhao S K, Li Y, Zhao Q L. Study on the secondary electron emission coefficient of metal doped tungsten based alloy cathodes. *Intense Laser and Particle Beam*, 2014, 26 (12): 123006.

CHAPTER 5

Secondary Electron Emission Properties of Insulators and Their Regulation Techniques

Yongning He and Yahui Cai

5.1 METHODS FOR MEASURING THE SECONDARY ELECTRON EMISSION PROPERTIES OF INSULATORS

Upon irradiation by an electron beam, the surface of an insulator with poor electrical conductivity not only exhibits the phenomenon of secondary electron emission but also forms a charge accumulation [1,2] on the surface. The existence of the accumulated charge affects the measurement of the secondary electron emission characteristics of the insulator. Accordingly, the existing methods of measuring the secondary electron emission properties of insulators can be classified into four categories based on the method used to eliminate the accumulated charge: the discharge method [3–7], the compensation method [8–11], the impulse method [12–21], and the neutralization method [22–31].

5.1.1 Discharge Method

The fundamental principle of the discharge method is to utilize the leakage current to discharge the accumulated charge. This can be achieved by preparing thin layers, subjecting the sample to heating, or preparing micrographic metal channels. The preparation of thin layers and sample heating can effectively increase the body leakage current, while the preparation of micrographic metal channels can effectively increase the surface leakage current. In the case of discharging the accumulated charge through the preparation of thin layers, the measurement process is analogous to that of the secondary electron emission yield of metallic materials. However, it should be noted that the thickness of the insulator material must be less than 100 nm and must be prepared on a conductive substrate. Figure 5.1a depicts a schematic diagram of a measurement procedure utilizing heating to discharge accumulated charge. The procedure for conducting a secondary electron emission yield

DOI: 10.1201/9781003562863-5

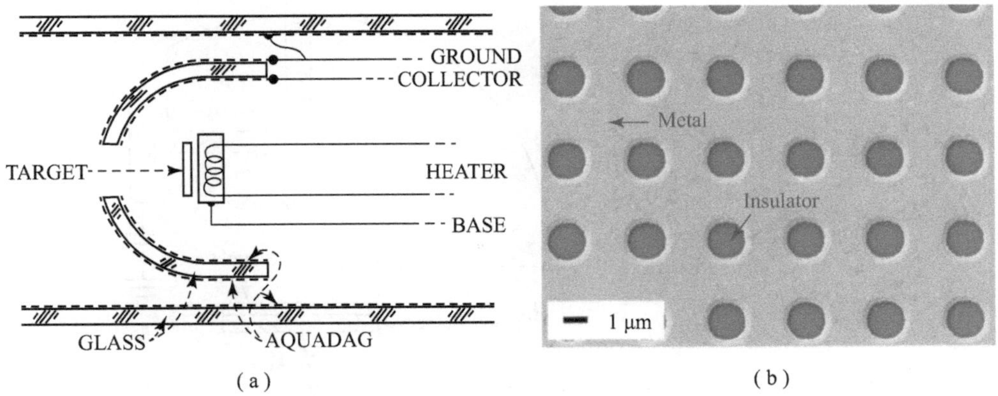

FIGURE 5.1 Working principle of different discharge methods. (a) Sample heating. (b) Micrographics channel.

test is as follows: (1) Measure the total electron beam current I_{PE}; (2) Measure the secondary electron beam current I_{SE} or sample current I_S; (3) Heat the sample to discharge the accumulated charge; (4) Change the incident electron energy E_{PE}, and repeat steps (2) and (3), at which SEY (σ) can be expressed as:

$$\sigma = \frac{I_{SE}}{I_{PE}} = \frac{I_{PE} - I_S}{I_{PE}} = \frac{I_{SE}}{I_{SE} + I_S}. \tag{5.1}$$

Accumulated charge is discharged by preparing micrographic metal channels on the insulator surface with the sample structure shown in Figure 5.1b, and the secondary electron emission yield test flow is as follows: (1) Measure the total electron beam current I_{PE}; (2) Measure the sample current I_{SM} in the pure metal region; (3) Measure the sample current $I_{SM\text{-}I}$ in the region of the periodic structure; (4) Vary the incident electron energy E_{PE}, repeat (2) and (3), at which σ can be expressed as:

$$\sigma = \left[\frac{I_{PE} - I_{SM\text{-}I}}{I_{PE}} - (1 - A_I)\frac{I_{PE} - I_{SM}}{I_{PE}}\right]\frac{1}{A_I}, \tag{5.2}$$

where A_I is the area proportion of the insulator in the periodic structure.

5.1.2 Compensation Method

The principle of the compensation method is to detect the surface potential of an insulator after the accumulation of charge by capacitance or Kelvin probe and to apply a bias voltage accordingly to eliminate the influence of the surface potential on the incident electrons. Figure 5.2a illustrates the compensation schematic for probing the surface potential by capacitance, and the measurement flow is as follows: (1) Measure the total electron beam current I_{PE}; (2) Measure the sample current I_S and the insulator surface potential V_S; (3) Adjust the sample base reversal voltage V_R such that $V_R + V_{s=}0\,\text{V}$; (4) Vary the incident electron energy E_{PE}, and repeat (2) and (3), at which σ is calculated by (5.1) is calculated.

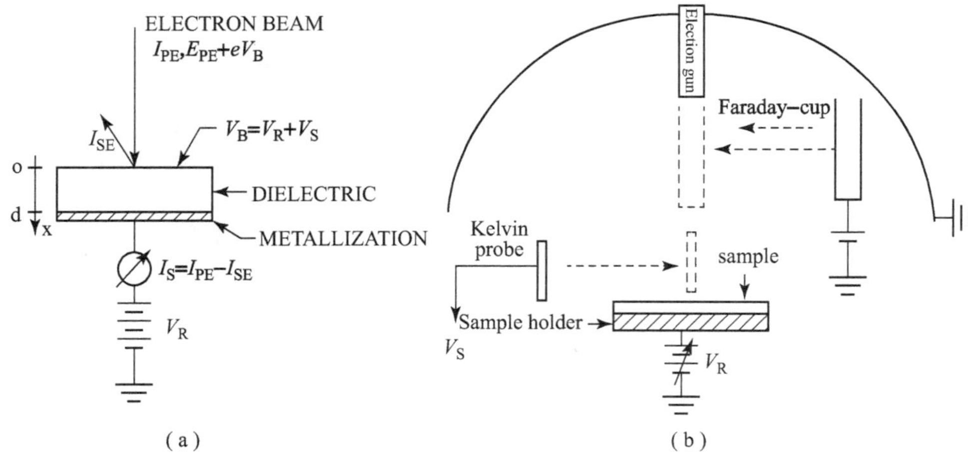

FIGURE 5.2 Working principles of different compensation methods. (a) Capacitance detection. (b) Kelvin probe detection.

Figure 5.2b illustrates the compensation schematic for probing the surface potential by a Kelvin probe with the following measurement flow: (1) Measure the total electron beam current I_{SE}; (2) Measure the secondary electron beam current I_{SE}; (3) Move the Kelvin probe to probe the surface potential V_s of the insulator; (4) Adjust the sample base reversal voltage V_R, so that $V_R + V_{s=}0$ V; (5) Change the incident electron energy E_{PE}, and repeat (2) to (4), at which σ is calculated from (5.1).

5.1.3 Pulse Method

The pulse method is used to attenuate the effect of charge accumulation on SEY measurements by emitting a short-pulse electron beam or by moving the measurement position to ensure that the level of potential accumulation on the insulator surface during the measurement process is low enough to be negligible with respect to the incident electron energy. Figure 5.3a illustrates the measurement schematic for emitting a short-pulse electron beam, with the following measurement flow:

1. Fire a single short-pulse electron beam;
2. Measure the secondary electron beam current I_{SE} and the sample current I_S; and
3. Vary the incident electron energy E_{PE}, and repeat (1) and (2), at which point σ is calculated from (5.1).

Figure 5.3b illustrates the schematic diagram of compensation by moving the measurement position with the following measurement flow: (1) Emit a single short-pulse electron beam; (2) Measure the secondary electron beam current I_{SE} and the sample current I_S; (3) Move the measurement position and change the incident electron energy E_{PE}, and repeat (1) and (2), at which point σ is calculated from (5.1).

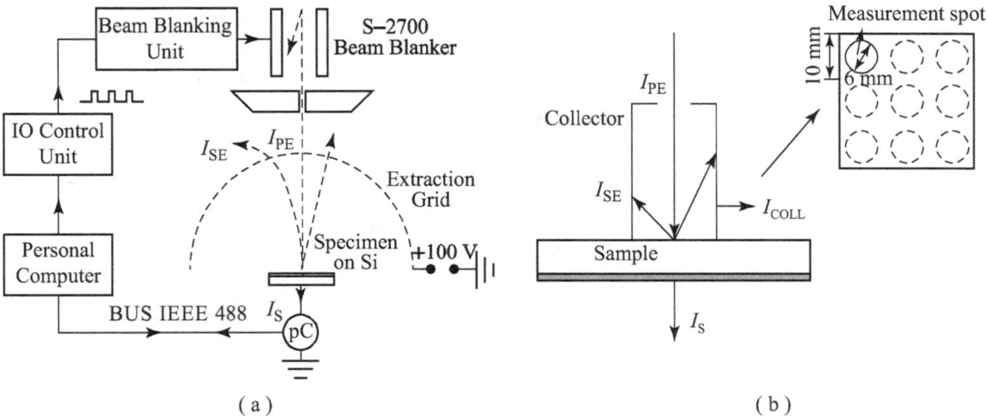

FIGURE 5.3 Working principle of different pulse methods. (a) Short pulse. (b) Shift measurement.

5.1.4 Neutralization Method

Based on the pulse method, in the neutralization method, the neutralization source is introduced to neutralize the accumulated charge; the neutralization source mainly includes the electron gun or applying the collection pole bias voltage, by which the electron gun neutralization is divided into single-gun neutralization and double-gun neutralization. The principle of single-gun neutralization is shown in Figure 5.4a, which is mainly used to neutralize the accumulated positive charge; the principle of double-gun neutralization is shown in Figure 5.4b, which can be used to neutralize the accumulated positive and negative charges. The measurement flow of the electron gun neutralization method is as follows: (1) Emit a single short-pulse electron beam; (2) Measure the secondary electron beam current I_{SE} and sample current I_S; (3) Turn on the neutralization gun to neutralize the accumulated charges; (4) Change the incident electron energy E_{PE} and repeat (1) to (3), at which σ is calculated from (5.1). The principle of collection pole bias neutralization is shown in Figure 5.4c, which can be used to neutralize the accumulated positive and negative charges, and the measurement flow is as follows: (1) Adjust the collection pole bias to positive voltage and emit a single short-pulse electron beam; (2) Measure the secondary electron beam current I_{SE} and the sample current I_S; (3) Adjust the collection pole bias to a certain negative bias for neutralization of the accumulated charges; (4) Change the incident electron energy E_{PE}, and repeat (1) to (3), at which σ is calculated from (5.1). It can be found that the neutralization method introduces the neutralization process again under the premise that the incident electrons are short pulses, which reduces the influence of the accumulated charge and further improves the accuracy of the SEY measurement.

5.1.5 Summary of the Section

By statistically examining the existing insulator SEY measurement methods, as shown in Table 5.1, the upper right corner of the literature is labeled as the year when the test method was proposed, and it can be seen that the pulse method and the neutralization method

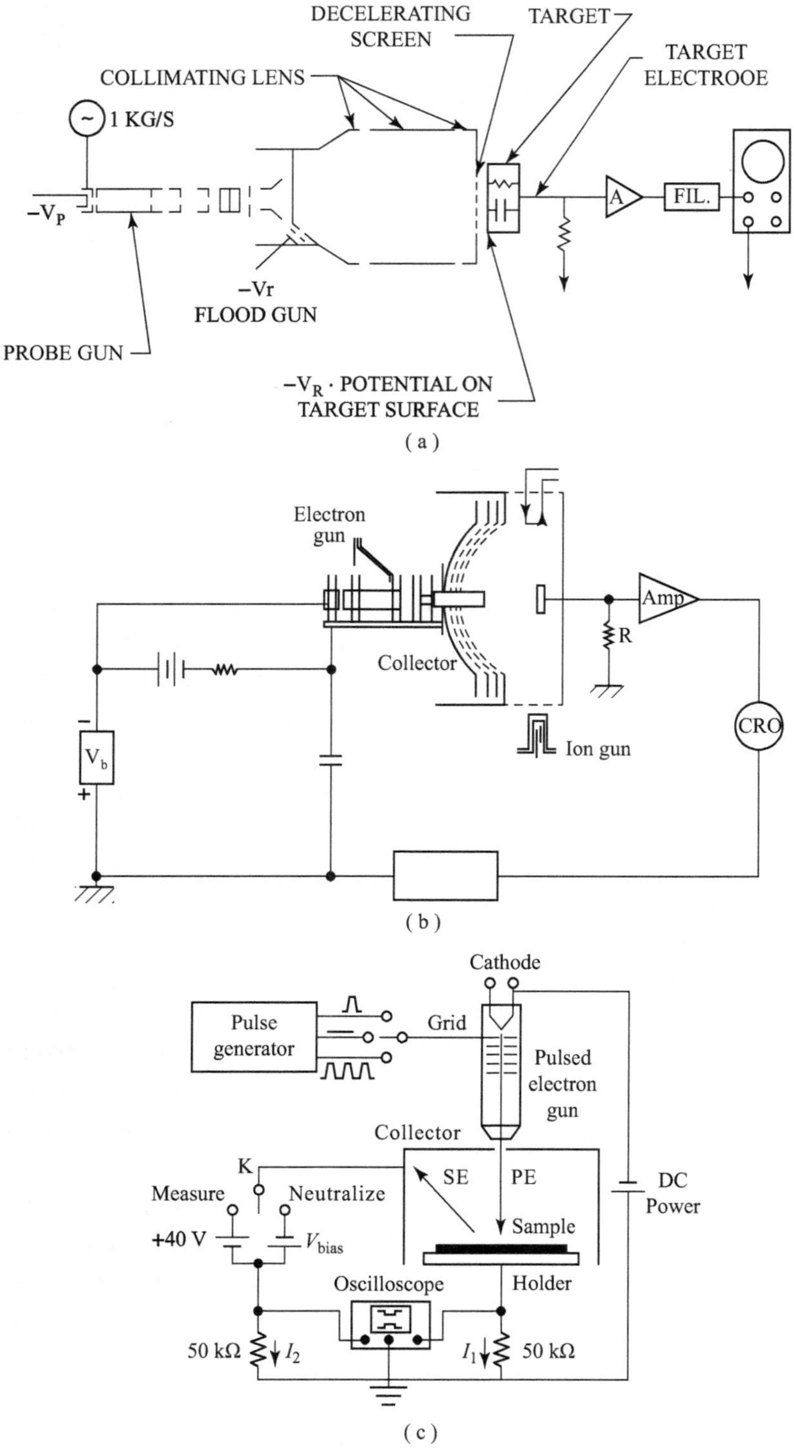

FIGURE 5.4 Working principles of different neutralization methods. (a) Single-gun neutralization. (b) Double-gun neutralization. (c) Bias neutralization.

TABLE 5.1 Classification and Year Distribution Statistics of Measurement Methods for Secondary Electron Emission Characteristics of Insulators

Measurement Method	Literature
Discharge	[3](1953), [4](1957), [5](2001), [6](2006), [7](2020)
Compensation	[8](1985), [9](2009), [10](2011), [11](2012)
Pulse	[12](1948), [13](1973), [14](1977), [15](1998), [16](2003), [17](2008), [18](2010), [19](1998), [20](2010), [21](2013)
Neutralization	[22](1966), [23](1976), [24](1983), [25](1985), [26](2003), [27](2014), [28](2014), [29](2016), [30](2017), [31](2022)

dominate among the charge elimination methods and have a long duration of practical application, which is related to the fact that it can effectively attenuate the influence of accumulated charge on SEY measurement and improve the accuracy of SEY measurement.

5.2 CHARACTERIZATION OF SECONDARY ELECTRON EMISSION FROM TYPICAL INSULATORS

The common insulator SEYs obtained by testing using the above insulator secondary electron emission characterization methods are counted and are mainly classified into inorganic ceramic and organic polymer types. In the energy range counted ($E_P < 5$ keV), the depth of electron incidence in insulator materials is less than 2 μm, so samples with thickness less than 2 μm are regarded as thin films, and samples with thickness greater than 2 μm are regarded as blocks, and SEY curves of some insulator materials are summarized for block materials and thin film materials, respectively.

5.2.1 Inorganic Ceramic

5.2.1.1 Silicon Dioxide (SiO_2)

The SEY of 3 μm SiO_2 grown by magnetron sputtering is measured in Yong [15] by using the pulse method, in Cai [31] using the neutralization method for 1 mm quartz crystal SiO_2, and in Ooka [32] using the pulse method for quartz crystal SiO_2. The statistics of the SEY curves of bulk SiO_2 are shown in Figure 5.5a, with σ_m of 2.1–4.4, E_{Pm} of 300–500 eV, and E_{P2} of 2–5 keV.

The SEY of 300 nm SiO_2 is measured in Cai [7] using the discharge method, in Yong [15] using the pulse method for 1 μm SiO_2 grown by hydrothermal method, in Cai [31] using the neutralization method for 300 nm SiO_2, and in Yi [33] using the pulse method for 55 nm SiO_2. The SEY statistics curve of the thin-film SiO_2 are shown in Figure 5.5b, with σ_m of 3.3–4.3, E_{Pm} of 300–500 eV, and E_{P2} of 3–6 keV.

5.2.1.2 Aluminum oxide (Al_2O_3)

The SEY of 2 mm polycrystalline Al_2O_3 is measured in Belhaj [10] by the compensation method, in Cai [31] by the neutralization method for 2 mm sapphire Al_2O_3, in Wang [34] by the neutralization method for bulk Al_2O_3, and in Cazaux [34] by the impulse method for hp-phase sapphire Al_2O_3.The SEY statistics curve of bulk Al_2O_3 are shown in Figure 5.6a, with σ_m of 2.4–6.4, E_{Pm} of 200–600 eV, and E_{P2} of 3–7 keV.

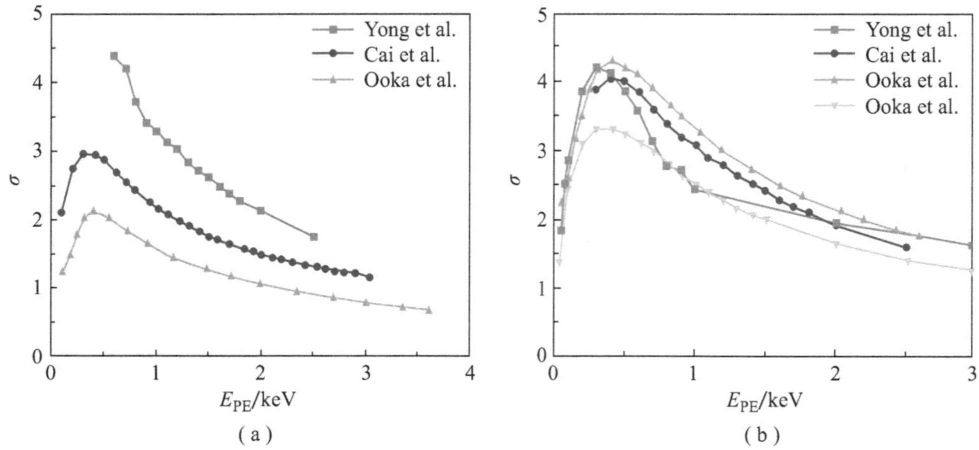

FIGURE 5.5 SEY curves of SiO_2 bulk and thin film. (a) SiO_2 bulk. (b) SiO_2 thin-film.

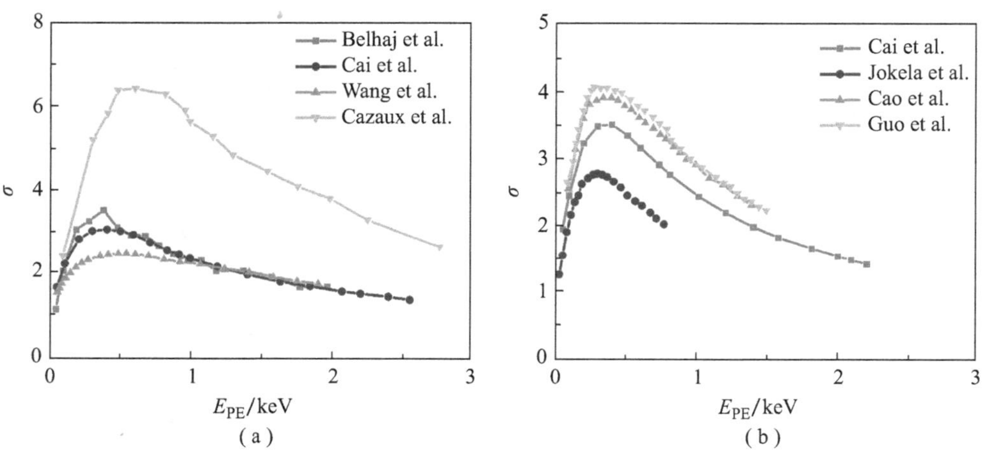

FIGURE 5.6 SEY curves of Al_2O_3 bulk and thin film. (a) SiO_2 bulk. (b) SiO_2 thin-film.

The SEY of 34 nm Al_2O_3 grown by atomic layer deposition (ALD) is measured in Cai [31] using the neutralization method, in Jokela [35] using the compensation method for 4 nm Al_2O_3 grown by ALD, in Cao [35] using the pulse method for 20 nm Al_2O_3, in Guo [36] using the pulse method for 30 nm Al_2O_3 grown by ALD. The SEY statistics curves of thin-film Al_2O_3 are shown in Figure 5.6b, with σ_m of 2.8–4.1, E_{Pm} of 300–500 eV, and E_{P2} of 3–6 keV.

5.2.1.3 Magnesium Oxide (MgO)

The SEY of single-crystal MgO is measured in Johson [12] using the discharge method and in Whetten [37,38] using the pulse method. The SEY statistics curves of bulk MgO are shown in Figure 5.7a, with σ_m of 7.2–23.3, E_{Pm} of 800–1,200 eV, and E_{P2} of 8–30 keV.

The SEY of 37 nm MgO grown by ALD measured in Cai [31] using the neutralization method, in Jokela [35] using the compensation method for 4 nm MgO grown by ALD, in Cao [39] using the pulsed method for 20 nm MgO, in Whetten [38] using the pulse method

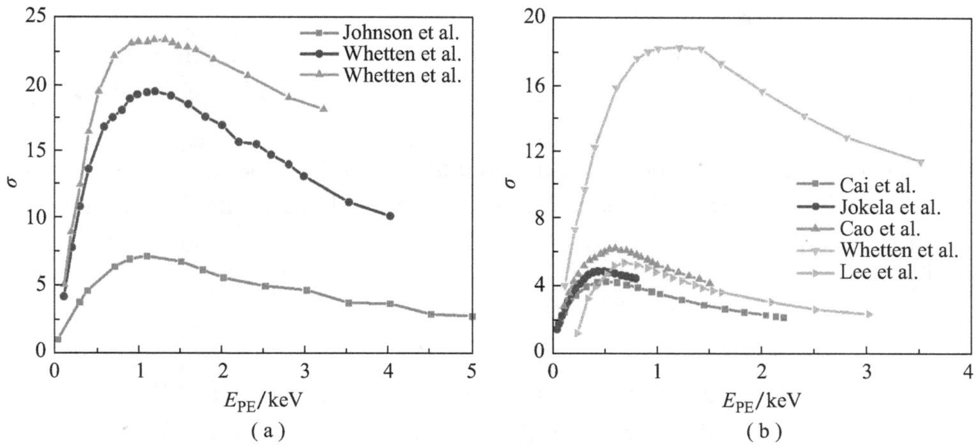

FIGURE 5.7 SEY curves of MgO bulk and thin film. (a) SiO_2 bulk. (b) SiO_2 thin-film.

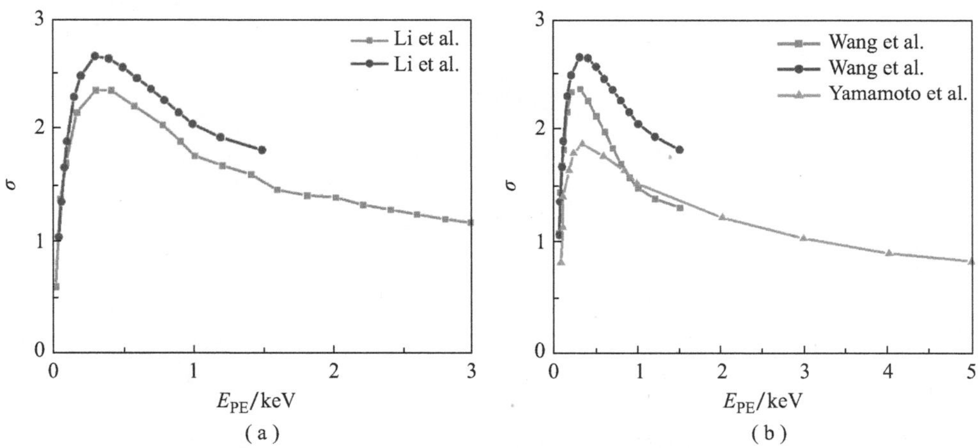

FIGURE 5.8 SEY curves of bulk and thin film ferrite. (a) Bulk ferrite. (b) Thin film ferrite.

for 80 nm MgO grown by evaporation oxidation, and in Lee [38,40] using the compensation method for 30 nm MgO. The SEY statistics curves for thin film MgO are shown in Figure 5.7b, with σ_m of 4.2–18.2, E_{Pm} of 300–1,500 eV, and E_{P2} of 3–20 keV.

5.2.1.4 Ferrite

The SEY of 2 mm block ferrite is measured in Li [41,42] by the neutralization method, and the SEY statistics curves of block ferrite are shown in Figure 5.8a, with σ_m of 2.4–2.7, E_{Pm} of 300–500 eV, and E_{P2} of 4–7 keV.

The SEY of 50 nm ferrite grown by magnetron sputtering is measured in Wang [43] using the discharge method, and in Yamamoto [44] using the discharge method for the thin-film ferrite. The SEY statistics curves of thin-film ferrite are shown in Figure 5.8b, with σ_m of 1.9–2.7, E_{Pm} of 300–500 eV, and E_{P2} of 3–7 keV.

5.2.1.5 Aluminum Nitride (AlN)

The SEY of bulk AlN is measured in Michizono [45] by the pulse method, and the results are shown in Figure 5.9a, with σ_m of about 2.5, E_{Pm} of about 1,000 eV, and E_{P2} of about 5.5 keV.

The SEY of 40 nm AlN is measured in Buzulutskov [46] by the discharge method, and in Lau [47] by the pulse method for 10 nm ALD-grown AlN, and the SEY statistics curves of thin-film AlN are shown in Figure 5.9b, with σ_m of 2.5–4.1, E_{Pm} of 300–1,200 eV, and E_{P2} of 3–5 keV.

5.2.1.6 Boron Nitride (BN)

The SEY of magnetron sputter-grown thin films of BN is measured in Villemant [48] using the discharge method, and the results are shown in Figure 5.10, with σ_m of bout 2.0, E_{Pm} of about 300 eV, and E_{P2} of about 2 keV.

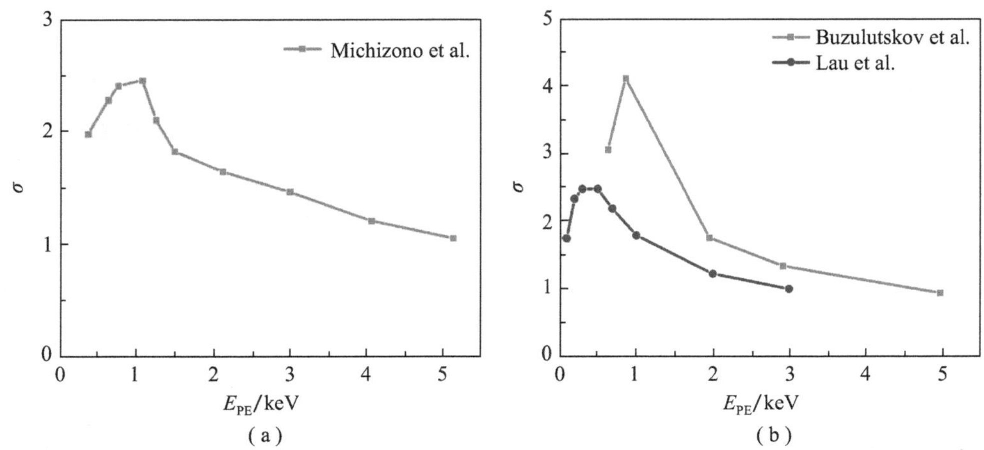

FIGURE 5.9 SEY curves of AlN bulk and thin film. (a) AIN bulk. (b) AIN thin film.

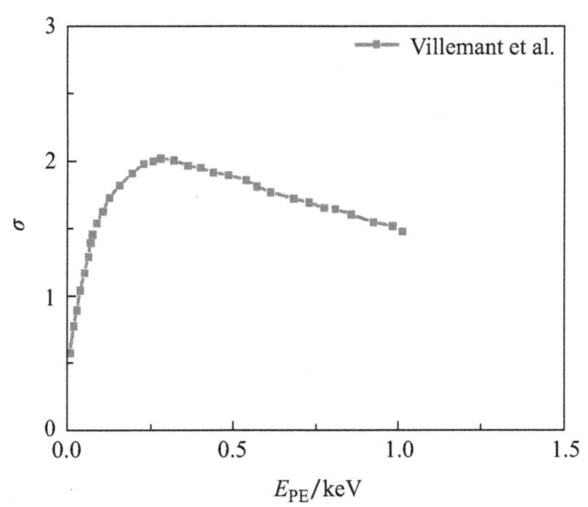

FIGURE 5.10 SEY curve of BN thin film.

5.2.1.7 Silicon Nitride (Si_3N_4)

The SEY of 10 nm Si_3N_4 grown by low-pressure chemical vapor deposition (LPCVD) is measured in Fijol [48,49] using the discharge method, and the results are shown in Figure 5.11, with σ_m of about 3.0, E_{Pm} of about 300 eV, and E_{P2} of about 4 keV.

5.2.2 Organic Polymers

5.2.2.1 Polytetrafluoroethylene (PTFE)

The SEY of bulk PTFE is measured in Song [21], Willis [50], Hazelton [51], and Yang [52] by the pulse method and in Cai [31] by the neutralization method for bulk PTFE, and the SEY statistics curves of bulk PTFE are shown in Figure 5.12, with σ_m of 1.5–3, E_{Pm} of 200–1,000 eV, and E_{P2} of 1.5–4 keV.

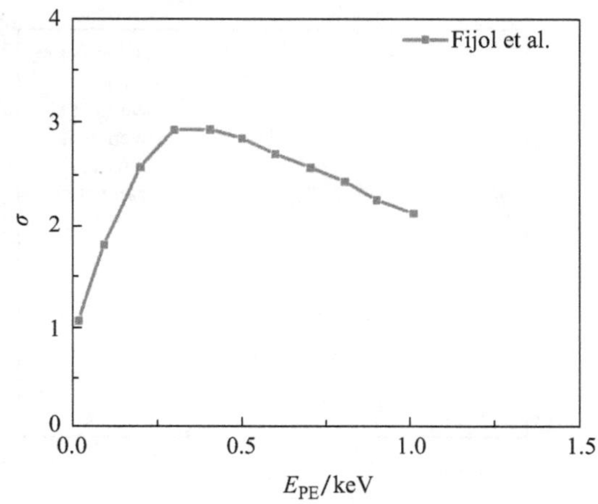

FIGURE 5.11 SEY curve of Si_3N_4 thin film.

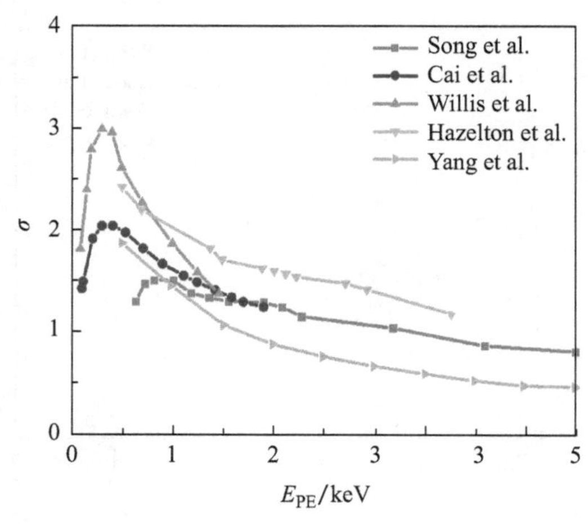

FIGURE 5.12 SEY curves of PTFE bulk.

5.2.2.2 Polyimide (PI)

The SEY of bulk PTFE is measured in Balcon [11] and Yang [53] by the compensation method, Chen [20] and Wu [54] by the pulse method for bulk PTFE, and Weng [28] and Wen [55] by the neutralization method for bulk PTFE, and the SEY statistics curves of the bulk PTFE are shown in Figure 5.13 shown, with σ_m of 1.8–2.6, E_{Pm} of 100–600 eV, and E_{P2} of 0.7–3 keV.

5.2.2.3 Polyethylene (PE)

The SEY of bulk PE is measured in Song [21], Zhou [56], and Wang [57] by pulse method and in Cazaux [58] by neutralization method for bulk PE, and the SEY statistics curves of bulk PE are shown in Figure 5.14, with σ_m of 2.2–3.0, E_{Pm} of 300–1,000 eV, and E_{P2} of 1.5–6 keV.

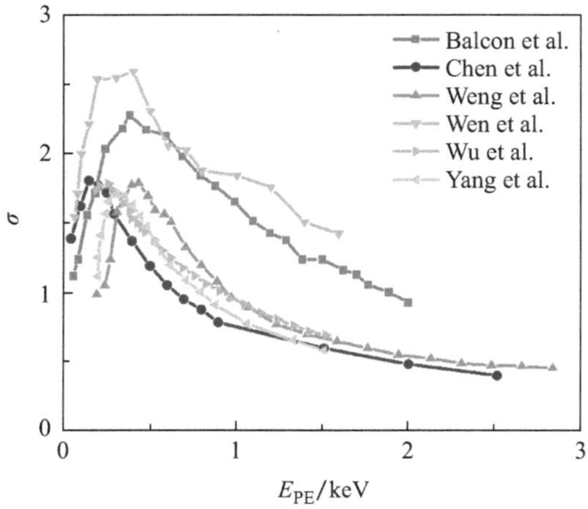

FIGURE 5.13 SEY curve of PI bulk.

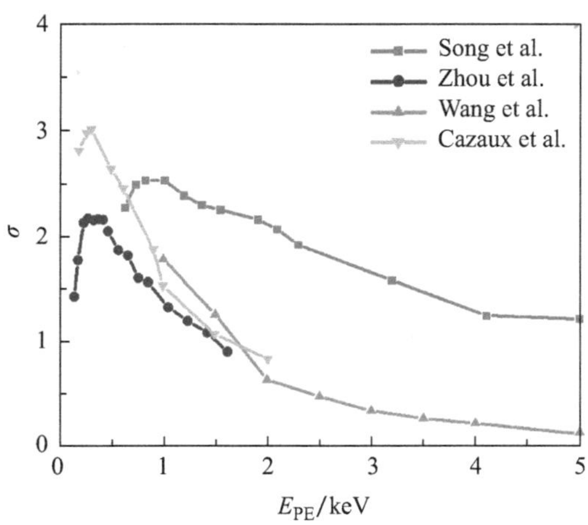

FIGURE 5.14 SEY curve of PE bulk.

5.2.2.4 Polymethyl Methacrylate (PMMA)

The SEY of bulk PMMA is measured in Huo [59] and Boubaya [60] using the pulse method and in Cazaux [58] using the neutralization method, and the SEY statistics curves of bulk PE are shown in Figure 5.15, with σ_m of 2.2–2.4, E_{Pm} of 400–900 eV, and E_{P2} of 1.2–4 keV.

5.2.3 Summary of the Section

The characteristic parameters, mainly including σ_m, E_{Pm}, and E_{P2}, of the above insulator secondary electron emission characteristics are statistically analyzed, as shown in Table 5.2. It can be found that there is a certain difference in the secondary electron

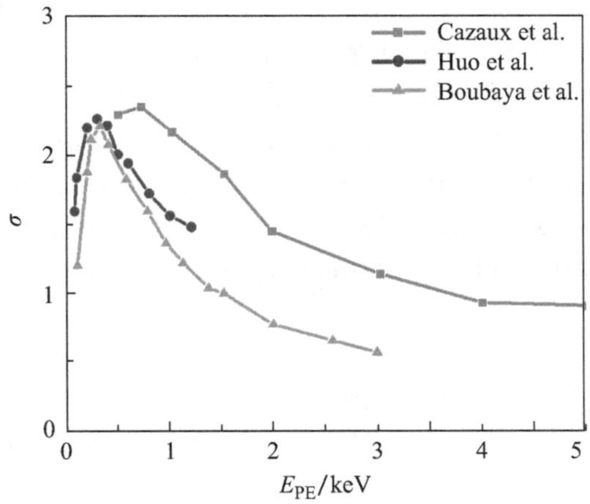

FIGURE 5.15 SEY curve of PMMA bulk.

TABLE 5.2 Statistics of Characteristic Parameters for Secondary Electron Emission from Insulators

Material	Film Layer Type	σ_m	E_{Pm}/eV	E_{P2}/eV
SiO_2	Bulk	2.1–4.4	300–500	2,000–5,000
	Thin film	3.3–4.3	300–500	3,000–6,000
Al_2O_3	Bulk	2.4–6.4	200–600	3,000–7,000
	Thin film	2.8–4.4	300–500	3,000–6,000
MgO	Bulk	7.2–23.3	800–1,200	8,000–30,000
	Thin film	4.2–18.2	300–1,500	3,000–20,000
Ferrite	Bulk	2.4–2.7	300–500	4,000–7,000
	Thin film	1.9–2.7	300–500	3,000–7,000
AlN	Bulk	~2.5	~1,000	~5,500
	Thin film	2.5–4.1	300–1,200	3,000–5,000
BN	Thin film	~2	~300	~2,000
Si_3N_4	Thin film	~3	~300	~4,000
PTFE	Bulk	1.5–3	200–1,000	1,500–4,000
PI	Bulk	1.8–2.6	100–600	700–3,000
PE	Bulk	2.2–3	300–1,000	1,500–6,000
PMMA	Bulk	2.2–2.4	400–900	1,200–4,000

emission characteristics between the same materials, among which there is a significant difference in the secondary electron emission between the bulk material of inorganic ceramics and the thin film material. The difference in secondary electron emission ability between different materials is large, among which the secondary electron emission of inorganic ceramics is significantly stronger than that of organic polymers, especially the strong secondary electron emission ability of MgO and the weaker secondary electron emission ability of PI and PTFE.

5.3 STUDY ON SECONDARY ELECTRON EMISSION MODULATION TECHNIQUES FOR INSULATORS

5.3.1 Secondary Electron Emission Suppression Modulation Techniques

The study of insulator secondary electron emission suppression technology and regulation theory is of great significance for suppressing the undesirable effects induced by secondary electrons in high-power microwave and radio frequency components, spacecraft and their components, and high-voltage vacuum insulating components. The current techniques that can achieve insulator secondary electron emission suppression include surface roughening treatment, surface low TEEY coating preparation, and elemental doping.

5.3.1.1 Surface Roughening

Surface roughening can be achieved by plasma etching, laser etching, machining, chemical etching, physical vapor deposition, and 3D printing processes, where plasma etching, laser etching, machining, and chemical etching processes are subtractive etching processes, and physical vapor deposition and 3D printing processes are additive manufacturing processes, which are all effective in increasing the surface roughness and thus achieving secondary electron emission suppression.

Plasma treatment is the reaction of reactive gases with materials to achieve surface cleaning and roughening. When the cover glass and polyimide were treated with oxygen plasma, their surface morphology is shown in Figure 5.16 [54,61]. It can be observed that as the oxygen plasma treatment dose increases, the roughness of the cover glass and polyimide increases. Figure 5.17 shows the SEY curves before and after oxygen plasma treatment, and it can be found that the SEY gradually decreases with the increase of the treatment dose, in which the σ_m of the coverslip decreases from 8 to 2, and the σ_m of the PI decreases from 1.7 to 0.8.

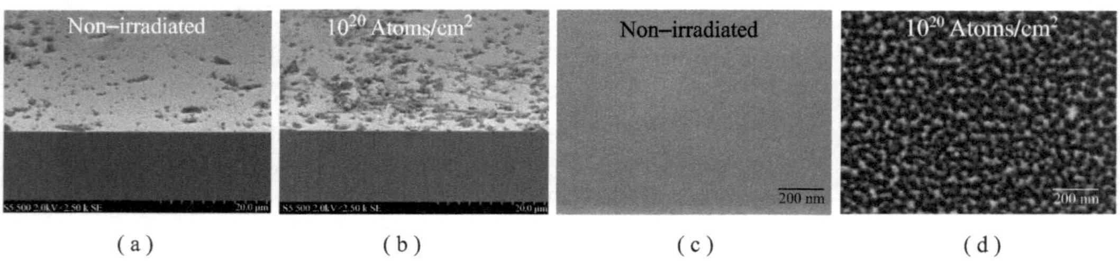

FIGURE 5.16 SEM images of samples before and after oxygen plasma treatment. (a) Before coverslip treatment. (b) After coverslip treatment. (c) Before polyimide treatment. (d) After polyimide treatment.

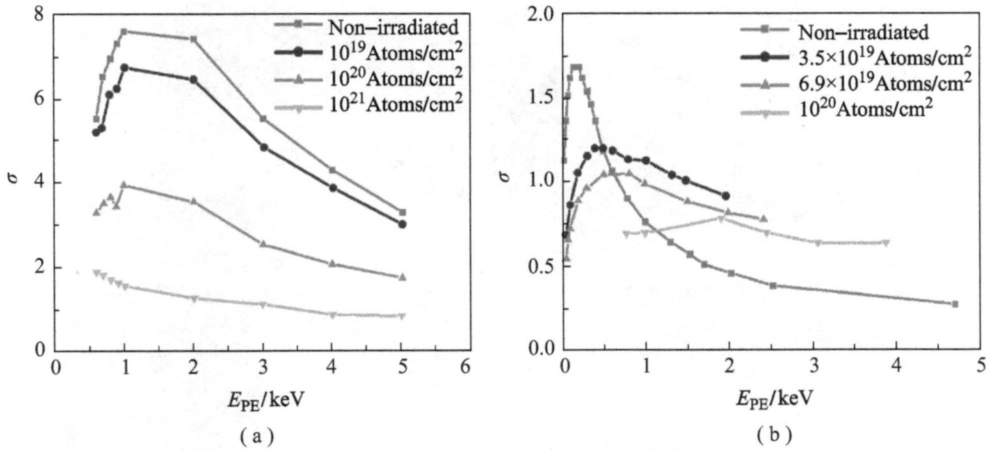

FIGURE 5.17 SEY curves of samples before and after oxygen plasma treatment. (a) SEY before and after coverslip treatment. (b) SEY before and after polyimide treatment.

It is also possible to carry out the treatment with other gaseous plasma such as the treatment of epoxy (EP) and high-density polyethylene (PE) with a hybrid F_2/N_2 plasma. It was found that after 30 minutes of treatment, the organic roughness was the largest, and SEY was the lowest, and the σ_m of EP and PE were reduced from 2.4 and 2.2 to 2, respectively, and all of them could effectively improve the antistatic discharge capability.

Laser etching is the process of inducing physical or chemical reactions in materials by thermal action to produce regular or irregular surface structures and to achieve the goal of suppressing secondary electron emission from insulating materials. When a regular microstructure is prepared on the surface of a ferrite medium by laser etching, its surface structure is shown in Figure 5.18. By measuring its SEY, as shown in Figure 5.19, it can be observed that the SEY of laser-etched samples with regular microstructures is significantly reduced with σ_m from 2.6 to 1.6, and after the samples with regular microstructures are placed in the ring-shaped dielectric-metal system reactor, their micro-discharge threshold is increased by 8 times [41].

The goal of mechanical processing is to reduce the TEEY of insulators by processing the insulator surface into a surface with a random rough or regular structure through physical methods such as cutting or grinding. For example, a triangular groove structure is prepared on the PMMA surface by machining, the structure of which is shown in Figure 5.20, and the SEY of the grooved sample is significantly lower than that of the untreated sample, its σ_m is reduced from 2.3 to 1.7, and the threshold of flashover discharge is almost doubled [59].

The goal of chemical etching is to reduce the TEEY of insulators by processing the insulator surface into a surface with a random rough or regular structure through chemical reaction. For example, PMMA is doped with SiO_2 particles of different sizes and then prepared into a sample, and after the sample is prepared and molded, the surface structure with different size microstructures can be obtained by etching SiO_2 with HF, as shown in Figure 5.21 [62].

FIGURE 5.18 Morphology of the sample before and after laser etching. (a) Untreated plan view. (b) 3D view of untreated sample. (c) Microstructure plan view. (d) 3D view of microstructure.

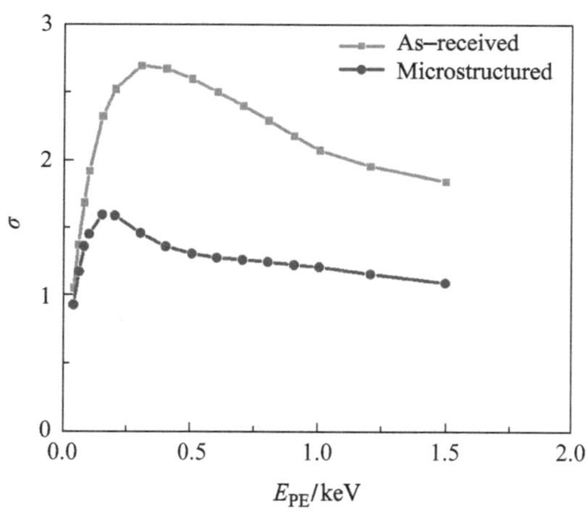

FIGURE 5.19 SEY curves of samples before and after laser etching.

Low-pressure evaporation can achieve nanostructure preparation on the surface, as shown in Figure 5.22a. 3D printing achieves surface addition by melting and curing and can also achieve the preparation of various surface structures, as shown in Figure 5.22b. At present, these two methods are less widely used in suppressing secondary electron emission on the surface of insulators.

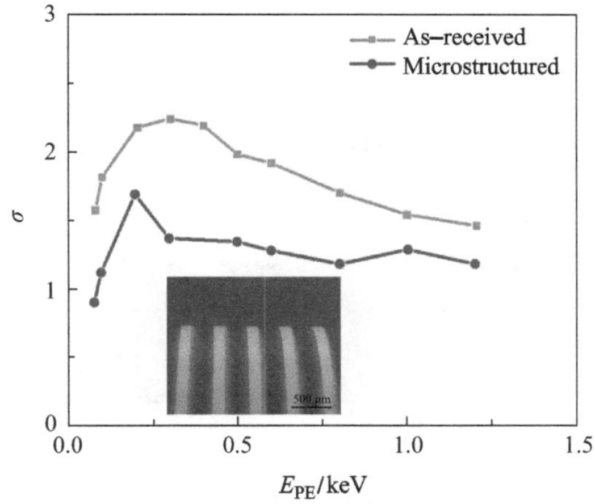

FIGURE 5.20 SEY curves of samples before and after machining.

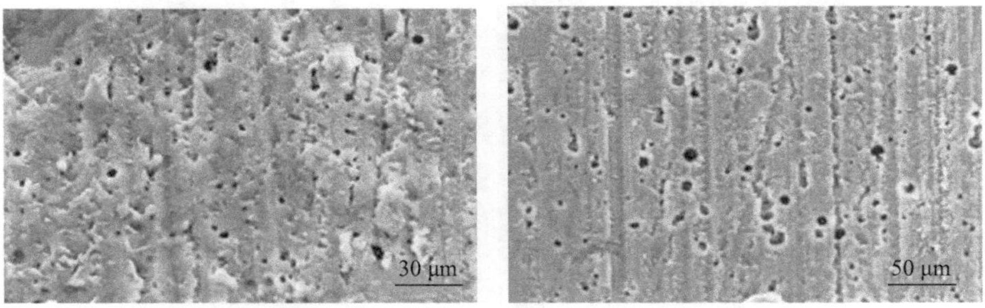

FIGURE 5.21 Surface microstructure prepared by chemical corrosion.

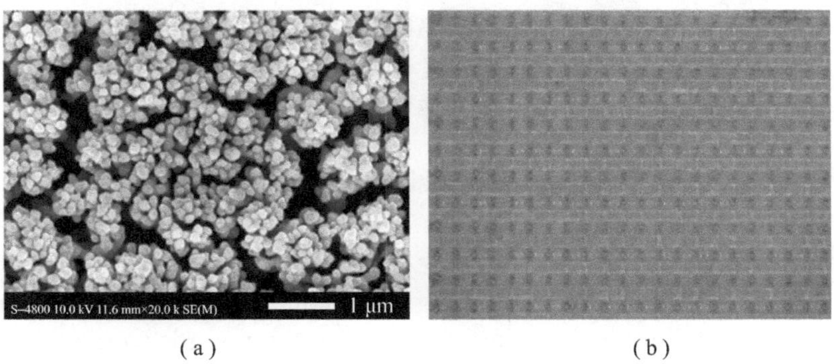

FIGURE 5.22 Surface morphology of low-pressure evaporation and 3D printed samples. (a) Low-pressure evaporated Ag film. (b) 3D printed PTFE film.

5.3.1.2 Low-SEY Coating

Surface low-SEY coating preparation can be realized by plating low-SEY films, such as by plating TiN films on Al_2O_3, which can effectively reduce the SEY of the Al_2O_3 substrate, as shown in Figure 5.23, whose σ_m is reduced from 2.5 to 1.9 [34], and the inset of which shows the surface morphology of TiN films.

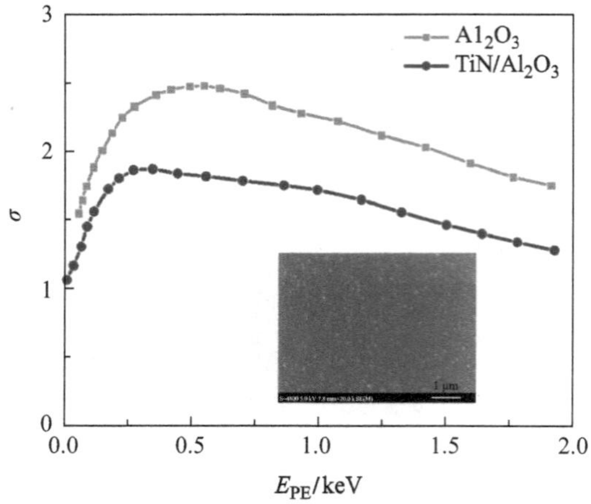

FIGURE 5.23 SEY curves of pure Al_2O_3 and Al_2O_3 coated with TiN films.

FIGURE 5.24 Surface morphology and elemental distribution of Cr_2O_3-doped PIs.

5.3.1.3 Elemental Doping

Elemental doping is a method of regulating the secondary electron emission characteristics of insulators by doping elemental elements or compounds to modify the physical properties of the material. For example, Cr_2O_3 particles are doped into PI, the structure of which is shown in Figure 5.24. The SEY is tested as shown in Figure 5.25, and the SEY of the doped PI film decreases, and its σ_m decreases from 2.6 to about 1, resulting in a 50% increase in the flash discharge threshold of the device [53].

5.3.1.4 Composite Process

More effective secondary electron emission suppression can be achieved by combining two or more of the above secondary electron emission suppression methods. For example, combining laser processing to prepare microstructures and TiN thin films can

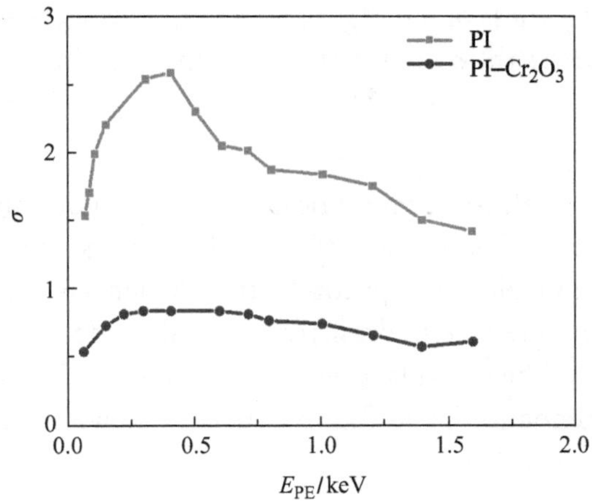

FIGURE 5.25 SEY curves of pure and Cr_2O_3-doped PIs.

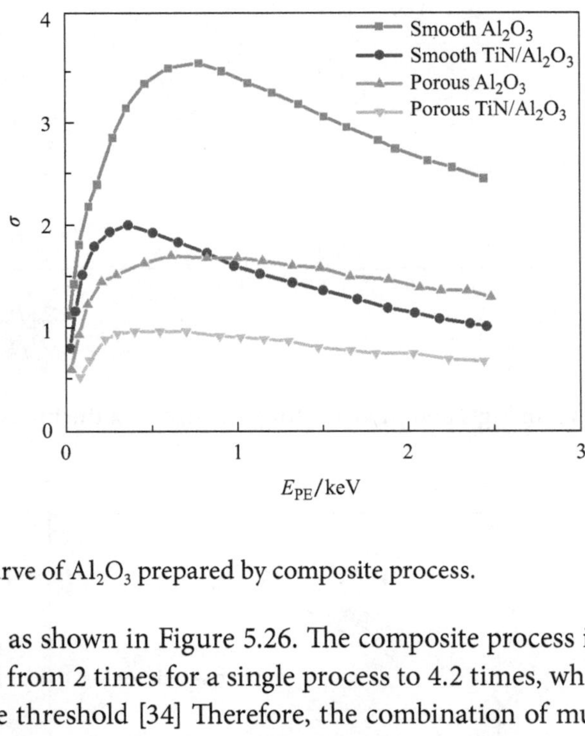

FIGURE 5.26 SEY curve of Al_2O_3 prepared by composite process.

further reduce SEY, as shown in Figure 5.26. The composite process increases the micro-discharge threshold from 2 times for a single process to 4.2 times, which greatly improves the micro-discharge threshold [34] Therefore, the combination of multiple processes can further improve the suppression of secondary electron emission in materials, reduce the TEEY of materials, and reduce the occurrence of adverse effects caused by secondary electrons (Figure 5.26).

5.3.2 Secondary Electron Emission Enhancement Modulation Technique

1. **High SEY Coating**

   Surface high TEEY coating preparation can be achieved by plating high TEEY films such as Al_2O_3 and MgO films. The microchannel plates are coated with MgO and Al_2O_3 coating by atomic layer deposition (ALD), and it is found that SEY increased with increasing thickness of the MgO and Al_2O_3 coating, as shown in Figure 5.27, in which Al_2O_3 enhances the σ_m of the substrate up to 3, whereas MgO enhances the σ_m up to 5 [35].

2. Elemental Admixture

The application of electron multiplication requires not only a high electron emission coefficient but also good conductivity and no charge accumulation. Research has shown that doping insulators with conductive elements (such as B, Si, Al, Ti, Au, Zn, and Zr) can not only reduce the resistivity but also effectively improve the electron emission capability. The secondary electron emission of MgO thin films doped with Al is significantly enhanced, and the SEY curve is shown in Figure 5.28 with increasing σ_m from 9.7 to 10.4 [63].

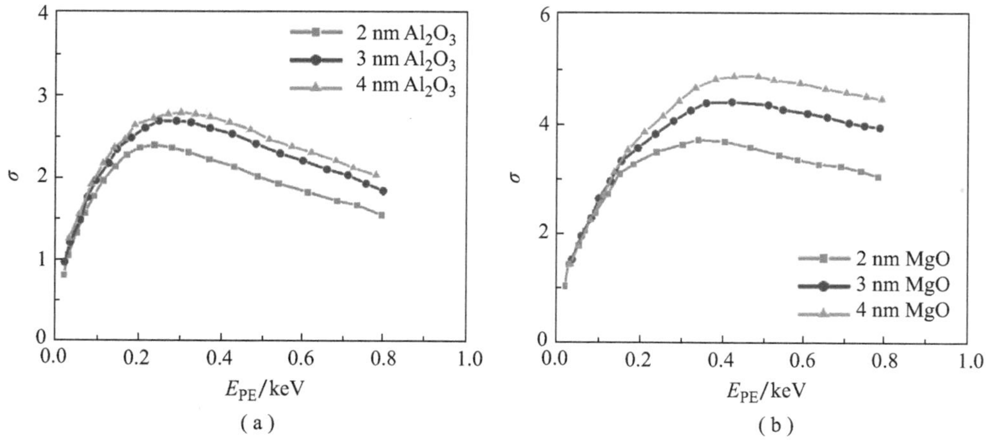

FIGURE 5.27 SEY curves of MgO and Al_2O_3 coating with different thicknesses. (a) MgO. (b) Al_2O_3.

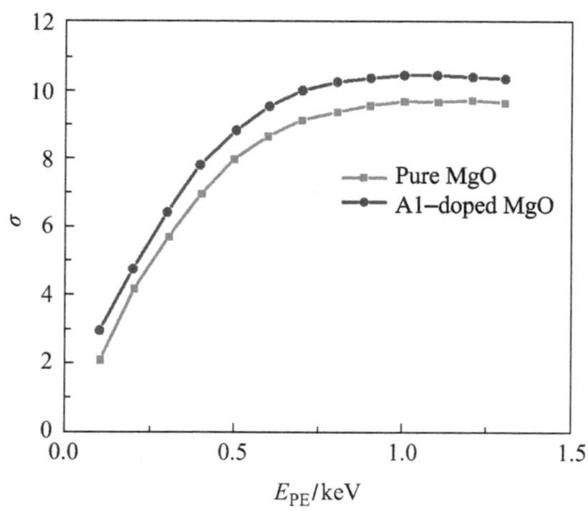

FIGURE 5.28 SEY curves of elemental Al-doped MgO.

5.4 SUMMARY OF THE CHAPTER

This chapter provides a systematic summarization and analysis of insulator secondary electron emission property testing methods, SEY of typical insulator materials, and insulator material secondary electron emission modulation techniques.

Based on the summarization and analysis of the insulator secondary electron emission property test methods, it can be found that the pulse method and the neutralization method are the most mature, and they are most widely used in the actual measurement of insulator secondary electron emission properties. However, there is no uniform standard for evaluating the degree of charge neutralization and the accuracy of SEY measurement, and it is of great significance for the development of insulator SEY test methods if a more systematic measurement evaluation standard can be established.

Based on the summarization and comparative analysis of the secondary electron emission properties of typical insulator materials, it can be found that oxide insulators have a higher SEY, while the SEY of organic materials is relatively low. However, from the summary, it can be found that the SEY of the same material has a large difference in the measured value between different literatures, so the study of the source of the difference in the electron emission properties of different materials is of great significance to the understanding of the electron emission properties of insulator materials and their influencing factors and also has a very great reference value for the application of the secondary electron emission of insulator materials.

Based on the summarization and analysis of the secondary electron emission control technology of insulator materials, it can be found that the surface microstructure and surface coating play a major role in the control of secondary electron emission of materials. Among them, the preparation of surface coating can not only inhibit the secondary electron emission but also enhance the secondary electron emission. Surface coating is an important means to regulate secondary electron emission in applications, so the study of surface coating with different electron emission characteristics has a very important application value. The research on surface microstructure mainly focuses on the inhibition of secondary electron emission, so exploring the mechanism and application scenarios of the enhancement of secondary electron emission by surface microstructure is a very worthwhile research direction.

REFERENCES

1. Zarbout K, Ahmed A S, Moya G, et al. Stability of trapped charges in sapphires and alumina ceramics: evaluation by secondary electron emission. *Journal of Applied Physics*, 2008, 103(5): 6334.
2. Wang D, Cai Y H, Mao Z S, et al. Evolution of dielectric surface potential induced by electron beam radiation. *Chinese Science Bulletin*, 2022, 67(2): 212–220.
3. Johnson J B, McKay K G. Secondary electron emission of crystalline MgO. *Physical Review*, 1953, 91(3): 582.
4. Whetten N R, Laponsky A B. Secondary electron emission of single crystals of MgO. *Journal of Applied Physics*, 1957, 28(4): 515–516.
5. Lee J, Jeong T, Yu S G, et al. Thickness effect on secondary electron emission of MgO layers. *Applied Surface Science*, 2001, 174(1): 62–69.
6. Ikeda S, Oohashi Y, Shimoyama H. Measurement of electron emission yield of insulators without charging using SEM. *IEEJ Transactions on Electronics, Information and Systems*, 2006, 126(12): 1418–1423.

7. Cai Y, Wang D, Ye M, et al. A new method for measuring total electron emission yield of insulators. *Review of Scientific Instruments*, 2020, 91(9): 095111.
8. Von Seggern H. Charging dynamics of dielectrics irradiated by low energy electrons. *IEEE Transactions on Nuclear Science*, 1985, 32(4): 1503–1511.
9. Belhaj M, Tondu T, Inguimbert V. Experimental investigation of the effect of the internal space charge accumulation on the electron emission yield of insulators submitted to e-irradiation: application to polycrystalline MgO. *Journal of Physics D-Applied Physics*, 2009, 42(14): 145306
10. Belhaj M, Tondu T, Inguimbert V. Effect of the incident electron fluence on the electron emission yield of polycrystalline Al_2O_3. *Applied Surface Science*, 2011, 257(10): 4593–4596.
11. Balcon N, Payan D, Belhaj M, et al. Secondary electron emission on space materials: Evaluation of the total secondary electron yield from surface potential measurements. *IEEE Transactions on Plasma Science*, 2011, 40(2): 282–290.
12. Johnson J B. Secondary electron emission from targets of barium-strontium oxide. *Physical Review*, 1948, 73(9): 1058–1073.
13. Henrich V E. Fast, accurate secondary electron yield measurements at low primary energies. *Review of Scientific Instruments*, 1973, 44(4): 456–462.
14. Kishimoto Y, Hayashi T, Hashimoto M, et al. Secondary electron emission from polymers and its application to the flexible channel electron multiplier. *Journal of Applied Polymer Science*, 1977, 21(10): 2721–2733.
15. Yong Y C, Thong J T L, Phang J C H. Determination of secondary electron yield from insulators due to a low-kV electron beam. *Journal of Applied Physics*, 1998, 84(8): 4543–4548.
16. Fakhfakh S, Jbara O, Belhaj M, et al. An experimental approach for dynamic investigation of the trapping properties of glass-ceramic under electron beam irradiation from a scanning electron microscope. *European Physical Journal-Applied Physics*, 2003, 21(2): 137–146.
17. Xie A G, Guo S L, Li C Q, et al. Device for measuring secondary electron emission yield of insulator. *High Power Laser and Particle Beams*, 2007, 19(1): 134–138.
18. Fakhfakh S, Jbara O, Rondot S, et al. An experimental approach for measuring surface potential and second crossover energy in insulators. *Journal of Physics D-Applied Physics*, 2008, 41(10): 105402.
19. Fakhfakh S, Jbara O, Rondot S, et al. Analysis of electrical charging and discharging kinetics of different glasses under electron irradiation in a scanning electron microscope. *Journal of Applied Physics*, 2010, 108(9): 093705.
20. Chen Y, Kouno T, Toyoda K, et al. Total electron emission yield measurement of insulator by a scanning small detector. *Applied Physics Letters*, 2011, 99(15): 152101.
21. Song B P, Shen W, Mu H B, et al. Measurements of secondary electron emission from dielectric window materials. *IEEE Transactions on Plasma Science*, 2013, 41(8): 2117–2122.
22. Handel K J, Jensen A S, Siedband M P. A two electron gun technique for the measurement of secondary emission characteristics of a variety of materials. *IEEE Transactions on Electron Devices*, 1966, ED-13(6): 525–528.
23. Authinarayanan A, Dudding R W. Changes in secondary electron yield from reduced lead glasses. *Advances in Imaging and Electron Physics*, 1976, 40(A): 167–181.
24. Sun Y N, Zhang J Q, Cheng Y F. Device for measuring secondary electron emission yield. *Vacuum and Cryogenics*, 1983, 2(4): 1–4.
25. Lu X J, Yu Z N, Zheng D X. Design of the about devices for measuring the secondary emission coefficient of MgO films. *Vacuum Electronics*, 2000, 228(5): 5–8.
26. Hopman H J, Alberda H, Attema I, et al. Measuring the secondary electron emission characteristic of insulators. *Journal of Electron Spectroscopy and Related Phenomena*, 2003, 131–132: 51–60.
27. Weng M, Cao M, Zhao H J, et al. Secondary electron yield measurement of insulator based on negatively biased collector. *Chinese Journal of Vacuum Science and Technology*, 2014, 34(11): 1262–1266.

28. Weng M, Cao M, Zhao H J, et al. Note: a simple charge neutralization method for measuring the secondary electron yield of insulators. *Review of Scientific Instruments*, 2014, 85(3): 036108.
29. Wu Q. *Research on the Measurement Device of Secondary Electron Emission Coefficient*. Chengdu: University of Electronic Science and Technology of China, 2016.
30. Li H L. *Design of Hardware and Software for Measurement and Control System of Secondary Electron Emission Coefficient*. Chengdu: University of Electronic Science and Technology of China, 2017.
31. Cai Y, Wang D, Qi K, et al. Measurement of total electron emission yield of insulators based on self-terminating charge neutralization. *Review of Scientific Instruments*, 2022, 93(5): 055103.
32. Ooka K, Dunn B, Mackenzie J D. Secondary electron emission from glass and its practical applications. *Journal of Non-Crystalline Solids*, 1973, 12(1): 1–17.
33. Yi W, Jeong T, Yu S G, et al. Study of the secondary-electron emission from thermally grown SiO_2 films on Si. *Thin Solid Films*, 2001, 397(1–2): 170–175.
34. Wang D, Mao Z, Ye Z, et al. Ultralow electron emission yield achieved on alumina ceramic surfaces and its application in multipactor suppression. *Journal of Physics D-Applied Physics*, 2022, 55(45): 455301.
35. Jokela S J, Veryovkin I V, Zinovev A V, et al. Secondary electron yield of emissive materials for large-area micro-channel plate detectors: surface composition and film thickness dependencies. *Physics Procedia*, 2012, 37: 740–747.
36. Guo J, Wang D, Xu Y, et al. Secondary electron emission characteristics of Al_2O_3 coatings prepared by atomic layer deposition. *AIP Advances*, 2019, 9(9): 095303.
37. Whetten N R, Laponsky A B. Secondary electron emission of single crystals of MgO. *Journal of Applied Physics*, 1957, 28(4): 515–516.
38. Whetten N R, Laponsky A B. Secondary electron emission of MgO thin films. *Journal of Applied Physics*, 1959, 30(3): 432–435.
39. Cao W, Wang B, Yang Y, et al. Secondary electron emission characteristics of the Al_2O_3/MgO double-layer structure prepared by atomic layer deposition. *Ceramics International*, 2021, 47(7): 9866–9872.
40. Lee J, Jeong T, Yu S G, et al. Thickness effect on secondary electron emission of MgO layers. *Applied Surface Science*, 2001, 174(1): 62–69.
41. Li Y, Cui W Z, He Y N, et al. Enhanced dynamics simulation and threshold analysis of multipaction in the ferrite microwave component. *Physics of Plasmas*, 2017, 24(2): 023505.
42. Li Y, Wang D, Yu M, et al. Experimental verification of multipactor discharge dynamics between ferrite dielectric and metal. *IEEE Transactions on Electron Devices*, 2018, 65(10): 4592–4599.
43. Wang D. Characteristics and Controllable Modulation Methods of Space Materials' Secondary Electron Emission. Xi'an: Xi'an Jiaotong University, 2019.
44. Yamamoto K, Kamiya J, Ogiwara N, et al. Estimation of secondary electron effect in the J-PARC rapid cycling synchrotron after first study. *Applied Surface Science*, 2009, 256(4): 958–961.
45. Michizono S, Saito Y, Yamaguchi S, et al. Dielectric materials for use as output window in high-power klystrons. *IEEE Transactions on Electrical Insulation*, 1993, 28(4): 692–699.
46. Buzulutskov A, Bondar A, Mironenko L, et al. Study of AlN coatings for microstrip gas chambers. *Nuclear Instruments and Methods in Physics Research Section A: Accelerators, Spectrometers, Detectors and Associated Equipment*, 1998, 409(1–3): 33–36.
47. Lau S P, Huang L, Yu S F, et al. Enhanced secondary electron emission from group III nitride/ZnO coaxial nanorod heterostructures. *Small*, 2006, 2(6): 736–740.
48. Villemant M, Belhaj M, Sarrailh P, et al. Measurements of electron emission under electron impact on BN sample for incident electron energy between 10 eV and 1000 eV. *Europhysics Letters*, 2019, 127(2): 23001.

49. Prodanović V, Chan H W, Graaf H V D, et al. Ultra-thin alumina and silicon nitride MEMS fabricated membranes for the electron multiplication. *Nanotechnology*, 2018, 29(15): 155703.
50. Willis R F, Skinner D K. Secondary electron emission yield behaviour of polymers. *Solid State Communications*, 1973, 13(6): 685–688.
51. Hazelton R C, Yadlowsky E J, Churchill R J, et al. Effect of material parameters on the charging characteristics of irradiated dielectrics. *IEEE Transactions on Nuclear Science*, 1981, 28(6): 4541–4546.
52. Yang K Y, Hoffman R W. Electron yields and escape depths from Kapton and Teflon. *Surface and Interface Analysis*, 1987, 10(2–3): 121–125.
53. Yang X, Sun G, Zhou R, et al. Ultralow secondary electron emission and improved vacuum surface insulation of polyimide with scalable nanocomposite coating. *Applied Surface Science*, 2022, 592: 153221.
54. Wu J, Miyahara A, Khan A R, et al. Effects of ultraviolet irradiation and atomic oxygen erosion on total electron emission yield of polyimide. *IEEE Transactions on Plasma Science*, 2013, 42(1): 191–198.
55. Wen K, Liu S, Yan B, et al. New neutralization method for measuring the secondary electron yield of insulative material. *Radiation Detection Technology and Methods*, 2020, 4(3): 319–326.
56. Zhou R D, Sun G Y, Song B P, et al. Mechanism of F_2/N_2 fluorination mitigating vacuum flashover of polymers. *Journal of Physics D: Applied Physics*, 2019, 52(37): 375304.
57. Wang W, Li S, Min D. Enhanced flashover strength in polyethylene nanodielectrics by secondary electron emission modification. *AIP Advances*, 2016, 6(4): 045022.
58. Cazaux J. A new model of dependence of secondary electron emission yield on primary electron energy for application to polymers. *Journal of Physics D: Applied Physics*, 2005, 38(14): 2433.
59. Huo Y, Liu W, Ke C, et al. Sharp improvement of flashover strength from composite micro-textured surfaces. *Journal of Applied Physics*, 2017, 122(11): 115105.
60. Boubaya M, Blaise G. Charging regime of PMMA studied by secondary electron emission. *The European Physical Journal-Applied Physics*, 2007, 37(1): 79–86.
61. Nitta K, Miyazaki E. Morioka C. Secondary electron emission measurements of atomic-oxygen-iirradiated cover glasses. *Proceedings of the 3rd AIAA Atmospheric Space Environments Conference*. AIAA, New York, 2011: 3826.
62. Yankun H, Wenyuan L, Changfeng K, et al. Construction of micro-structure on polymer insulators and their surface flashover characteristics. *High Power Laser and Particle Beams*, 2018, 30(3): 035006-1–035006-7.
63. Gao B, Li J, Hu W, et al. Analysis of secondary electron emission properties of MgO/Au composite film with an Al-doped MgO surface layer. *AIP Advances*, 2018, 8(11): 115031.

CHAPTER 6

Charging Effect and Dynamic Secondary Electrons Emission

Guobao Feng and Wanzhao Cui

6.1 CHARGING PROPERTIES OF THE DIELECTRIC MATERIALS AFTER ELECTRON-BEAM IRRADIATION

Dielectric microwave components and the phenomenon of secondary electron multiplication in them have received more attention with the increase in high power and integration requirements of microwave devices [1–3]. However, the imbalance between incident and emitted electrons in dielectric materials exposed to electron irradiation can lead to internal charge deposition, unlike in metallic materials with higher electrical conductivity [4]. This irradiation-induced dielectric charging can alter the spatial electric field of the microwave component, while having a more complex effect on the secondary electron multiplication phenomenon of the microwave dielectric component.

The dielectric sample will produce surface charging by the incident electrons. For a sample with a grounded lower surface, the charge will gradually diminish as the incident electrons stop and the charge continues to leak, which is the discharge process of the sample. For dielectric materials with low conductivity, the net charge from the imbalance of incident and emitted electrons does not conduct out of the sample very smoothly. Therefore, the ability of internal charge leakage becomes an important factor affecting the charged state of the sample [5].

In previous experimental measurements, studies have focused more on the variation of each current flowing through the sample during the sample discharge phase, while it is difficult to directly measure the relevant micro-physical quantities such as sample potential, internal electric field and charge distribution. In addition, numerical studies of the transient discharge characteristics of electron-beam irradiated dielectric samples are still lacking.

In this chapter, a more complete numerical physical model is established by using numerical simulation methods to study the transient discharge properties of the dielectric samples after electron-beam irradiation, where the interaction between the incident

DOI: 10.1201/9781003562863-6

electrons and the sample is treated by the Monte Carlo (MC) method considering of the elastic and inelastic scattering collision process between the electron and the material, while the leakage transport process of the charge inside the sample is treated by the finite-difference time-domain (FDTD) method, which considers the complex processes of charge migration, diffusion, and trapping [6]. The microscopic transient processes such as charge, potential, and electric field inside the dielectric sample during the transient discharge after irradiation have been simulated. The distribution of each microscopic quantity charged in the dielectric sample during the transient discharge process and its influence by the material sample parameters including sample thickness, electron mobility, and defect-trapped density are investigated. The method and results of this numerical study are of great importance for the analysis of the charge during the secondary electron multiplication of dielectric microwave components and for the in-depth study of the multipactor process of dielectric microwave components.

6.1.1 Charge Transport Theory and Model

The charge deposited in the sample during scattering undergoes charge transport in the presence of a built-in field and a charge concentration gradient. Microscopically, in the vast majority of cases, the number of incident electrons is not equal to the number of secondary electrons emitted, leaving the sample with a net residual charge. From the secondary electron yield (SEY) curve, it is known that when the incident electron energy is greater than the second critical energy, the SEY is less than 1. This means that the secondary electrons emitted are less than the incoming irradiated electrons, resulting in a net negative charge accumulation in the sample. Microscopically, the free electrons deposited in the sample migrate and diffuse under the action of the electric and charge gradient fields generated by the accumulated electrostatic charges in the sample. In addition, during the electron migration and diffusion, some of the free electrons are captured by the trapping center due to defects in the material itself.

As shown in Figure 6.1, the charging process of electron non-penetrating irradiated dielectric film samples requires consideration of the scattering process of incident electrons

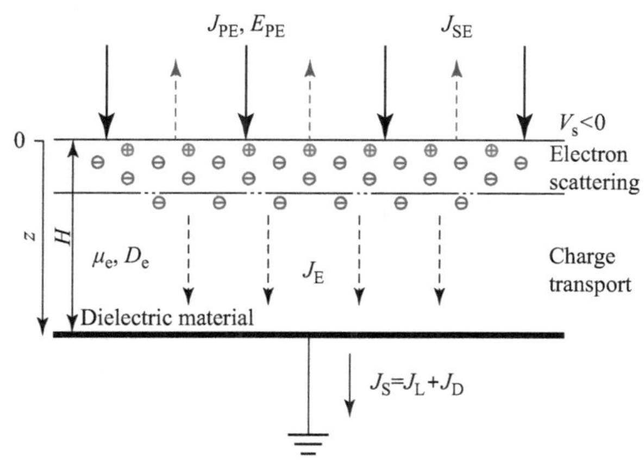

FIGURE 6.1 Schematic diagram of the electron-beam irradiated dielectric sample charging process.

and the transport process of internal charges. When the electrons with the energy of E_{PE} enter the sample, they will be elastically and inelastically scattered with the internal atoms of the sample and excited to produce inner secondary electron and hole pairs, and a part of the inner secondary electrons will leave the surface and be emitted as secondary electrons, forming the secondary electron current J_{SE}, and a part of the charge inside the sample will be transported to the bottom of the sample under the action of the built-in field and the charge concentration gradient, forming the transport current J_E. In the process of electron transport to the bottom of the sample, some of the free electrons will be captured by the defects inside the sample to form trapped electrons due to the existence of defects inside the sample. The free charge that is not captured reaches the bottom of the sample and forms the leakage current J_D. In addition, according to the current continuity principle, the displacement current J_D is generated at the bottom of the metal grounded plate due to the induced electric field. For grounded metal lower plate, the actual measured sample current J_S is the sum of the leakage current J_L and the displacement current J_D, i.e., $J_S = J_D + J_L$.

We will study the widely concerned phenomenon of negative charging of dielectric samples for conditions where the incident electron energy is greater than the second critical energy, which is commonly used in experimental environments. In the case of negatively charging, since the number of holes inside the sample is smaller than that of electrons and holes are bound near the shallow surface layer, the scattering accumulation of equivalent negative charges (net electrons) and the transport migration can be considered when simulating the sample charging process.

During the transport process, the free electrons $n_F(z,t)$ and trapped electrons $n_{trap}(z,t)$ inside the sample satisfy the current continuity equation, charge transport equation, capture equation and Poisson's equation, i.e.

$$\frac{\partial\left[n_F(z,t)+n_{trap}(z,t)\right]}{\partial t}=\frac{\nabla\cdot \boldsymbol{J}_E(z,t)}{e}, \tag{6.1}$$

$$\boldsymbol{J}_E(z,t)=-e\mu_e n_F(z,t)\nabla V(z,t)+eD_e n_F(z,t), \tag{6.2}$$

$$\frac{\partial n_{trap}(z,t)}{\partial t}=\frac{n_F(z,t)}{\tau_e}\frac{N_T-n_{trap}(z,t)}{N_T}, \tag{6.3}$$

$$\nabla^2 V(z,t)=\frac{e\left[n_F(z,t)+n_{trap}(z,t)\right]}{\varepsilon}, \tag{6.4}$$

where

N_T is the trapped density (cm^{-3}) of the sample material;

τ_e is the trapping time constant (s);

e is the individual electron power;

ε is the relative permittivity;

μ_e and D_e are the electron mobility and diffusion coefficient of the dielectric material, respectively, and satisfy the Nernst–Einstein equation $D_e = \mu_e k_B T/e$, where k_B is the Boltzmann constant, and the temperature $T=300$ K.

In addition, in charged simulations, the transient total charge density $Q(t_i)$ within the sample can be found by time domain charge conservation or by integrating of the space charge density in the z-direction, i.e.

$$Q(t_i)=\int_0^{t_i} J_D \, dt=\int_0^{t_i} \varepsilon \frac{\partial\left[\nabla V(z,t)\big|_{z=H}\right]}{\partial t} dt$$

or

$$Q(t_i)=\int_0^{H} e\left[n_F(z,t)+n_{trap}(z,t)\right]dz. \tag{6.5}$$

In the numerical solution of the transport process, for the above-mentioned set of charge transport equations, the finite-difference time-domain (FDTD) method is used in this book. Let the upper surface $z=0$, and the sample is discretized with the mesh in the depth direction facing downward as the positive direction. The relationship between the total number of longitudinal meshes N and the mesh spacing d and the sample thickness H is $H=d\cdot N$. Using the FDTD solution, the electron density relation at moments t and $t+\Delta t$ in the depth direction j can be obtained as

$$\begin{aligned}\frac{n_j(t+\Delta t)-n_j(t)}{\Delta t}&=D_e\frac{n_{j-1}(t)+n_{j+1}(t)-2n_j(t)}{d^2}-\mu_e n_j(t)\frac{V_{j-1}(t)+V_{j+1}(t)-2V_j(t)}{d^2}\\&-\mu_e\frac{\left(n_{j+1}(t)-n_{j+1}(t)\right)\left(V_{j+1}(t)-V_{j-1}(t)\right)}{4d^2}-\frac{n_F(z,t)}{\tau_e}\frac{N_T-n_{trap}(z,t)}{N_T}\end{aligned} \tag{6.6}$$

The above equation holds only for the electron density relation inside the sample, while for the upper and lower surfaces of the sample, special treatment is required for the case of the upper and lower surfaces of the sample since $j-1$ and $j+1$ will no longer exist. For the upper surface with $j=0$, according to the natural spline boundary condition and using Poisson's equation, the electron density relation at the upper surface in the depth direction at moments t and $t+\Delta t$ satisfies

$$\begin{aligned}\frac{n_j(t+\Delta t)-n_j(t)}{\Delta t}&=\mu_e n_j(t)\rho_j(z,t)/\varepsilon_0\varepsilon_r-\mu_e\frac{\left(n_{j+1}(t)-n_{j+1}(t)\right)\left(V_{j+1}(t)-V_{j-1}(t)\right)}{d^2}\\&-\frac{n_F(z,t)}{\tau_e}\frac{N_T-n_{trap}(z,t)}{N_T}\end{aligned} \tag{6.7}$$

For the lower surface of $j=N$, the grounding of the bottom plate makes the potential there 0 ($V_N(t)=0$), so the electron density relation at the lower surface in the depth direction at the moments t and $t+\Delta t$ satisfies

$$\frac{n_j(t+\Delta t)-n_j(t)}{\Delta t}=\mu_e n_j(t)\rho_j(z,t)/\varepsilon_0\varepsilon_r-\mu_e\frac{\left(n_j(t)-n_{j-1}(t)\right)\left(0-V_{j-1}(t)\right)}{d^2} \\ -\frac{n_F(z,t)}{\tau_e}\frac{N_T-n_{trap}(z,t)}{N_T} \tag{6.8}$$

To solve the Poisson equation for the internal potential, the one-dimensional coordinate form is

$$\frac{\partial^2 V}{\partial z^2}=\frac{-e\left(-n_F(z,t)-n_{trap}(z,t)\right)}{\varepsilon_0\varepsilon_r} \tag{6.9}$$

Then, according to the one-dimensional difference form of the second-order differential:

$$\left.\frac{\partial^2 V}{\partial z^2}\right|_j=\frac{1}{d^2}\left(V_{j+1}-2V_j+V_{j-1}\right) \tag{6.10}$$

Substituting them into the above equation, Poisson's equation can be written as

$$V_{j+1}-2V_j+V_{j-1}=-d^2\frac{e\left(-n_F(j,t)-n_{trap}(j,t)\right)}{\varepsilon_0\varepsilon_r} \tag{6.11}$$

Similarly, it can be seen that the above equation still does not fit for the case of the upper surface of the sample. Moreover, considering the boundary condition of Poisson's equation at the upper surface, a continuous boundary condition of the electric displacement vector should be used due to the different dielectric constants, i.e.,

$$\varepsilon_0\frac{\partial V_0}{\partial z}=\varepsilon_1\frac{\partial V_1}{\partial z}, \tag{6.12}$$

where V_0 and V_1 are the potentials in vacuum and in the sample, respectively; ε_0 and ε_1 are the dielectric constants in vacuum and in the sample, respectively. Considering that the potential at infinity in vacuum is 0, the potential at the surface can be approximated as

$$V_0=V_1+\frac{d^2 e\left(-n_F(j,t)-n_{trap}(j,t)\right)}{2(1+\varepsilon_r)\varepsilon_0} \tag{6.13}$$

In the numerical simulations in this chapter, the Poisson equations for solving the sample and spatial potentials are accelerated using a successive over-relaxation iteration algorithm for the calculations.

6.1.2 Process of Dynamic Charging Simulation

The above simulation describes that when the sample is continuously irradiated with electrons, the charge is deposited during the scattering process and migrated during the transport process, and the total charge of the sample shows a gradual accumulation of charging process. Then, when the electron irradiation stops after a period of time, there is no net charge entering from the outside, and the internal charge will be gradually transported to the bottom of the sample to leak and decrease, forming a discharge process.

A block diagram of the simulation flow of the charging dynamic process of the sample before and after the electron irradiation period is shown in Figure 6.2. The specific process is as follows:

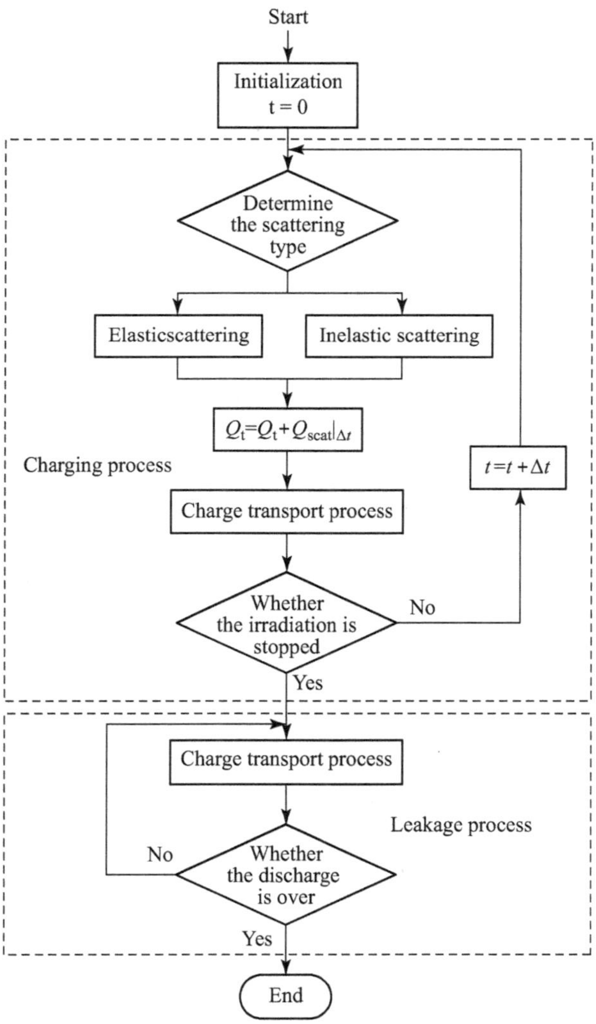

FIGURE 6.2 Block diagram of the numerical simulation flow of sample charging.

6.1.2.1 Charging Process

1. **Charge accumulation during scattering:** During the scattering process, the internal charge eventually generates a large amount of deposited charge $dQ_{scat}|_{\Delta t}$ under the effect of elastic and inelastic scattering because the incident electrons are more than the emitted secondary electrons.

2. **Charge transport**: As the scattered charge is gradually deposited, the total charge increases, $Q_t = Q_t + dQ_{scat}|_{\Delta t}$. The next step of the charge transport process begins, causing the charge distribution to change.

3. **Cyclic iteration of the time domain**: The charge in the sample, after the transport process, goes once again to the next superposition of the scattering deposited charge and continues cyclically until the charging reaches saturation or the electron radiation stops.

6.1.2.2 Leakage Process

When the incident electrons cease to irradiate, the charge inside the sample no longer has an applied deposited charge and $dQ_{scat}|_{\Delta t} = 0$ at this point. During the transport process, the internal charge is gradually released downwards until all the internal free electrons have been completely released, the charge inside the sample is essentially unchanged and the discharge process finally ends.

6.1.3 Charged Distribution Characteristics of Electron-Irradiated Medium

In this section, a polymer material with low electron mobility and strong charging, Polymethyl Methacrylate (PMMA), is chosen as the sample material. In the numerical simulation of this chapter, the parameters are set as shown in Table 6.1.

The irradiation of the incident electrons charges the sample and a charging process takes place, when the irradiation is stopped the internal charge (mainly free electrons) continues to be transported and leaks out under the effect of the built-in field and gradient field, and a leakage process takes place after the irradiation is stopped. As the electrons continue to irradiate the dielectric sample to saturate its charge, the currents in and out of the sample

TABLE 6.1 Default Parameters for the Simulation of Charged Transient Processes

Parameter Name	Parameter Value	Unit
Irradiation beam current density J_{PE}	20	$nA \cdot cm^{-2}$
Incident electron energy E_{PE}	5	keV
Electron mobility μ_e	1.0×10^{-10}	$cm^2 V^{-1} S^{-1}$
Sample thickness H	20	μm
Sample trapped density N_T	1.0×10^{15}	cm^{-3}

reach equilibrium and the charge state within the sample no longer changes. Therefore, in this book it is assumed that the interruption of incident electrons occurs after the sample charge has reached saturation. In the actual simulation, considering that the induced current J_D of the charged saturation process is only infinitely close to zero, it is assumed that the charged equilibrium is reached when the induced current reaches a small threshold, i.e. $J_D < 0.005 J_{PE}$.

6.1.3.1 Transient Characteristics of Sample Charging

During the continuous irradiation of the sample by incident electrons, electrons with a certain energy will elastically and inelastically scatter with the material inside the sample. Some of these electrons will overcome the surface potential to form secondary electrons, while the remainder will eventually be deposited inside the sample. The imbalance in the number of incident and outgoing electrons leads to the gradual accumulation of a net charge inside the sample, which makes the sample electrically charged.

Figure 6.3 illustrates the secondary electron yield curves of the outgoing PMMA material for electron energies ranging from 100 eV to 10 keV. This figure compares the results of the simulations presented in this paper with those of the experiments conducted by Boubaya M. in (Boubaya M, *J. Appl. Phys.*, 2007) [7]. In this case, both the simulations and the experiments are conducted under the assumption of an uncharged state for the sample. The simulation is a simulation that considers only the scattering process and does not consider the effect of charge. In the experiment conducted by Boubaya et al., the measured SEY values represent those observed at the commencement of irradiation, which can be considered to represent the uncharged case. During our experiments, we employed a novel electrical neutralization methodology with the objective of eliminating the influence of dielectric charge on SEY. In the case of secondary electron yields $\delta > 1$, the sample

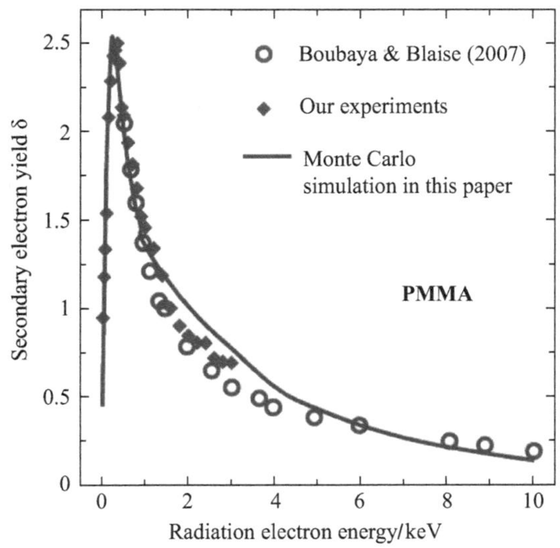

FIGURE 6.3 Comparison of simulated and experimental secondary electron yield curves for PMMA materials.

is positively charged as a whole, since the number of outgoing secondary electrons exceeds the number of incoming electrons. Conversely, when the secondary electron yield $\delta < 1$, the number of emitted electrons is less than the number of incident electrons, resulting in a negatively charged sample. Given the heightened interest in negatively charged samples due to their more pronounced charging effect, this paper will concentrate on the charging characteristics of these samples in the context of negative charge. The energy of the incident electrons is also set to be within the range of 4–10 keV. The reliability of the model can be further verified by comparing the results of the simulations in this paper with the experimental measurements in Figure 6.3.

6.1.3.2 Transient Charging Process

When subjected to a negative charge, the accumulation of negative net charge within the sample will be accelerated by the continuous irradiation of incident electrons. Concurrently, the surface potential will undergo a corresponding change. Furthermore, the internal charges will undergo transport processes and undergo changes under the influence of the self-built electric field.

6.1.3.2.1 Internal Distribution The accumulation of negative net charge within the sample will be accelerated by the continuous irradiation of incident electrons. In accordance with the theoretical model presented in Chapter 2 of this paper, the free electrons deposited within the sample undergo a charge transport process, which includes migration, diffusion, and trapping.

Figure 6.4 illustrates the charge distribution within the sample in the depth direction at different moments under the default conditions of incident electrons and sample parameters as outlined in this chapter. The total number of electrons within the sample, including both free and trapped electrons, is represented by the equation $n_q = n_F + n_T$. From the distribution of total electrons in Figure 6.4a, it can be observed that as the irradiation continues (from 40.5 to 121.5 s), the negative internal charge gradually extends from the scattering region in the surface layer to the bottom of the sample in the depth direction. The position corresponding to the peak of the internal charge density remains relatively stable, exhibiting only slight movement toward the surface layer. This movement remains within the scattering range of the electrons. This is due to the fact that in the process of charging the sample by continuous electron irradiation, the charge accumulation in the scattering region is larger than the charge leakage process, and thus the net charge is still dominated by a large amount of scattering accumulation. The slight leftward shift of the peak in the figure is a consequence of the narrowing of the scattering region, which is a result of the enhancement of the negative surface charge.

The total number of electrons within the sample can be divided into two categories: free electrons and trapped electrons. The former are electrons that are not captured by the sample defects, while the latter are electrons that are captured by the defects. Figure 6.4b and c illustrate the distribution of free electrons and trapped electrons within the sample at different moments of electron irradiation. In accordance with the capture model presented in this paper, if the capture center is not occupied, there exists a certain probability that the

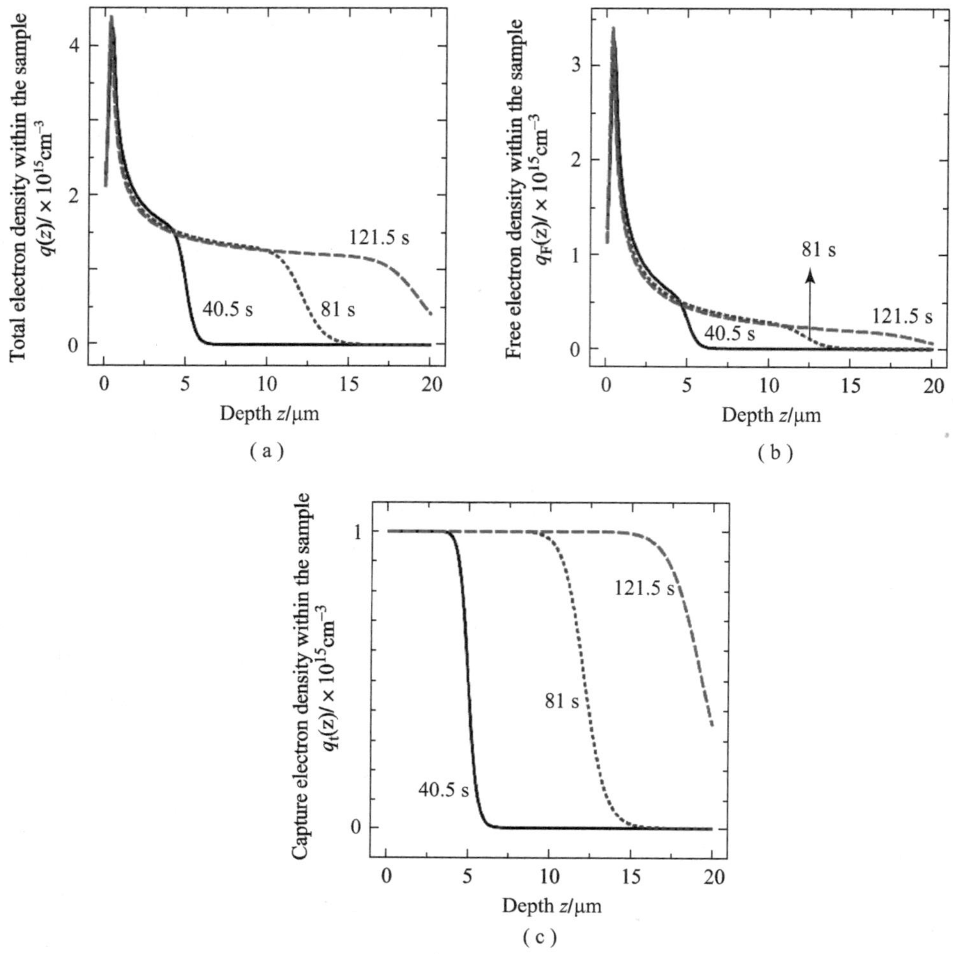

FIGURE 6.4 Charge distribution in the sample at different moments of the discharge process (a) total negative charge distribution, (b) free electron distribution, and (c) capture charge distribution.

free electrons will be captured by the trap at that specific location. Figure 6.4b depicts the distribution of free electrons at different moments of irradiation (40.5, 81, and 121.5 s). This distribution is analogous to the distribution of the total electron density. This is because, when the internal capture centers are filled (as illustrated in Figure 6.4c), the free electrons are no longer captured, and all the excess free electrons are in the form of free electrons. Moreover, as free electrons are transported down the pole plate, the capture center will be gradually filled.

The net charge deposited by the sample generates a built-in electric field, which is directly related to the charge distribution inside the sample. As illustrated in Figure 6.5a, the electric field distribution within the sample (e.g., at 40.5 s) demonstrates a gradual increase in the depth direction as the net charge accumulates. As the internal net charge gradually decreases in the depth direction, the internal electric field also gradually tends to be constant. During the continuation of electron irradiation (from 40.5 to 121.5 s), the electric field inside the sample further enhances along the depth direction as the internal

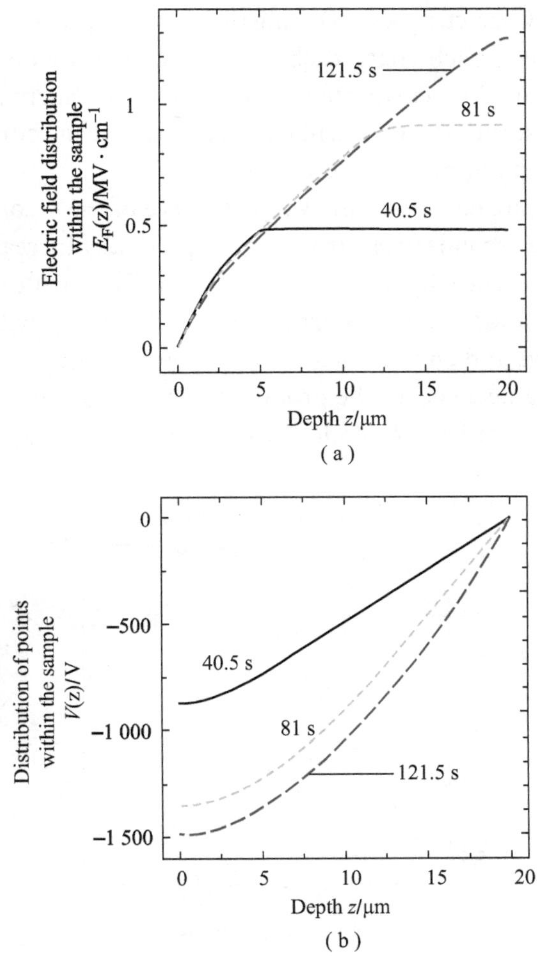

FIGURE 6.5 Distribution of charged quantitatively within the sample during charging transients (a) internal electric field distribution and (b) internal potential distribution.

charge accumulates and pushes toward the depth direction. The maximum field strength inside the sample is observed at the bottom of the sample. In the simulations presented in this paper, the maximum field strength is approximately 1 MV/cm, which is insufficient to induce an electrical breakdown in PMMA polymer materials with robust electrical insulation properties.

The internal electric field induced by the internal charge acts as an integral in the depth direction, thereby causing a certain potential difference to form at different depths within the sample. In the model established in this paper, the sample is placed on a platform with the lower pole plate grounded, thus presenting a negative potential overall under the negatively charged state. Figure 6.5b illustrates the distribution of potentials within the sample during the charging process at different moments. The negative potential within the sample is gradually augmented from the lower polar plate ($z = 20\ \mu$m) to the sample surface ($z=0$). Furthermore, as the electron irradiation persists, the negative potential within the sample is further augmented.

6.1.3.2.2 Time-Varying Characteristics During the transient process of continuous electron irradiation, the change in each charged characteristic quantity on the inner surface and within the sample results in a corresponding change in the current flowing through the sample. Figure 5.6 illustrates the transient characteristics of the currents flowing through the sample during the charging process.

As illustrated in Figure 6.6a, the time-varying curve of the secondary electron current density reveals that the secondary electron current gradually increases with the continuation of irradiation and eventually reaches a stable value. This is due to the fact that prior to reaching charged saturation, continued irradiation of the charge within the sample results in an increase in charge and enhancement of the negative potential on the surface, which subsequently leads to a decrease in the grounding energy of the irradiated electrons due to the repulsion of the negative potential. Consequently, while the incident electrons enter

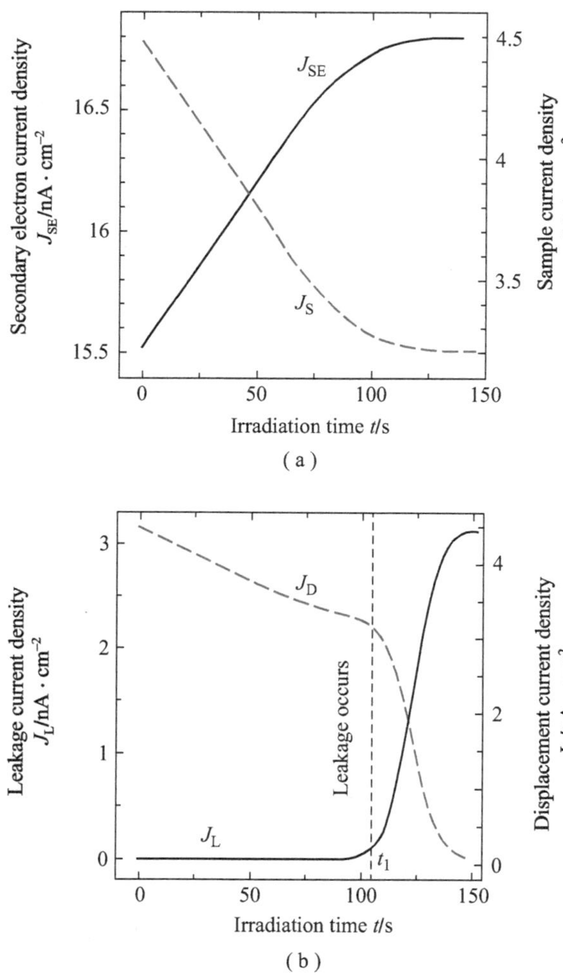

FIGURE 6.6 Transient variation curves of various current densities flowing through the sample during the charging process (a) secondary electron current and sample current and (b) leakage current and displacement current.

a more superficial region, the secondary electrons produced are more likely to be ejected. Concurrently, the secondary electron yield curve (Figure 6.1) indicates that in the negatively charged irradiation energy range, the corresponding change in the yield of outgoing secondary electrons increases with the decrease in the energy of grounding until the charge reaches the saturated surface potential, at which point no further change occurs.

In addition to the secondary electrons exiting from the upper surface, the internally deposited free electrons will also be transported in the depth direction under the action of the built-in electric field, causing the dielectric sample to become more negatively charged. This process is known as charge leakage, and it occurs when the electrons reach the bottom of the sample. Figure 6.6b depicts the transient variation curve of the leakage current. In the absence of free electrons transported to the sample's bottom, the leakage current is zero. At the moment of $t_1 = 107$ s, the free electrons are transported to the bottom of the sample, resulting in the gradual appearance of the leakage current at the bottom of the sample and an increase until it reaches the equilibrium of the total current in and out of the sample. This represents the charged saturation state of the sample.

The presence of displacement currents in the dielectric sample, placed on a metal pole plate, allows for the gradual measurement of the sample current at the lower pole plate at the beginning of the irradiation, rather than after the leakage current has appeared. Figures 6.6a and b illustrate the transient variation of the currents flowing through the sample during charging. In accordance with Maxwell's equations, the sample current J_S, the leakage current J_L, and the displacement current J_D satisfy the following equation:

$$J_S = J_L + J_D, \tag{6.14}$$

and, according to the theory of total charge conservation, the incident current J_{PE}, secondary electron current J_{SE} and sample current J_S satisfy the following equation:

$$J_{PE} = J_{SE} + J_S. \tag{6.15}$$

As illustrated in Figure 6.6b, the sample current will gradually diminish from the early irradiation stage due to the presence of displacement current, which is accompanied by a decrease in displacement current and the emergence of leakage current. Ultimately, it will reach a stable state of charged saturation. The change in displacement current (as illustrated in Figure 6.6b) can be divided into two stages: $t < t_1$, prior to the emergence of leakage current, and $t > t_1$, subsequent to the appearance of leakage current. Prior to the onset of leakage current, the reduction in displacement current is primarily attributed to the rise in secondary current. The advent of leakage current, in turn, precipitates a further acceleration in the decline of displacement current. Once the charge has reached saturation, the internal electric quantity remains unchanged, resulting in a decrease in the displacement current to zero.

Figure 6.7 depicts the numerical simulation results of the transient change process of the secondary electron current I_{SE} and the displacement current I_D generated by the induced electric field out of the sample surface under the continuous irradiation of incident

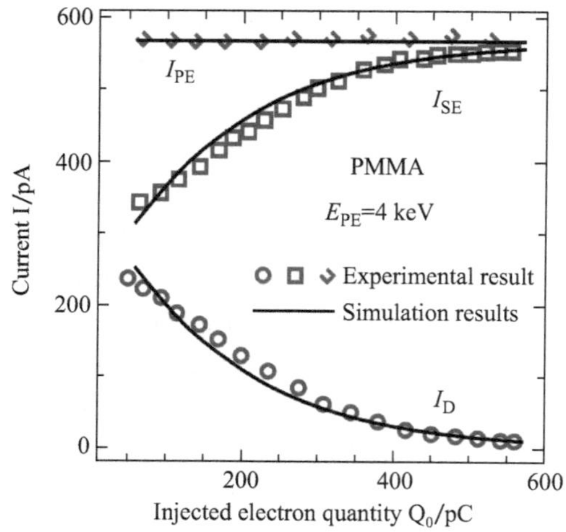

FIGURE 6.7 Comparison between simulation and experiment of transient processes of various currents flowing through the sample.

electrons, in comparison with the experiments conducted by the French group of G. Blaise [8]. Prior to the commencement of the experiments, the PMMA samples were subjected to a series of preparatory steps, including surface flattening, annealing, surface cleaning, and drying. They were then placed on a grounded metal substrate. The relevant parameters (incident electron energy $E_{PE} = 4\,\text{keV}$, incident current $I_{PE} = 570\,\text{pA}$, PMMA sample thickness $H = 2\,\text{mm}$, dielectric constant $\varepsilon = 2.6\varepsilon_0$, etc.) in the simulation were set to be identical to those in the experimental conditions. As illustrated in Figure 6.7, the simulation outcomes align with the experimental findings. With continuous irradiation of incident electrons, the negative charge within the sample increases, the surface negative potential becomes more pronounced, the incident electrons on the ground energy diminish, and the secondary electron yield gradually rises. The displacement current tends to 0, and the sample charge gradually reaches a steady state. In this case, the displacement current does not exhibit clear segmental changes. This is primarily because in the case of thick samples, the leakage current value is relatively small, and its effect on the displacement current is not significant. The congruence between the simulated transient processes of secondary electron current and displacement current and the experimental outcomes serves to substantiate the soundness of the numerical model in simulating the dynamic properties of dielectric materials charged in this study.

During the charging of an electron-irradiated sample, the current into the sample is greater than the current out of the sample. This results in a continuous increase in the negative charge accumulated within the sample and a corresponding gradual increase in the resulting negative surface potential. Figures 6.8 and 6.9 illustrate the transient time-varying curves of the surface potential, V_S, and the total charge, Q, within the sample during the transient process of electron irradiation charging. Both the negative surface potential and the total negative charge of the sample continue to increase with irradiation and eventually reach a steady state value with the saturation of the charge.

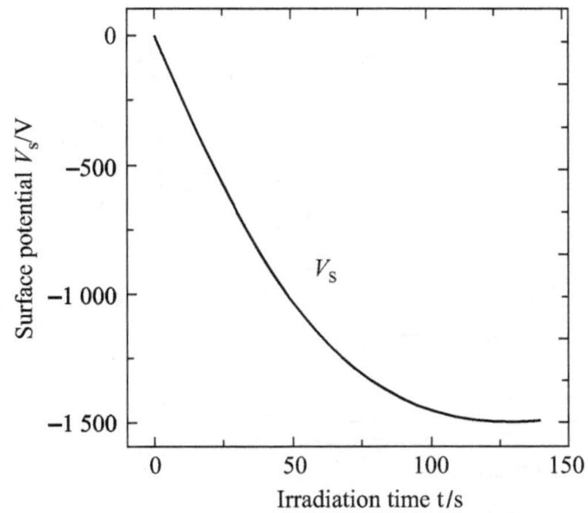

FIGURE 6.8 Transient curve of surface potential.

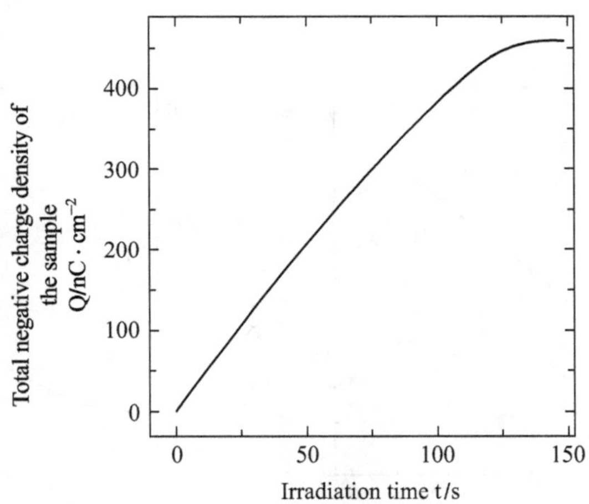

FIGURE 6.9 Transient curve of total negative charge of the sample.

6.1.3.3 Transient Process of Sample Leakage Discharge

6.1.3.3.1 Internal Distribution After the incident electron irradiation stops, the sample has no more scattered charge deposition, while the charge transport process continues due to the presence of the built-in field and internal charges [9]. Figure 6.10 shows the distribution of the total and free electron densities within the sample at different moments of the discharge. The migration and diffusion processes do not occur because the trapped charge is bound by the defect center. Therefore, the charge transport and leakage in the discharge process are mainly free electron change processes. And the change law of total charge distribution in the discharge process is basically consistent with the change law of free electrons. As shown in Figure 6.10b, the internal free electrons gradually decrease with the leakage from the 10 s to the 80 s process. Due to the weakening of the scattering effect, the peak value of the internal charge gradually moves toward the depth of the sample.

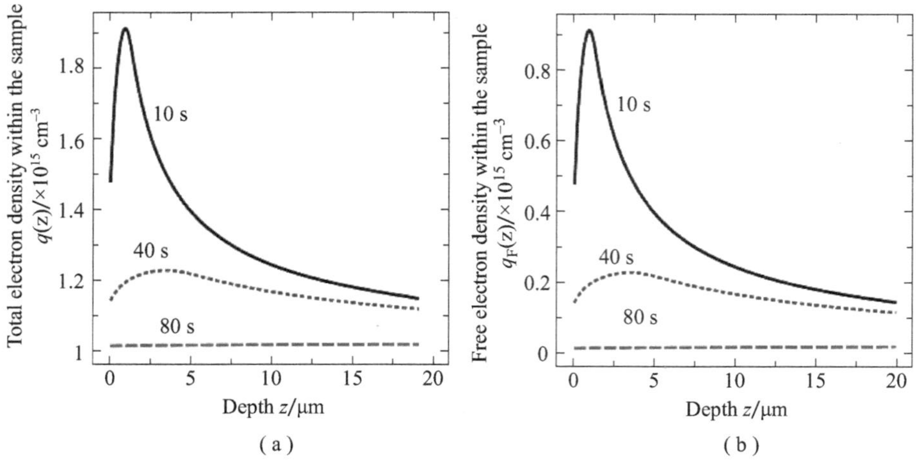

FIGURE 6.10 Distribution of electric field and potential in the sample at different moments of leakage discharge process (a) total electrical density and (b) free electrical density.

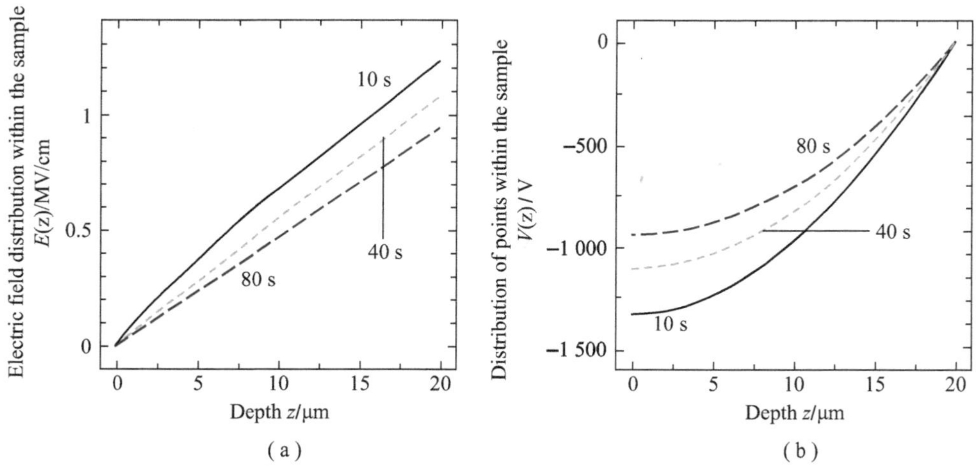

FIGURE 6.11 Distribution of electric field (a) and potential (b) in the sample at different moments of leakage discharge process.

With the complete leakage of free electrons, the total electrons inside the sample eventually consist almost entirely of trapped electrons.

The gradual leakage of the internal free electrons also causes a corresponding weakening of the internal electric field and potential. Figure 6.11 shows the distribution of the internal electric field $E_F(z)$ and potential $V(z)$ of the sample at different moments in the leakage discharge process. During the leakage discharge process, the internal electric field $E_F(z)$ and potential $V(z)$ show an overall weakening process from the moment $t=10\,\text{s}$ to the moment $t=80\,\text{s}$. However, again, due to the presence of the internal trapped charge, a certain residual of the internal electric field and potential still exists.

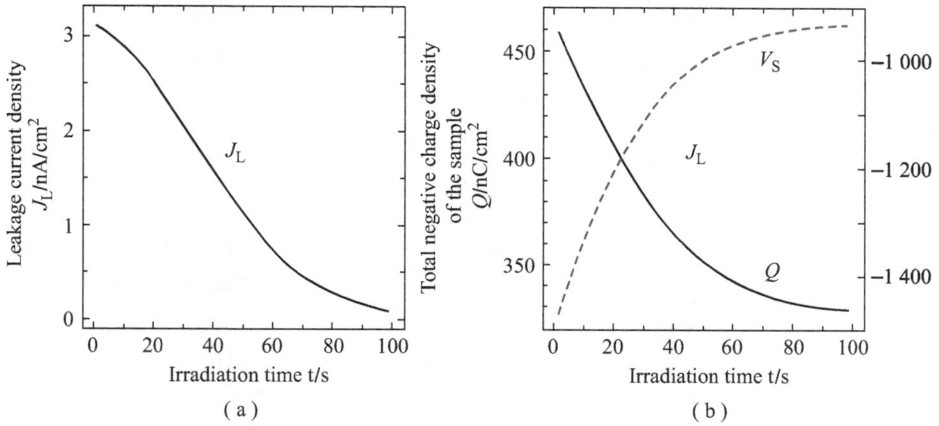

FIGURE 6.12 Transient variation curves of leakage current (a) and total power at the bottom of the sample (b) and surface potential during the discharge process.

6.1.3.3.2 Time-Varying Characteristics During the leakage process, the incident electron current and secondary electron current no longer exist due to the cessation of incident electrons, $J_{PE} = 0$, $J_{SE} = 0$. At this point, only the leakage current J_L of the sample needs to be considered. Meanwhile, with the gradual leakage of charge, the total negative charge and surface potential within the sample will gradually weaken. Figure 6.12 shows the transient process of leakage current (a) and surface potential V_S and total charge Q (b) during the leakage process.

The transient discharge process analyzed in this section occurs after the irradiated sample is saturated with charge, which makes the study in this book more relevant to the actual situation than the study of the transient discharge process in the artificially constructed identical charged state. This is because various material parameters can affect the transient discharge process in addition to the charged process. Although the increase in material mobility reduces the amount of sample released in the discharge phase, it also results in a faster internal charge release because a large electron mobility implies a large electron migration rate, and samples with larger mobility exhibit weaker charge strength during the discharge process, so that materials with larger electron mobility or metal-doped dielectric materials can be used to avoid the effects of charging. The charge trapped by material defects can have a very significant effect on the discharge process. Since the trapped charge is difficult to escape, the final remaining charge of the transient discharge process consists mainly of trapped charge, and the magnitude of the trapped density directly determines the discharge process and the final charged state. Therefore, materials with smaller defects are more conducive to charge leakage and charge escape.

In this section, the transient discharge process in dielectric materials after electron irradiation is studied using numerical simulation. By establishing a numerical physical model of the interaction of electrons with the sample material and the internal charge transport, the transient processes of each charged microscopic quantity during the transient discharge of the sample after saturation charging are studied and analyzed. During the charging process, the charge distribution within the sample gradually extends from the scattering region to the bottom of the sample as the electrons continue to irradiate. This results in a slight shift of the corresponding peak position of the charge concentration toward the surface. The negative potential within the sample gradually diminishes from the surface to the deeper layers of the sample and is enhanced with the continuation of irradiation. As the leakage current emerges and intensifies, the charge increment of the sample gradually diminishes until it reaches a steady-state equilibrium of charge. Moreover, the displacement current at the bottom of the sample decreases at a faster rate as the leakage current emerges. During the transient discharge process, the maximum peak of the charge inside the sample near the surface gradually moves toward the bottom of the sample and gradually decreases, and the total charge and surface potential inside the sample gradually decrease to a final state value directly related to the trapped density. The increase of the trapped density within the sample hinders the discharge from the sample, and the final state of the discharge is directly related to the trapped density. In general, although each parameter of the dielectric material affects the discharge process, the specific effects vary such that the mobility changes the discharge time more significantly; the material thickness reverses the trend of the initial charge of the discharge; and the trapped density affects the discharge ratio by an exponential order of magnitude.

6.2 DYNAMIC CHARACTERISTICS OF THE SECONDARY ELECTRON EMISSION OF THE CHARGED DIELECTRIC

In contrast to the stable secondary electron emission properties of metals, the secondary electron emission of dielectric materials has complex dynamic characteristics due to the influence of the charged state. Many studies have been conducted at home and abroad on the secondary electron emission and charged properties of dielectric materials. Previous studies have mainly used experimental methods to measure the SEY variation of the dielectric and to make some qualitative analysis of the charged state of the dielectric. In the numerical simulation studies of dielectric irradiation charging, the secondary electron emission process is usually simplified. The intrinsic correlation and microscopic mechanism of the secondary electron emission and charged state of dielectric materials still need to be further explored.

In this section, the intrinsic relationship and mechanism between secondary electron emission and charged state of dielectric materials under electron irradiation are investigated by using numerical simulation. The charged equilibrium model of electron irradiation is proposed for the first time, and the dynamic characteristics of the secondary electron emission and the corresponding intrinsic mechanism are studied from the perspective of the charged equilibrium model. In this section, the numerical model of the secondary electron emission and charging process of the electron-irradiated dielectric material is obtained by combining the MC method and the finite-difference time-domain method. The elastic and inelastic scattering are considered for the collision process between the electron and the material and the Mott model and the fast Penn-secondary electron model are used to simulate, respectively, and the transport process of the scattered charge includes the process of charge transfer, diffusion, and capture. The transient changes, microscopic distributions, and steady-state characteristics of the secondary electron emission and charged state in each charged equilibrium mode are analyzed. The research in this section provides important theoretical exploration and engineering guidance for the mechanistic study and suppression of dielectric secondary electrons.

6.2.1 Numerical Model for Charged Equilibrium

Based on the above numerical simulation method, the charged and secondary electron emission dynamics of electron-beam irradiated dielectric materials are simulated in this section [10]. In order to better represent the charged and secondary electron dynamic processes of the dielectric, polymeric materials with low electron mobility and commonly used in the space environment, here polymethyl methacrylate (PMMA), are selected for the numerical simulations in this section. As in most irradiated environments, the boundary conditions for the numerical simulations in this section are grounded at the bottom of the dielectric sample. Considering that the negatively charged state of the dielectric material by irradiation can be much larger than the positively charged state, the simulated electron irradiation energy in this book is set to be greater than the second crossover energy of the SEY curve. In this case, the SEY is less than 1, resulting in a negatively charged dielectric sample.

Figure 6.13 shows the process of each current transient variation with time for three different conditions simulated in this section, and the simulated parameter settings corresponding to Figure 6.13a–c, respectively, are shown in Table 6.2.

The three conditions of Figure 6.13a–c are simulated for sample thicknesses of 50, 5, and 25 μm, respectively. In general, the secondary electron current J_{SE} shows a different degree of increase with the continuation of electron irradiation, while the sample current J_S decreases to different degrees at the bottom of the material. According to the current continuity principle, the sample current here includes the leakage current J_L formed by the charge migration to the bottom and the displacement current J_D caused by the changing electric field, and satisfies $J_S = J_D + J_L$.

The main reason for the rising trend of the secondary electron current: after the electrons larger than the second cross-energy are incident into the dielectric material, SEY is less than 1, the emitted electrons are less than the incident electrons, the negative charge is accumulated inside the sample, and the surface shows a negative potential, which reduces the landing energy of the incident electrons and increases the SEY; and as the incidence continues, the surface negative potential will continue to increase and the secondary electron production will continue to increase accordingly. Eventually, the electrons emitted from the surface and the electrons emitted from the bottom reach a charged dynamic equilibrium with the incident electrons.

From a macroscopic point of view, the dynamic process of secondary electron emission mainly depends on the charged equilibrium process of the sample. Figure 6.13a–c correspond to different charged equilibrium processes, and the right figure corresponds to their equilibrium schematics, respectively. In (a), the secondary electron current increases significantly and the leakage current changes very little, and the internal charge equilibrium is mainly dominated by the increase of the secondary electron current, which is the

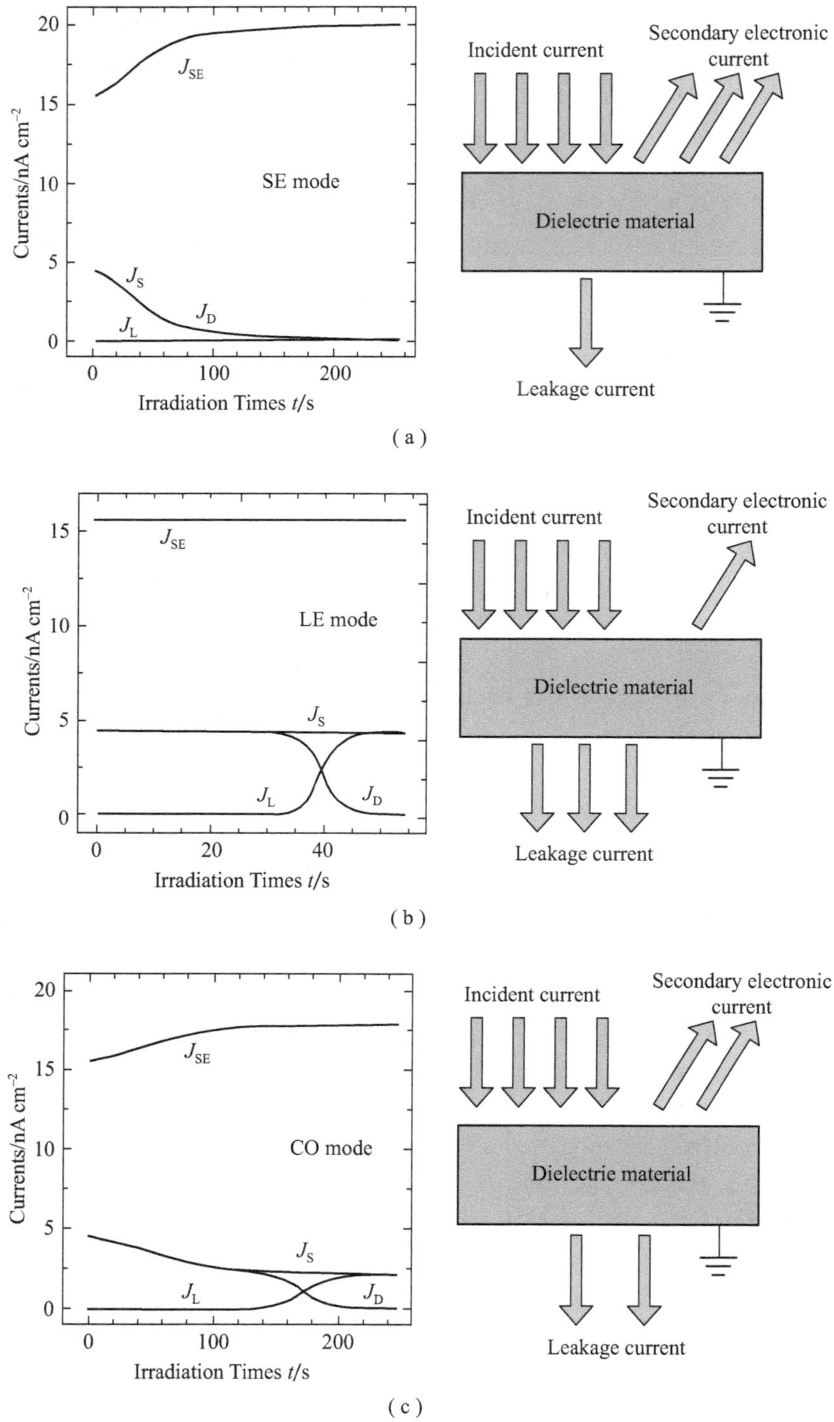

FIGURE 6.13 Current transient relations of three charged equilibrium modes: (a) secondary electron equilibrium mode (SE), (b) leakage current equilibrium mode (LE), and (c) common mode (CO).

TABLE 6.2 Numerical Simulation-Related Parameter Settings

Parameter	(a)	(b)	(c)
Energy/keV	5	5	5
Beam/($nA \cdot cm^{-2}$)	20	20	20
Mobility/($cm^2 \cdot V^{-1} \cdot S^{-1}$)	10^{-10}	10^{-10}	10^{-10}
Thickness/μm	50	5	25
Trapped density/cm^{-3}	10^{15}	10^{15}	10^{15}

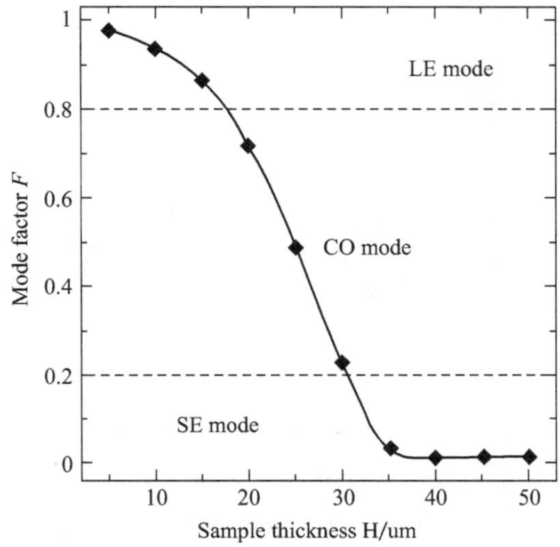

FIGURE 6.14 Effect of sample thickness on the charged equilibrium mode.

secondary electron equilibrium mode (SE); in (b), the leakage current shows significant changes and the secondary electron current changes very little, which is the leakage current equilibrium mode (LE); in (c), both changes are comparable, which is the common mode (CO) (Figure 6.14).

To express the degree of charged equilibrium mode more intuitively, the equilibrium mode factor (mode factor) F is defined as the ratio of the equilibrium state leakage current value $J_L(t=\infty)$ to the initial sample current $J_S(t=0)$, i.e.

$$F = \frac{J_L(t=\infty)}{J_S(t=0)} \tag{6.16}$$

Here, the mode factor F directly expresses the ratio achieved by the charge leakage. When $F \in (0,0.2]$, it is in SE mode; when $F \in [0.8,1)$, it is in the LE mode; and when $F \in (0.2,0.8)$, it is in CO mode. The variation curves of the mode factor F are shown in Figure 4.7 for different material sample thicknesses ($H = 5 \sim 50\,\mu m$). As the sample thickness increases, the charged equilibrium mode shifts from the LE mode to the SE mode.

6.2.2 Effect of Sample Parameter on Equilibrium Mode

For the same material sample, the most direct reference parameter for simulation is the sample thickness because the material properties are not easily changed. In this book, the sample thickness is used as a parameter variable to simulate and analyze the dynamic properties of secondary electrons and other related charged transient properties under different charged equilibrium modes.

Figure 6.15 also shows the relationship between the SEY emitted from the dielectric surface and the induced displacement current at the bottom of the sample with time for

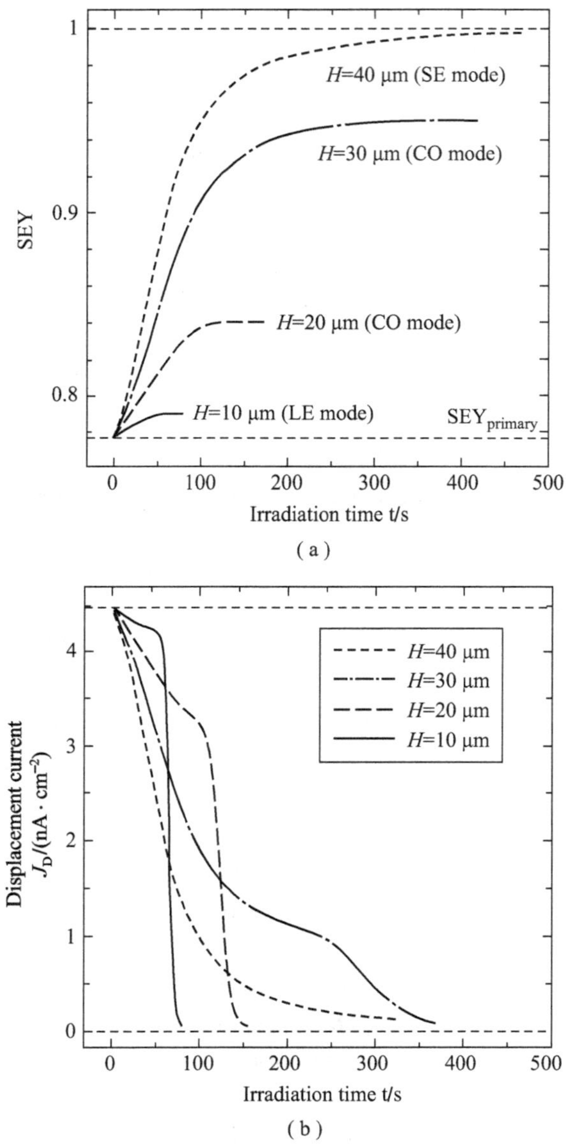

FIGURE 6.15 Secondary electron emission and displacement current transient curves at different sample thicknesses (a) SEY and (b) displacement current.

different sample thicknesses of 10, 20, 30, and 40 μm, respectively. In Figure 6.15a, for relatively thin samples (e.g., 10 μm), corresponding to the LE mode, the SEY grows slowly with small variation, while for thick samples (e.g., 40 μm), corresponding to the SE mode, the SEY grows fast with large variation and is closer to 1. In Figure 6.15b, the displacement current transient curve can be divided into two segments, whose inflection point is caused by the leakage current. It can be seen that increasing the sample thickness delays the appearance of the displacement current inflection point. For the SE mode (e.g., 40 μm), the inflection point of the displacement current is almost invisible.

The charged state of the sample in different modes is determined by the internal charge distribution. Figure 6.16 shows the distribution of the internal charge of the sample after reaching equilibrium for different thickness cases. The horizontal coordinates are the relative depth z_r (the ratio of the absolute depth to sample thickness z/H), the sample surface on the left and the grounded bottom of the sample on the right. For thinner samples (e.g., 10 μm), corresponding to the LE mode, the charge distribution is more inclined toward the interior. In the case of the SE mode (e.g., 40 μm), the charge is not transported to the bottom of the sample, and a larger ratio of the charge is distributed on the sample surface.

The final steady-state related characteristic quantities resulting from the transient process also change under different charged equilibrium modes. Figure 6.17 shows the variation curves of the sample surface potential V_s, *SEY*, total charged quantity *Q*, and equilibrium time constant T_c at steady state for different equilibrium modes due to different thicknesses. In Figure 6.17a, the *SEY* gradually increases and approaches 1 as the sample thickness increases and the equilibrium mode shifts to the SE mode. In addition, the SE mode implies a large secondary electron change, which corresponds to a stronger surface potential.

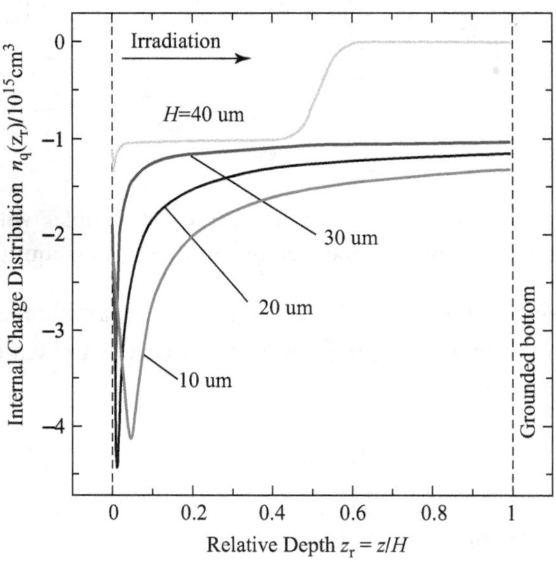

FIGURE 6.16 Internal charge distribution at different sample thicknesses.

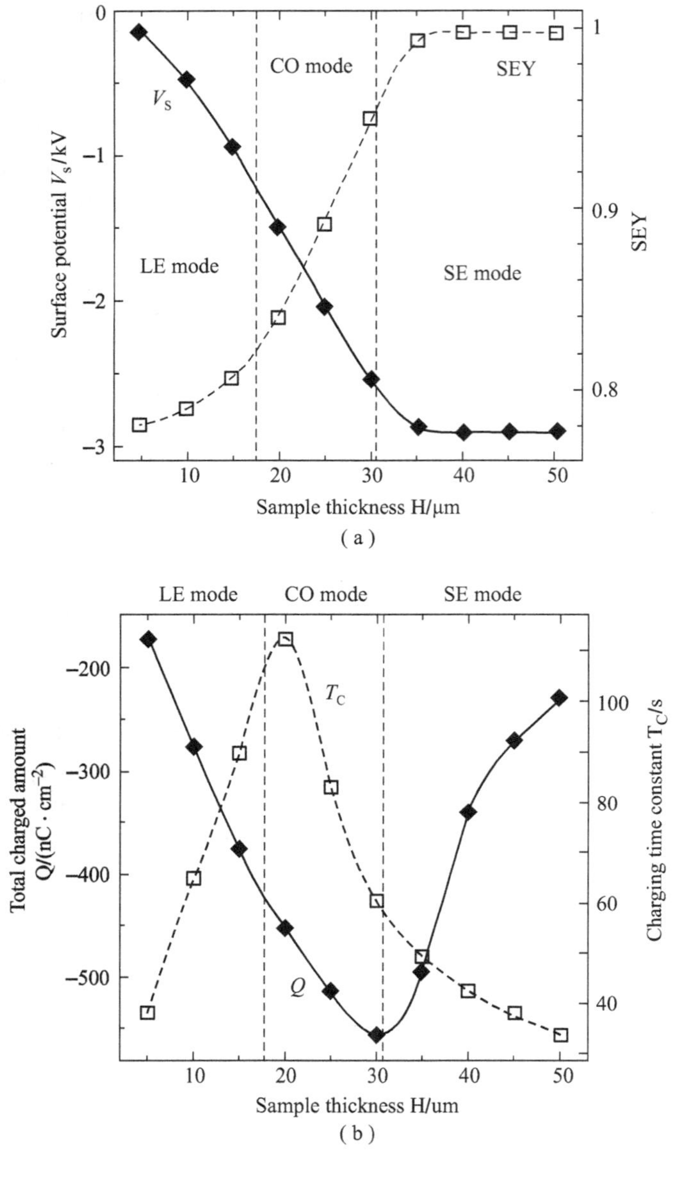

FIGURE 6.17 Dynamic correlation characteristic curves of samples with different thicknesses (a) surface potential V_s and SEY and (b) total charged amount Q and equilibrium time constant T_c.

To describe the speed of the equilibrium process as the sample reaches charged equilibrium, the charged time constant T_c is defined in this book as the time for the displacement current to decrease from its initial value to $1/e$:

$$T_c = t\big|_{J_D=\frac{1}{e}J_D(t=0)}, \tag{6.17}$$

where e is the natural constant.

From Figure 6.17b, it can be seen that the equilibrium mode changes as the sample thickness increases, and the equilibrium time constant T_c shows an opposite trend in the middle CO mode zone. This is mainly due to the change in the dominant factor of the equilibrium, from leakage current dominated to secondary electron current dominated, and the internal charged equilibrium mechanism is changed. Similarly, for the total charge, the opposite trend of the total charged quantity occurs with the change of the equilibrium pattern. According to the previous results on the internal charge distribution, the charge tends to be more distributed on the surface in the SE mode and the enhancement of the surface negative potential slows down with the increase of the sample thickness, while the total charge shows a trend of first increasing and then decreasing. This trend can also be obtained by combining Figure 4.17b with the theoretical analysis above.

In fact, the charged equilibrium mode not only reflects the macroscopic current equilibrium mode, but also distinguishes the deeply charged and surface charged states of the dielectric material. The secondary electron SE mode usually corresponds to the surface charged state of the sample, while the leakage current LE mode corresponds to the deep-charged state. The equilibrium mode study in this book is mainly based on the change of sample thickness. In addition, material property parameters such as electron mobility and trapped density of the sample material as well as electron irradiation parameters (e.g., incident electron energy and beam density) can change the equilibrium mode [11].

In this section, the secondary electron emission and negative charging characteristics based on the charged equilibrium mode are studied by numerical simulation, and the transient variation law and microscopic influence mechanism of each characteristic parameter under different equilibrium modes are analyzed, and the following specific conclusions are obtained:

1. In the case of negative charging, with the continuous incidence of electrons, the SEY from the dielectric surface gradually increases, the surface negative potential is enhanced, and the displacement current at the bottom of the sample will show a fast decreasing inflection point with the appearance of the leakage current.

2. According to the change of the dielectric surface current, the charging process of the dielectric surface can be divided into SE mode, LE mode and CO mode; the sample presents the surface charged state in SE mode, while the deep charged state in LE mode.

3. As the LE mode shifts to the SE mode, the steady-state SEY increases and the surface negative potential increases; the total charge and the equilibrium time constant show the opposite trend due to the change of the equilibrium mode.

REFERENCES

1. Sorolla E, Belhaj M, Sombrin J, et al. New multipactor dynamics in presence of dielectrics. *Physics of Plasmas*, 2017, 24(10): 103508.
2. Sinitsyn O, Nusinovich G, Antonsen Jr T. Studies of multipactor in dielectric-loaded accelerator structures: Comparison of simulation results with experimental data. *AIP Conference Proceedings*, Annapolis, Maryland, USA. American Institute of Physics, 2010, 1299(1): 302–306.
3. Ye M, Li Y, He Y, et al. Study of multipactor suppression of microwave components using perforated waveguide technology for space applications. *Physics of Plasmas*, 2017, 24(5): 052109.

4. Ganachaud J P, Attard C, Renoud R. Study of the space charge induced by an electron beam in an insulating target. *Physica Status Solidi (B)*, 1997, 199(1): 175–184.
5. Zhang H B, Li W Q, Cao M. Space charge characteristics of an insulating thin film negatively charged by a low-energy electron beam. *Journal of Electron Microscopy*, 2012, 61(2): 85–97.
6. Feng G B, Wang F, Hu T C, et al. Characteristics of charge and discharge of PMMA samples due to electron irradiation. *Chinese Physics B*, 2015, 24(11): 117901.
7. Boubaya M, Blaise G. Charging regime of PMMA studied by secondary electron emission. *The European Physical Journal-Applied Physics*, 2007, 37(1): 79–86.
8. Blaise G, Pesty F, Garoche P. The secondary electron emission yield of muscovite mica: Charging kinetics and current density effects. *Journal of Applied Physics*, 2009, 105(3): 034101.
9. Zhang H B, Li W Q, Cao M. Leakage current simulation of insulating thin film irradiated by a nonpenetrating electron beam. *Chinese Physics Letters*, 2012, 29(4): 047901.
10. Feng G, Liu L, Cui W, et al. Transient characteristics of charging effects due to e-beam irradiation: A method of SEY-based charging balance mode. *IEEE Transactions on Plasma Science*, 2019, 47(8): 3783–3789.
11. Feng G B, Cao M, Yan L P, et al. Combined effects of sample parameters on polymer charging due to electron irradiation: A contour simulation. *Micron*, 2013, 52: 62–66.

CHAPTER 7

Application of Secondary Electron Emission

Wanzhao Cui, Guobao Feng, Fang Wang, and Guanghui Miao

THIS CHAPTER INTRODUCES THE application of secondary electron emission using the secondary electron multiplication phenomenon (i.e., multipactor) and electron beam micro-imaging of spacecraft microwave load under space high-power state as examples.

7.1 MICRO-DISCHARGING EFFECT

The micro-discharging effect, a vacuum resonant discharge phenomenon occurring between two metal surfaces or on a single dielectric surface, is usually formed by the secondary electrons generated by the collision of electrons excited by the RF electric field transmitted in the components and accelerated to obtain energy in the RF electric field on the surface. Typical multipactor phenomena include (1) double-surface multipactor in metal resonant structures and (2) multipactor on single dielectric surfaces. The conditions for multipactor between metal double surfaces are that the average free range of electrons must be greater than the gap distance between the two metal surfaces and that the average crossing time of electrons between the two surfaces must be an odd number of times the half-cycle of the RF electric field; the condition for multipactor on a single dielectric surface is that the DC electric field generated by the surface charge must be able to accelerate the electrons back to the dielectric surface, which can generate secondary electrons.

Multipactors in spacecraft load systems occur mainly in two-surface metallic structures. The process (Figure 7.1) is as follows: as seen in 0(I), the electrons entering between the two plates accelerate toward the upper plate during the positive half-cycle of the RF field. Assuming that the electron hits the upper plate when the electric field passes through the zero point, a secondary electron is created (II), and the electron will accelerate back to the lower plate in the negative half-cycle (III). This process is repeated continuously (IV–VI), due to secondary electrons, each impact will release more electrons than the original

DOI: 10.1201/9781003562863- 7

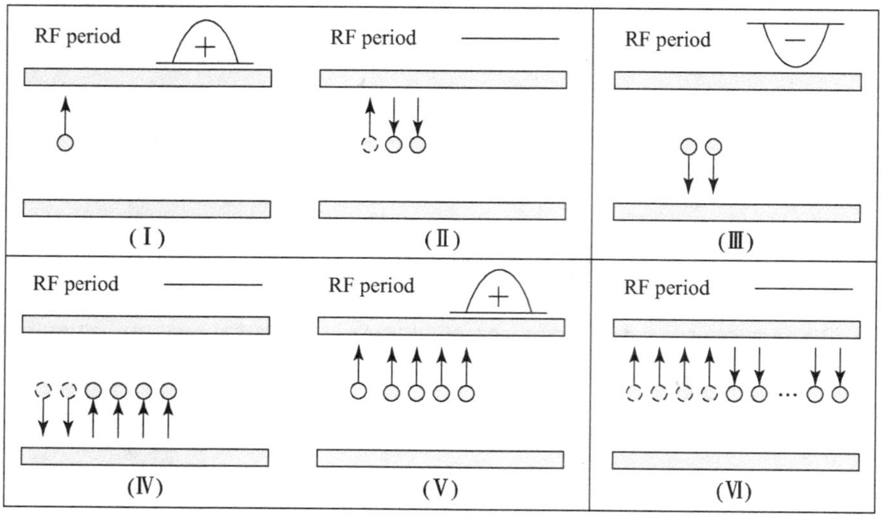

FIGURE 7.1 Electron "avalanche" process of electron secondary multiplication in a two-surface metal structure.

hit the plate, which will produce an avalanche phenomenon, until the appearance of steady state. At this point, the rate of electron generation is equal to the rate of loss.

7.1.1 Multipactor Simulation Method

Trajectory tracking is the most used multipactor simulation method, such as ANALYST, developed in the USA, MEST, developed at the University of Madrid, Spain, and Multp developed at the Moscow Institute of Engineering Physics. PIC is a simulation method for real time, dynamic tracking of macro-particles composed of a large number of particles, providing the most intuitive simulation of physical phenomena. It is currently the only simulation method capable of studying nonlinear effects of arbitrarily complex structures, multi-carriers, and multipactors. ESA has developed the simulation software FEST based on the simulation modules of FDTD and PIC. The Institute of Applied Physics of the Russian Academy of Sciences has also proposed a fast simulation method based on statistical theory, by which the general solution of the steady-state distribution of the secondary electron emission phase and the multipactor threshold growth can be obtained by the exact resolution of the probability density of the secondary electron arrival time. The method is fast, but it can only be used for calculating microwave components with a flat plate structure. Using simulation techniques and simulation software, the initial rate, direction, and scattering characteristics of the electrons have been added to the multipactor simulation study and have led to a reasonable interpretation of the experimental results.

The above gives an overview of the development of some current multipactor simulation models and briefly analyzes the current simulation methods that are more widely used and the algorithms of each software. Several multipactor simulation models are discussed in detail below.

7.1.1.1 Full-Wave Electromagnetic Simulation Tool (FEST)

The Full-Wave Electromagnetic Simulation Tool (FEST) was originally designed as an electromagnetic simulation tool to which multipactor modeling has been added in recent years. This model was first developed at ESA to calculate the field distribution inside microwave devices for the first time before multipactor simulations were performed. Compared to Microwave Studio, this tool calculates the fields more accurately and very quickly. Although FEST is capable of calculating rectangular, circular, and arbitrary shapes consisting of rectangles and circles, based on the authors' experience with multipactor simulations, only structural simulations can be done in FEST. Although FEST was initially developed for waveguide design, the driving force behind the tool would be to analyze waveguides using advanced techniques such as edge-integrated resonant mode expansion. Later, more multipactor components were added so that multipactor simulations could only be completed for structures selected in the FEST library to evaluate field information and simulate each electron trajectory separately, including secondary electrons.

For each collision, important parameters such as impact energy, angle of incidence, and impact position are recorded. Although secondary electrons are the main cause of breakdown, the convergence domain induced by the multipactor simulation in FEST varies with the number of initial particles used in the simulation. The number of initial particles emitted depends on the geometry of the individual device and its field distribution. FEST can simulate the scattering field so that more electrons accelerate along the curved path between the two walls rather than the straight path if the height-to-width ratio of the gap is greater than 1. This behavior greatly increases the breakdown threshold, because high power must drive the secondary electrons through the gap, and because more electrons are generated in the strong field area, which takes up a long passage time. FEST is also able to model the effect of space charge. The main reason for the change in the microwave components after the breakdown is the space charge generated during the breakdown. Propagation of the space charge generated field through the structure causes reflection of the incident power and some other nonlinear effects such as passive intermodulation (PIM) effects.

FEST calculates space charging effects through particle simulation models by inserting space charge into the mesh nodes to generate the field and adding it to the field calculation through FEST. The total field is used to calculate the force of electron motion, and this force is of great importance to simulate the effect of space charge, being able to correlate simulation results with experimental parameters, such as evaluating changes in scattering parameters before and after the breakdown of the test device and the generation of noise and harmonics. However, FEST can only simulate the breakdown of geometrical structures in its own library, and the convergence domain of the simulation results is determined to some extent by the number of initial particles. FEST also does not consider inelastic and elastic scattering electrons.

Algorithm: Particle simulation, Maxwell distribution.

Advantages:

The simulator can consider the effect of space charge and has refined edge field distribution.

Noise and harmonic analysis functions are available.

Disadvantages:

The simulator can only predict the multipactor of existing structures in the fest library.

Convergence depends on the number of primary particles selected.

Electrons with elastic scattering and inelastic scattering are not considered.

7.1.1.2 ESA Multipactor Calculator

The ESA multipactor calculator provides predictions of multipactor thresholds for different structures, such as parallel plates and square-axis transmission lines, and for different surface materials (silver and aluminum). The ESA curves are based on models developed by Woode and Petit. The Woode and Petit curves are the ESA design standard because they matched their experimental data plots and were accepted by the aerospace industry. However, the graphs used by ESA are based on the mean-field method and are therefore also conservative in their predictions of multipactor breakdown. In other words, ESA curves typically underestimate the power capacity handled by high-power RF devices, so researchers have developed more accurate multipactor models for the space industry.

7.1.1.3 Multipactor Simulation and Analysis Tool (MSAT)

In the present model, a three-dimensional self-consistent simulation method based on PIC technique with full-wave electromagnetic calculation method is proposed. The method can be used to describe the evolution of multipactors in passive microwave components with time and the prediction of multipactor thresholds. The electromagnetic field distribution in a specific structure is obtained by solving Maxwell's differential equations in a discrete mesh using an FDTD method. The charges and currents generated by the electron motion are added to the source terms of Maxwell's equations, while the field quantities are recalculated at each time step. Based on the laws of particle motion, the electron will remain in motion until the conditions for the end of the operation are met or the total computation time has elapsed, provided that the initial conditions for the electron and the field are defined and a suitable expression for the fundamental physical laws of electron motion and electromagnetic field evolution is given. The following steps are involved:

7.1.1.3.1 Initialization of Electrons and Fields First, as the basis of the whole simulation, a three-dimensional model of the passive device needs to be created, and an appropriate mesh is performed. For this simulation method, macro-particles are used to simulate the initial electrons and secondary electrons. To simulate the real environment, the initial electrons are randomly distributed in the air mesh. The kinetic energy of the initial electrons is taken from 0 to 10 eV, and the phase is taken from 0° to 180°. Since the initial electrons will disappear when they collide with the metal surface, the specific values of the initial electron energy and phase do not affect the multipactor breakdown threshold.

The factors that determine the multipactor breakdown threshold are the number and state of the emitted secondary electrons. The initial value of the field amplitude at the fixed mesh node is set to zero.

7.1.1.3.2 The Process of Self-Consistent Interaction of Electrons and Fields After the initialization of the electron and field distributions, a self-consistent evolution mechanism is proposed to regulate the motion of the electrons and the evolution of the electromagnetic field. First, the actual electromagnetic field distribution in the microwave component is obtained by numerically solving the difference form of Maxwell's equations with the FDTD method. If the mesh is reasonable, the computational accuracy and numerical efficiency can be guaranteed. The carrier signal is predefined while the mesh is excited at the transmitter wave port.

Then, the electron dynamics is determined by the relative Lorentz force equation. The differential form of the relative Lorentz force equation is

$$\frac{v^{n+1/2} - v^{n-1/2}}{\Delta t} = \frac{q}{m}\left(E^n + \frac{v^{n+\frac{1}{2}} - v^{n-\frac{1}{2}}}{2} \times B^n\right) \tag{7.1}$$

$$\frac{r^{n+1} - r^n}{\Delta t} = v^{n+1/2}, \tag{7.2}$$

where n is the nth time step; q is the electron charge; and m is the electron mass.

In particular, particle-to-mesh interpolation and mesh-to-particle interpolation are used in the present simulation model as key methods to combine electron motion with magnetic field evolution. On the one hand, the motion of electrons is driven by electromagnetic forces. The electromagnetic field force on the electron can be obtained by interpolating the electromagnetic field values at the mesh nodes to obtain the electromagnetic field values at the location of the electron. On the other hand, the charge and current generated by the electron motion as newly generated source terms are substituted into the EM field equations on the mesh nodes to recalculate for the EM field values by interpolating the charge and current from the electron location to the mesh nodes. Finally, the field components are updated at the end of each time step of the calculation. Considering the mutual repulsive Coulomb forces between electrons in this model improves the accuracy of the simulation of multipactor electrons.

7.1.1.3.3 Secondary Electron Emission In microwave devices, electrons are driven by electromagnetic field forces and move in the direction of the electromagnetic field lines, eventually colliding with the metal wall. When the collision occurs, the state of the impacted secondary electrons ultimately determines whether the multipactor electron avalanche effect occurs. In the present model, the SEY at the metal boundary is described by an empirical equation derived by Vaughan. Depending on the collision energy $E_\text{i} = mv^2/2$ and the collision angle θ, the SEY δ is calculated by the empirical equation proposed by Vaughan:

$$\delta = \delta_{\mathrm{max}0}\left(1+\frac{k_s\theta^2}{2\pi}\right)\left(we^{1-w}\right)^k \tag{7.3}$$

$$E_{\mathrm{max}} = E_{\mathrm{max}0}\left(1+\frac{k_s\theta^2}{2\pi}\right), \tag{7.4}$$

where $w = E_i / E_{\mathrm{max}}$, $k=0.62$ when $w < 1$, $k=0.25$ when $w > 1$; k_s is the roughness of the metal surface; $\delta_{\mathrm{max}0}$ and $E_{\mathrm{max}0}$ are the secondary electron emission parameters in the case of vertical incidence determined by the metal material.

7.1.1.3.4 Multipactor Breakdown Threshold Acquisition After several hundred-time steps calculations, the variation of the number of electrons with time is recorded for certain field strength conditions. When the multipactor avalanche effect occurs, the electrons move between the metal gaps in a resonant form and the number of electrons increases exponentially. Therefore, it is reasonable to use the power level that leads to the resonant steady state as the criterion for judging the multipactor breakdown threshold. The number of electrons remains fluctuatingly balanced as the electrons reach the resonant steady state.

Since the input power of microwave components can be changed by changing the input carrier signal, the power level of the multipactor breakdown threshold can be determined by comparing the evolution trend of the number of electrons with time at different power levels.

7.1.2 Multipactor Dynamic Process Analysis

The flat plate structure is a classic structure for studying secondary electron multiplication. In this book, a simple analysis of the electron trajectory based on the flat plate structure shown in Figure 7.2 is used as an analogy to the particle trajectory tracking process of multipactor simulation of special structure microwave components by simulation software. The electric field distribution between the flat plate structure is given by

$$E_y = E_0 \sin(\omega t + \varphi) \tag{7.5}$$

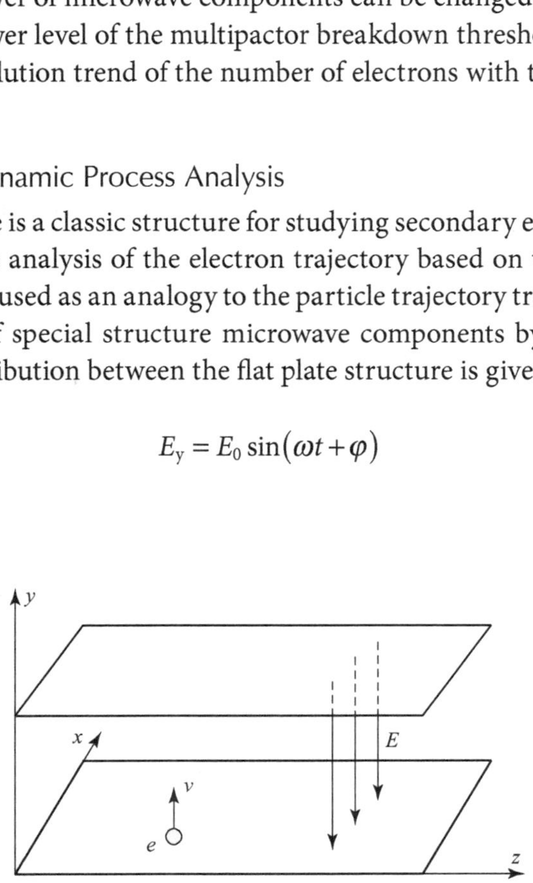

FIGURE 7.2 Classic flat plate structure.

The acceleration of an electron in an electromagnetic field and the force exerted on it can be obtained from Newton's law and Lorentz's law, i.e.,

$$\vec{a} = \frac{\vec{F}}{m} \tag{7.6}$$

$$\vec{F} = q\left[\vec{E} + \vec{v} \times \vec{B}\right] \tag{7.7}$$

For a flat plate structure with small gaps, the effect of the magnetic field on the electron motion is generally not considered, so the electron has acceleration only in the y-direction, and the velocity in the y-direction can be expressed as

$$v_y(t) = \int_0^t a_y \, dt = \frac{qE_0}{m} \int_0^t \sin(\omega t + \varphi) dt = \frac{qE_0}{m} \cos\varphi - \cos(\omega t + \varphi) + C \tag{7.8}$$

Let $t=0$. The constant C in equation (7.8) is

$$C = v_y(t=0) = v_{0y}$$

Thus, the velocity of the electron in the electromagnetic field between the flat plate structures can be expressed as

$$\left[v_x(t), v_y(t), v_z(t)\right] = \left[v_{0x}, \frac{qE_0}{m\omega}\left[\cos\varphi - \cos(\omega t + \varphi)\right] + v_{0y}, v_{0z}\right] \tag{7.9}$$

There is no electric field distribution in the xOz plane, and the velocity in these directions remains constant in the initial state, and for any moment t, the position of the electron can be expressed as

$$\begin{aligned} s_y(t) &= \int_0^t v_y(t) dt = \frac{qE_0}{m\omega} \int_0^t \cos\varphi dt - \frac{qE_0}{m\omega} \int_0^t \cos(\omega t + \varphi) dt + \int_0^t v_{0y}(t) dt \\ &= \frac{qE_0}{m\omega} t \cos\varphi - \frac{qE_0}{m\omega^2}\left[\sin(\omega t + \varphi) - \sin\varphi\right] + v_{0y} t + s_{0y}, \end{aligned} \tag{7.10}$$

where $\vec{s_0} = \left[s_{0x}, s_{0y}, s_{0z}\right]$ is the position of the electron at the moment t_0, so the position of the electron at any moment t can be expressed as

$$s_x = v_x t + s_{0x} \tag{7.11}$$

$$s_y = \frac{qE_0}{m\omega} t \cos\varphi - \frac{qE_0}{m\omega^2}\left[\sin(\omega t + \varphi) - \sin\varphi\right] + v_{0y} t + s_{0y} \tag{7.12}$$

$$s_z = v_z t + s_{0z} \tag{7.13}$$

Here is only a simple modeling for the electron trajectory in the flat plate structure, considering the narrow gap and ignoring the influence of the magnetic field on the electron motion. However, for higher-order multipactor, the electron motion time between the flat plates is longer, and the influence of the magnetic field must be considered, and the presence of the magnetic field in this case will have a greater impact on the electron trajectory and the angle of incidence of the electrons. In addition, this book only introduces the structure of a relatively simple flat plate structure, which can obtain relatively simple analytical results, while for complex microwave devices, the analytical calculation is relatively complex.

7.1.2.1 Threshold Voltage

The trajectory of electrons between flat plates is mainly derived from the above content, which is an essential part of the multipactor simulation process for tracking the trajectory of electrons. The ultimate purpose of the simulation is to predict the multipactor threshold, threshold voltage, or threshold power of microwave components in the process of designing microwave components; therefore, the threshold voltage of microwave components and the influencing factors of the multipactor threshold is needed to be simply analyzed and then to study the multipactor threshold of microwave components in a targeted manner.

Here is a simple analysis of the threshold voltage using a flat plate structure as an example. An expression for the velocity of the electron motion in the *y*-direction can be obtained in the form of

$$v_{0y}(t) = \frac{qE_0}{m\omega}\left[\cos\varphi - \cos(\omega t + \varphi)\right] + v_{0y} \tag{7.14}$$

When the occurrence of multipactor necessarily generates electronic resonance, the following condition is satisfied

$$t = (2n-1)\frac{1}{2f}, n = 1,2,3,\ldots \tag{7.15}$$

Then, we have

$$v_y = \frac{2qE_0}{m\omega}\cos\varphi + v_{0y} \tag{7.16}$$

When the resonance condition is satisfied, the trajectory of the electron satisfies

$$s_y = \frac{qE_0}{m\omega}\frac{2n-1}{2f}\cos\varphi - \frac{qE_0}{m\omega^2}\left[\sin\left(\frac{2n-1}{2f}t + \varphi\right) - \sin\varphi\right] + \frac{2n-1}{2f}v_{0y} + s_{0y} = d \tag{7.17}$$

This yields the transit equation, which is

$$d - s_{0y} = \frac{qE_0}{m\omega}\left[(2n-1)\pi \cos\varphi + 2\sin\varphi\right] + \frac{\pi}{\omega}(2n-1)\, v_{0y} \tag{7.18}$$

Accordingly, the electric field strength between the gaps can be obtained as

$$E_0 = \frac{m\omega^2\left(d - s_{0y}\right) - (2n-1)m\omega\pi\, v_{0y}}{q\left[(2n-1)\pi \cos\varphi + 2\sin\varphi\right]} \tag{7.19}$$

The final threshold voltage can be obtained as

$$V = E_0 d = \frac{m\omega^2 d\left(d - s_{0y}\right) - (2n-1)m\omega\pi\, v_{0y}}{q\left[(2n-1)\pi \cos\varphi + 2\sin\varphi\right]} \tag{7.20}$$

From the above threshold voltage equation, it can be seen that the threshold voltage V is closely related to the initial emission velocity v_{0y} of the electron, and the smaller the initial exit velocity v_{0y}, the larger the threshold voltage. In general, it is assumed that the electron energy is only kinetic energy, i.e., $E = 1/2mv^2$. Therefore, the electron emission velocity distribution indirectly reflects the energy distribution of electrons, and the threshold voltage is closely related to the emission velocity, which is also necessarily closely related to the energy distribution of the emitted electrons.

7.1.3 Effect of Secondary Electrons on Multipactor

Multipactor threshold is the maximum power that can be transmitted through microwave components without affecting the performance and operating stability of the microwave components. Multipactor experimental tests are conducted in a closed vacuum environment, so it is difficult to test and observe the phenomena during multipactor experiments. Numerical simulation is based on a large number of experimental phenomena, experimental results, and theoretical analysis and calculations, so the experimental phenomena are difficult to observe, which inevitably increases the difficulty of numerical simulation.

The secondary electron emission theory is the basis for the study of multipactor, which can also be expressed graphically as a secondary electron multiplication phenomenon, that is, the formation of multipactor is the effect of multiple secondary electron emission after the accumulation. A single secondary electron emission process involves a very complex physical process, for which in the microwave components, the initial electron is generally free electrons, ions, etc. in the microwave components. When the microwave components are put into the working state, the free electrons start to move in the internal existing RF field, collide with the microwave component wall, and produce secondary electrons, thus emitting the first secondary electron; the emitted secondary electrons will also be affected by the RF field and collide with the microwave component wall again to generate new secondary electrons; in this cycle, the micro-discharging effect is formed when the secondary electrons in the microwave components accumulate continuously and grow exponentially (see Figure 7.3). If the process is analyzed in detail, the angle, energy, and the location of the incident point of the incident electrons will have an impact on the secondary electron emission when the incident electrons into the microwave component wall; the field distribution, the location of the emission point, the emission angle, emission energy, etc. of the secondary electron emission in space determine the trajectory of the secondary electrons and the location of the secondary electron collision point with the microwave component wall, the collision angle and energy, which in turn determine the physical characteristics of the secondary electrons to produce new secondary electrons. This is a continuous cycle process, but in each individual cycle, the influence of the secondary electron emission coefficient and energy spectrum distribution of the material at the incident point and the angle on them must be considered, only if these quantities are known to predict the characteristics of the incident electrons to produce secondary electrons. The secondary electron emission is only a small cycle in the multipactor process, and the multipactor is the superposition effect of multiple secondary electron emission processes, so the study of the secondary electron emission characteristics is the basis for the in-depth study of multipactor.

7.1.3.1 Impedance Converters

Bilateral discharge is one of the classic problems in multipactor research. In bilateral discharge research, the impedance converter is a more classical and narrower microwave component. This part uses an impedance converter with a higher pass-band frequency, so that the overall structure is relatively small, which can effectively reduce the production cost and achieve the purpose of multipactor research.

7.1.3.1.1 Structure and Size The detailed process of microwave component design is not described in detail here, only the structural dimensions of the impedance converter are given directly, as shown in Figure 7.4, where the pitch of the narrowest zone is only 0.32 mm. The corresponding model structure is shown in Figures 7.4 and 7.5.

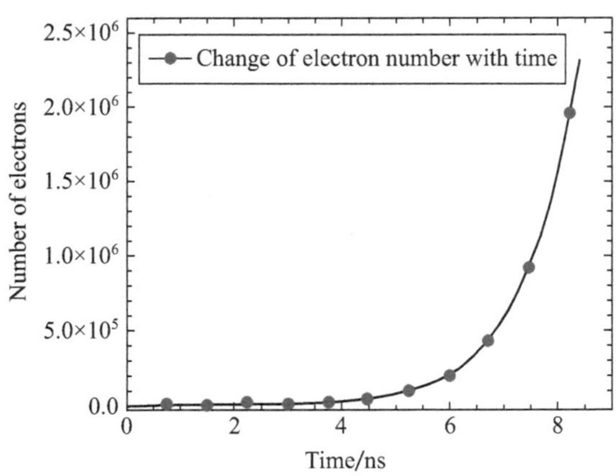

FIGURE 7.3 Exponential increase in the number of electrons during multipactor.

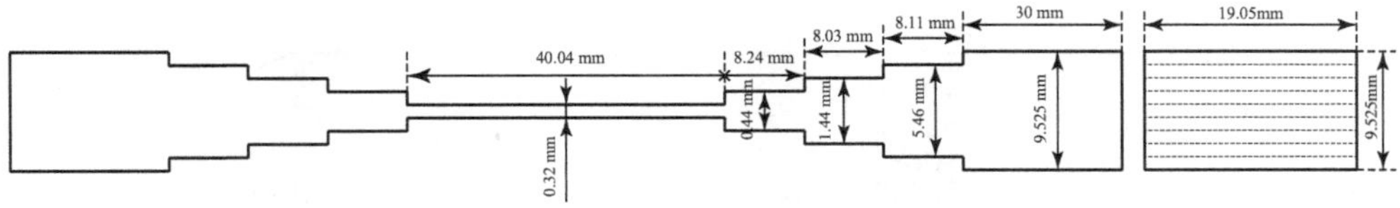

FIGURE 7.4 Diagram of impedance converter dimensions.

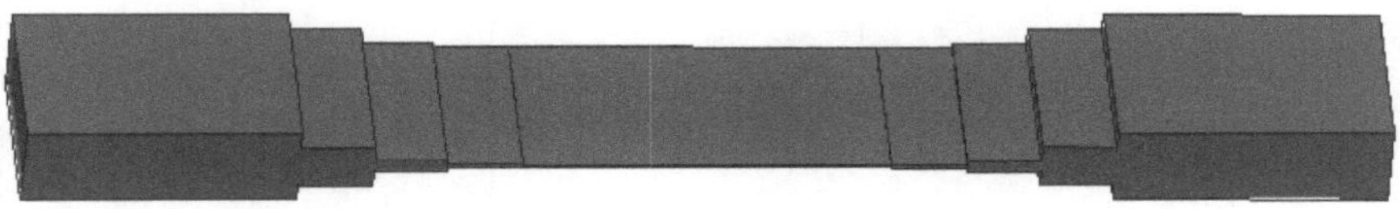

FIGURE 7.5 Diagram of impedance converter structure.

7.1.3.1.2 Electrical Performance Simulation The premise of the multipactor experiment is that the electrical properties of the microwave components are relatively good, the multipactor test system is very sensitive, and the reflected power has a serious impact on the multipactor test system, so it is necessary to ensure that the test microwave components themselves have good transmission characteristics in the test frequency range. Figure 7.6 shows the simulation results S_{11} and S_{21} of the impedance converter, and Figure 7.7 shows the corresponding field distribution of the impedance converter.

7.1.3.1.3 Threshold Simulation Analysis by Multipactor Joint Simulation Software One of the major differences between the simulation process of the software MSAT and the commercial software introduced earlier is the meshing. While most of the commercial software is adaptive meshing, MSAT also must perform manual meshing by targeting the structure of the microwave components, which can improve the speed and accuracy of the simulation, but the profiling efficiency is relatively low. The results of manual mesh generation using MSAT software are shown in Figure 7.8.

In addition, the software MSAT can observe the electron accumulation in the narrow gap of the microwave component, as shown in Figure 7.9. The trend of electron accumulation can be obtained, as shown in Figure 7.10, and the trend of electron growth for the predicted multipactor threshold, as shown in Figure 7.11.

To sum up, there are some algorithm differences between the software MSAT and current commercial software, but other multipactor simulation processes and judgment of multipactor are similar to commercial software. One significant advantage of the MSAT software is that the code is open source, allowing users to develop their own derived secondary electron emission models and incorporate them into the MSAT software for research purposes. This enhances the flexibility of research.

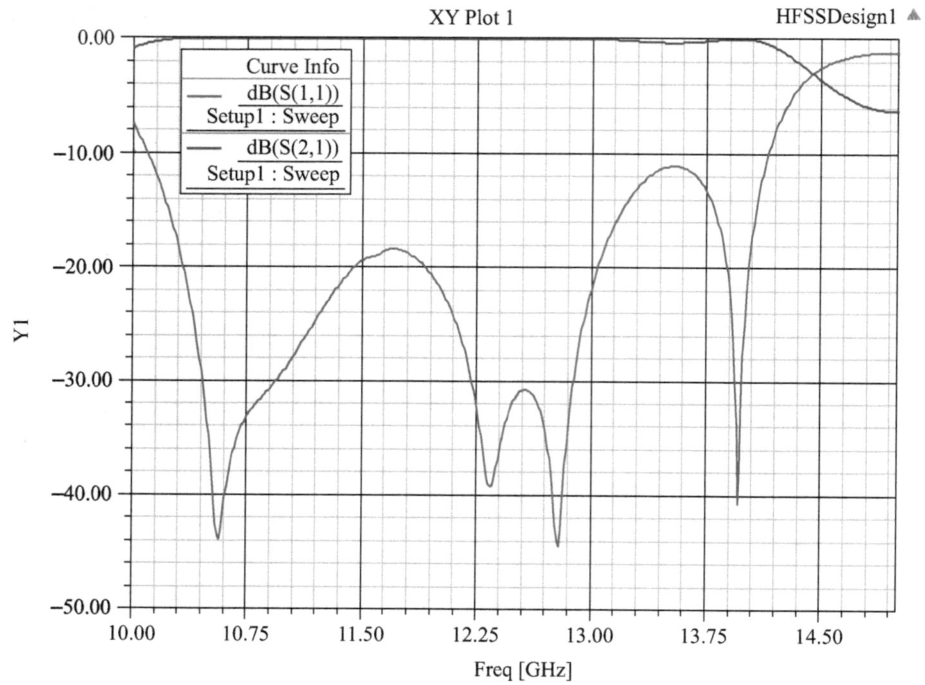

FIGURE 7.6 S_{11} and S_{21} amplitude and frequency characteristics simulation curve.

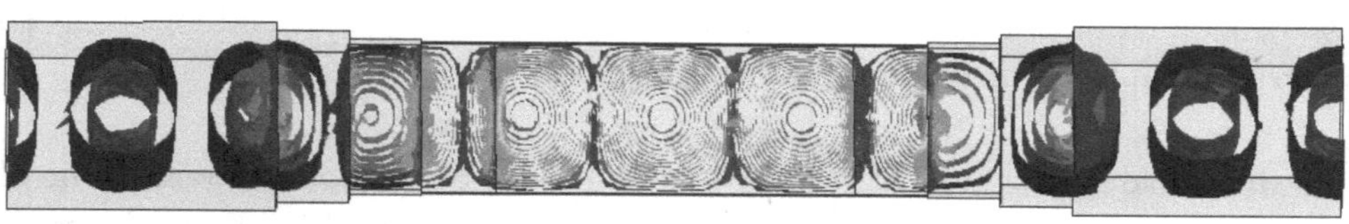

FIGURE 7.7 Field distribution.

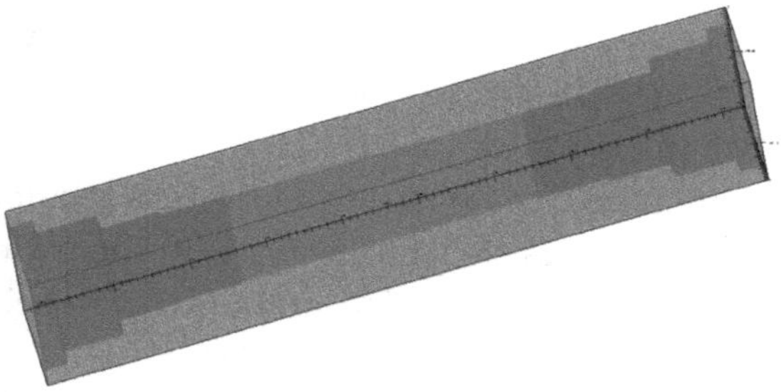

FIGURE 7.8 Meshing.

7.1.3.2 Metal Triplexer

Design a TM-mode metal triplexer with the following specifications:

Channel 1: 1.525 GHz ± 25 MHz

Channel 2: 1.635 GHz ± 25 MHz

FIGURE 7.9 Electronic accumulation.

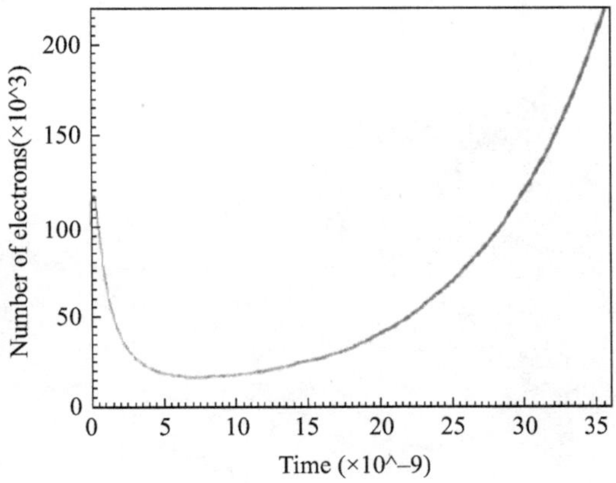

FIGURE 7.10 Growth curve of the number of electrons.

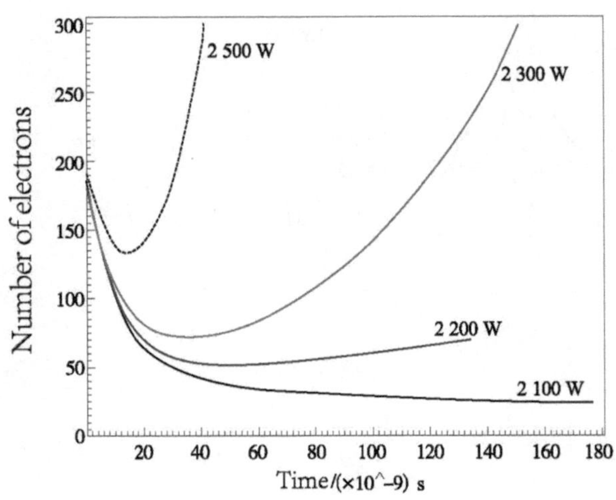

FIGURE 7.11 Electron growth curves at different input powers.

Channel 3: 1.745 GHz ± 25 MHz

Return loss: less than –20 dB

Insertion loss: less than 1.2 dB

Inter-channel isolation: more than 20 dB

The overall design model is shown in Figure 7.12.

The overall cuboidal triplexer for simulation has the *S*-parameters, as shown in Figure 7.13.

This means that the triplexer has good electrical performance. It can be seen from the simulation model that the top surface of the feeder post and the common end of the feeder are very close to the cover, which is the place where multipactor can easily occur.

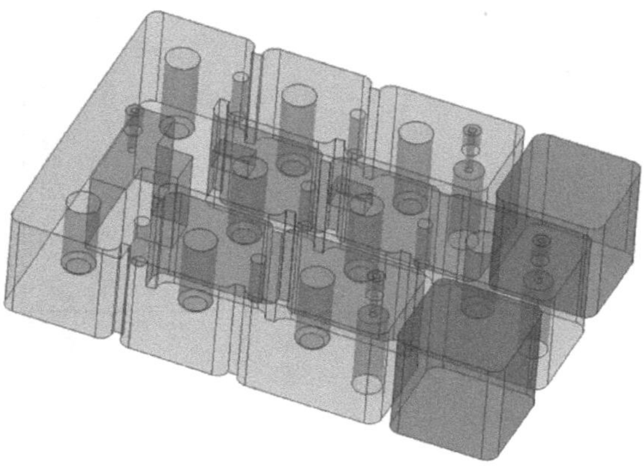

FIGURE 7.12 Overall simulation design model of metal triplexer.

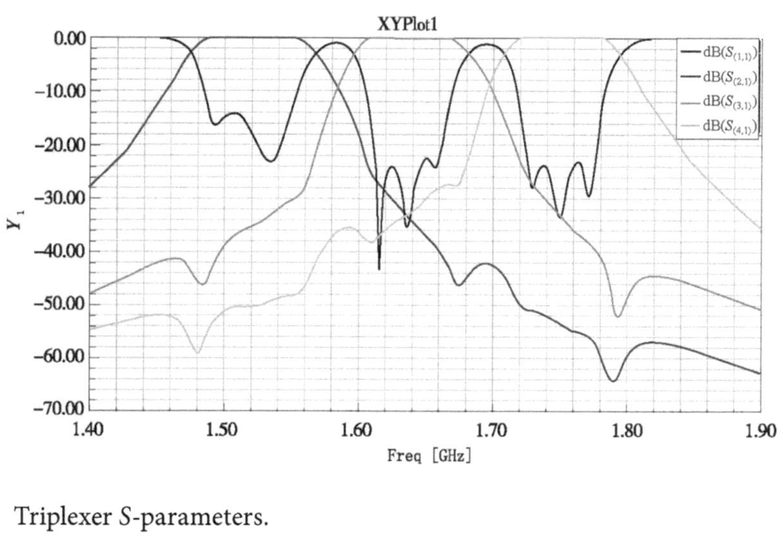

FIGURE 7.13 Triplexer *S*-parameters.

In order to improve the consistency between the multipactor threshold simulation results and the experimental tests, the secondary electron emission characteristics of the samples were obtained by testing the silver-plated aluminum alloy samples and ZNT-45 ceramic sample samples of the triplexer. The particle simulation method is used to simulate the multipactor process of the metal triplexer in dynamic real time, and the physical images of the multipactor formation process are revealed by obtaining the multipactor thresholds at three frequencies through the simulation of the initial formation process of the multipactor.

First, the multipactor formation process of the metal triplexer at 1.525 GHz is simulated, and the evolution of electrons formed by multipactor in the space of metal triplexer is shown in Figure 7.14. Multipactor occurs first in channel 1 at 1.525 GHz, and the accumulation of electrons near the common end multiplies with time or power, which is also a potential discharge area and needs attention in the design process. Through multiple simulations, the multipactor threshold at 1.525 GHz was obtained to be 205 W.

Then the multipactor formation process of the metal triplexer at 1.635 GHz is simulated, and the evolution of the electrons formed by the multipactor in the metal triplexer space is shown in Figure 7.15. It can be seen that the electron motion is confined to channel 2 at 1.635 GHz, and the multipactor occurs first at the common end when the power reaches 322 W, but there are also electrons gathered in channel 2, which is a potential discharge zone and needs attention during the design process. Through multiple simulations, the multipactor threshold at 1.635 GHz was obtained to be 322 W.

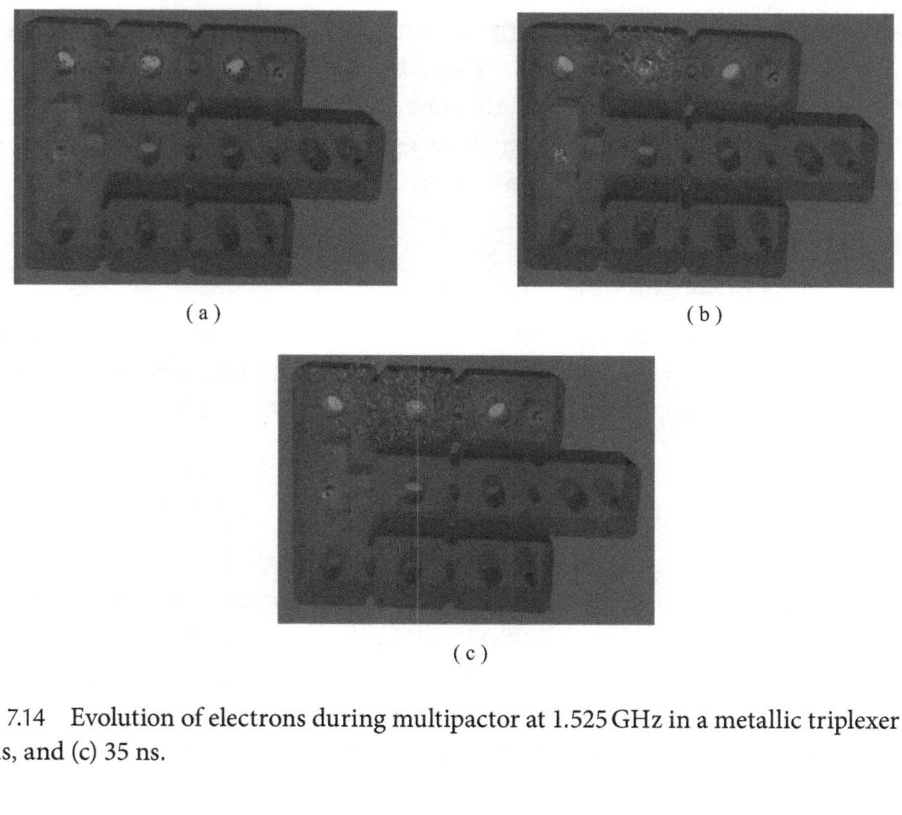

FIGURE 7.14 Evolution of electrons during multipactor at 1.525 GHz in a metallic triplexer (a) 2 ns, (b) 25 ns, and (c) 35 ns.

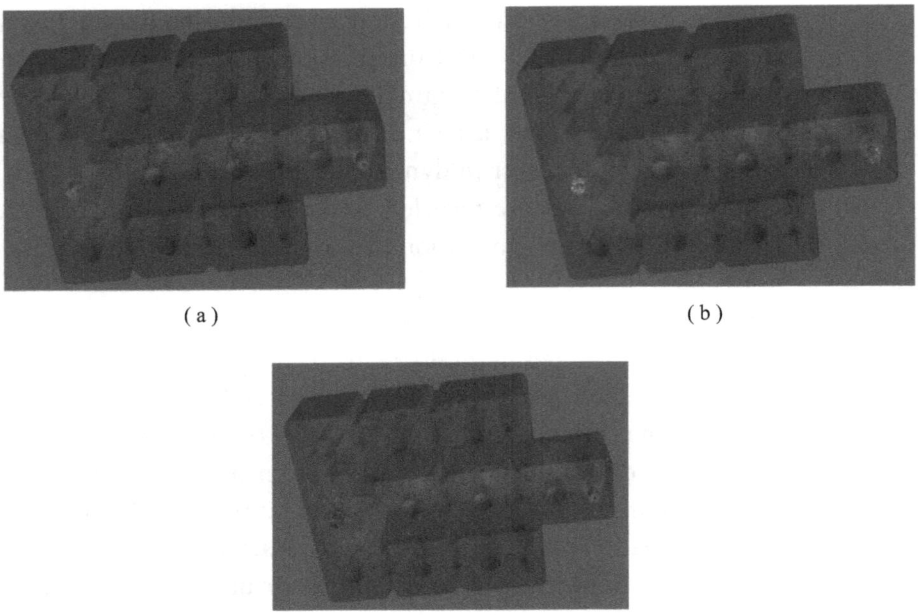

(a) (b) (c)

FIGURE 7.15 Evolution of electrons during multipactor at 1.635 GHz for metal triplexers (a) 2 ns, (b) 20 ns, and (c) 50 ns.

Finally, the formation process of multipactor in metal triplexer is simulated at 1.745 GHz, and the evolution of electrons formed by multipactor in the space of metal triplexer is shown in Figure 7.16. The electrons mainly move in channel 3 at 1.745 GHz, and multipactor occurs at the common end when the power reaches 582 W. Through several simulations, the multipactor threshold at 1.745 GHz frequency is obtained as 582 W.

7.2 SECONDARY ELECTRONS IN MICROSCOPIC IMAGING

7.2.1 Application of Secondary Electron Imaging

With the progress of secondary electron emission research, secondary electron emission has been widely used in various fields. At present, secondary electron emission is applied to multipliers for nuclear physics detection, image multipliers, SEM, and surface physics analysis instruments.

7.2.1.1 SEM

After the electron beam is incident on the material surface, the interaction of the electrons with the material substance will produce various signals carrying material information such as transmitted electrons, elastically scattered electrons, secondary electrons, backscattered electrons, energy loss electrons, absorption electrons, X-rays, oscillating electrons, and cathodoluminescence. Electron microscopy is to use this information to observe the topography, composition analysis and structure determination of the sample.

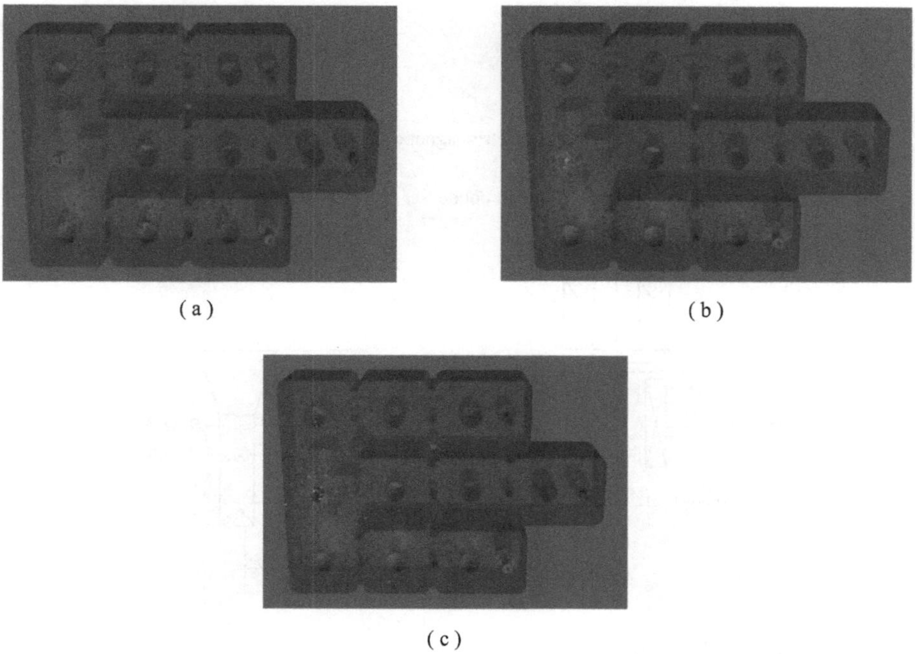

FIGURE 7.16 Evolution of electrons during multipactor at 1.745 GHz for metal triplexers (a) 2 ns, (b) 30 ns, and (c) 50 ns.

As one of the main types of electron microscopes, scanning electron microscopy (SEM) has been commonly used.

The principle of SEM is to use a very finely focused electron beam to scan on the surface of the sample to excite various physical information and to obtain the surface topography and other characteristics of the test sample by receiving, amplifying, and displaying the imaging of this information. The principle of SEM is shown in Figure 7.17. An electron beam of several tens of nanometers in diameter is emitted from the cathode electron gun, electrostatically focused by the gate, accelerated under a voltage of 2–30 kV, converged into an electron beam with a small aperture angle and a beam spot of 5–10 nm by passing through an electron optical system composed of two to three electromagnetic lenses, and finally focused on the sample surface. Under the action of the magnetic field generated by the scanning coil above the last lens, the electron beam performs a raster scan on the sample surface in a certain spatial order and frequency. The interaction between the incident electrons and the sample material excited a signal (such as secondary electrons) from the sample that is collected by the detector, which is amplified and used to modulate the brightness of the fluorescent screen. Since the current passing through the scanning coil is synchronized with the current through the corresponding deflection coil of the picture tube, the signal emitted from any point on the sample surface corresponds to the corresponding bright spot on the fluorescent screen of the picture tube one by one. The SEM adopts this point-by-point imaging method; the order of light imaging is from the top left to the bottom right, until the last line of the bottom right pixel is scanned to complete a

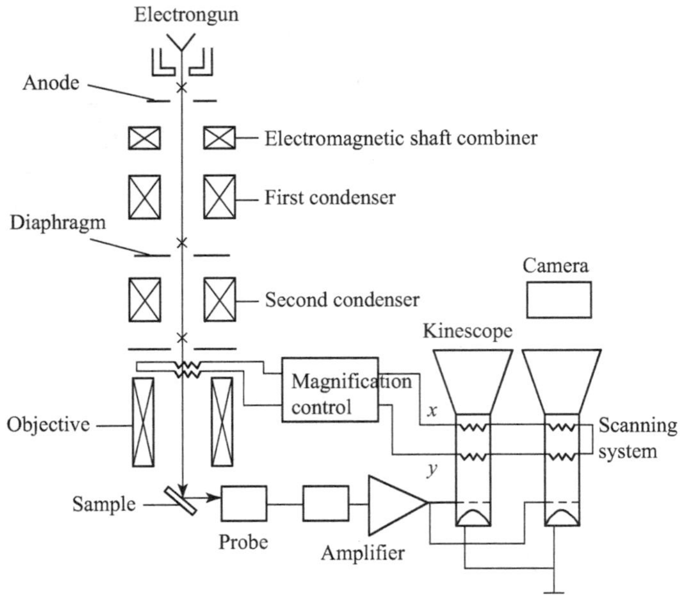

FIGURE 7.17 Schematic diagram of SEM structure.

frame, to get a secondary electron image reflecting the sample surface topography through the synchronous modulation of the tube brightness.

SEM has a large depth of field and high resolution, so that a very clear image of the sample structure can be obtained. The sample size of SEM can be very large (several millimeters×several millimeters) and the sample preparation is simple; it has a high magnification, continuously adjustable between 20,000 and 200,000 times, which is suitable for observation and analysis of various samples.

7.2.1.2 SEM Image Contrast

In SEM, the commonly used imaging signals are secondary electrons and backscattered electrons. The emission of secondary electrons is confined to a very small volume near the electron beam bombardment zone, mainly reflecting the state within the 10 nm layer of the specimen surface, and thus a relatively high-resolution image can be obtained. Several imaging contrast modes commonly used in SEMs include topographic contrast, composition contrast, and voltage contrast, which are briefly explained below.

7.2.1.2.1 Topographic Contrast The SEY is affected by the tilt angle between the incident electron beam and the sample surface, so the secondary electron image can reflect the topographic characteristics of the sample surface. This image contrast is the topographic contrast, which is one of the most used imaging modes.

When the sample with smooth and flat surface and no topographic features is placed obliquely, the secondary electron emission current is larger than when it is placed horizontally, but it only increases the brightness of the image without forming contrast.

For samples with certain surface topography, the topography can be seen as a number of convex peaks, steps, pits, and other details composed of different inclined surfaces, and the number of secondary electrons emitted from different parts of these topographic details is also different, thus generating the topographic contrast that reflects the undulation and geometry of the sample surface. The formation of the topographic contrast is mainly determined by the inclination angle of the sample surface with respect to the incident electron beam.

In the SEM, the direction of the incident electrons is fixed, but the unevenness of the specimen surface results in a different angle of the electron beam incidence on the specimen surface. As shown in Figure 7.18, let α be the angle between the incident electron beam and the normal to the sample surface, and for the two planes A and B in the sample, the corresponding incident angle α are different. For the smooth sample surface, the intrinsic SEY $\delta \propto 1/\cos\alpha$. The larger α is, the higher the intrinsic SEY δ and the higher the reflection coefficient η of the backscattered electrons.

Due to the low energy of secondary electrons, when detecting with the detector, the secondary electrons can be attracted by applying 250 V positive voltage to the collecting pole, and when the detector is placed above the sample, the detector can also receive part of the backscattered electrons, in which case, the secondary electron image also contains part of the backscattered electron information. Therefore, the number of secondary electrons and backscattered electrons received by the SEM detector varies in different topographic zones, and thus the brightness of the image varies. For example, in Figure 7.18, the incidence angle of zone A is smaller than that of zone B, and the secondary electrons and backscattered electrons received in zone A are smaller, so the image of zone A is darker than in zone B, thus showing the topographic contrast of the sample.

Therefore, the principle of topographic contrast is that the intensity of the secondary electron signal in the modulated image is a function of the inclination angle of the sample surface. The sample surface micro-zone topography difference is the micro-zone surface relative to the incident electron tilt angle is different. When the electron beam scans on the sample, the difference in the topography of any two points is manifested as a difference in signal intensity, thereby forming the surface topographic contrast in the image.

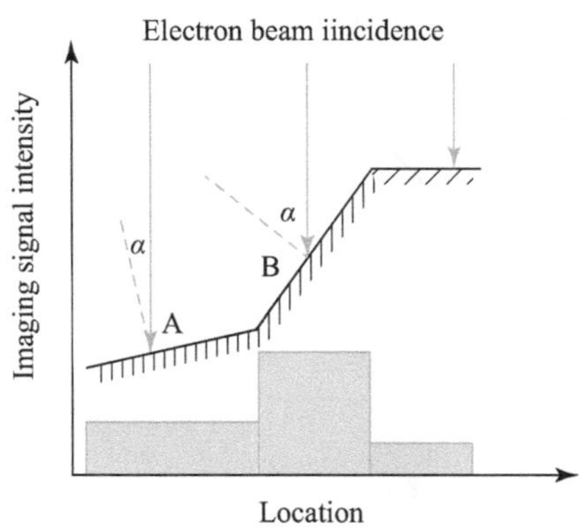

FIGURE 7.18 Correlation of secondary electron imaging signal intensity with angle of incidence.

Due to the high resolution of secondary electron image, no obvious shadow effect, large field depth, and strong stereoscopic sense, it is the main imaging method of SEM, which is especially suitable for the topographic observation of rough surface. In the manufacturing process of integrated circuit, the topographic contrast can be used for strip width and aperture analysis in photolithography, lateral analysis after etching, topographic analysis of circuit, and profile analysis of various film thicknesses and junction depths; topographic contrast can also be used to measure the critical size and perform plate alignment.

7.2.1.2.2 Atomic Number Contrast Atomic number contrast is also called composition contrast. A composition contrast image can be obtained by using physical signals that are sensitive to changes in the atomic number or chemical composition of the sample surface for imaging [1]. For samples with smooth surfaces and no topographic features, if the sample consists of a single element, the signal intensity generated when the electron beam is scanned to each point on the sample is the same and there is no contrast exists in the obtained image. However, if the sample consists of elements with different *Z*s, different signal intensities are generated over different elements, thus forming contrast.

Figure 7.19 shows the relationship between SEY and material *Z*. The differences in the number of extranuclear electrons and the ionization energy of substances with different *Z* result in a certain relationship between the SEY and *Z*. Moreover, the emission coefficient of backscattering varies significantly with *Z*, and the backscattered electrons also produce secondary electrons.

It can be seen from Figure 7.19 that the SEY generally increases as *Z* increases. When $Z < 20$, the SEY also increases with the increase of *Z*; when $Z > 20$, the SEY basically does not change with *Z*. Only the SEY of elements with small *Z* is related to the composition of the sample. Therefore, secondary electrons are usually used to observe the surface topography

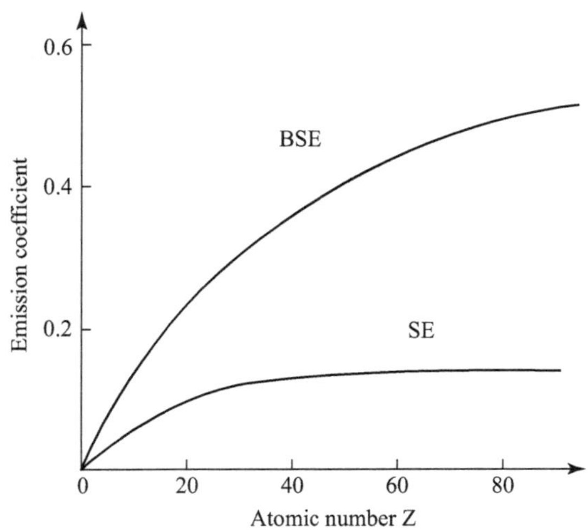

FIGURE 7.19 Relationship between secondary electron and backscattered electron yields and material *Z*.

rather than the composition distribution; only when Z is small or the difference is large, the Z contrast can be seen through the secondary electron energy.

The backscattered electron emission coefficient increases significantly with the increase of Z. During the analysis, the area with higher Z in the sample can emit more secondary electrons and backscattered electrons than the area with lower Z. Since the sensitivity of the backscattered electrons to Z is always much higher than that of the secondary electrons, the backscattered electrons are more widely used in the composition analysis. Under the same experimental conditions, the intensity of the backscattered electron signal increases with the increase of Z. In the area with a larger average Z on the sample surface, the intensity of the backscattered signal is higher. The corresponding area in the image shows a brighter contrast, while the area with a smaller average Z on the sample surface shows a darker contrast. Therefore, the difference of contrast in different zones reflects the difference in average Z in different zones of the sample, and the chemical composition distribution in the sample micro-zone can be qualitatively analyzed.

When Z is small or has a large difference, the secondary electron can also show a good Z contrast.

7.2.1.2.3 Charging Effect of Dielectric and Potential Contrast When the electron beam bombards the material surface, the SEY is 1 and the number of incident electrons and secondary electrons are equal only when the electron beam energy is the first critical energy or the second critical energy. If the electron beam energy does not meet this condition, for the non-conductive or poorly conductive and poorly grounded samples, the excess charge cannot be transferred, and the charge will accumulate on the surface of the dielectric sample, generating an electrostatic field that interferes with the incident electron beam and the emission of the secondary electron, and causing the charging effect. In the fields of SEM imaging and detection, electron beam probe micro-analysis, electron beam exposure technology, and other fields, sample charging will affect the accuracy of imaging, detection, and micro-machining. This charging effect can be suppressed in the low-energy electron beam device but cannot be eliminated. In addition, the charging effect of electron beam irradiation can form a special contrast and has a promising prospect. For example, the surface and buried layer microstructure of charged dielectric samples can be observed by SEM.

7.2.1.3 Dielectric Charging and Its Effects

In the early SEM research, it was found that the charging of the dielectric sample irradiated by the electron beam will cause image distortion, an amorphosis, formation of pseudo-image, and affect the accuracy of imaging and detection. The charging effect will have the following series of effects on the image [2].

The image contrast is abnormal: In SEM imaging, it is found that small changes in the surface potential of the insulation can also cause significant changes in image brightness. Under different incident electron energy conditions, since the secondary electron emission coefficient may be greater or less than 1, the polarity of the dielectric surface charge is different, divided into positive and negative charges, and the impact on the secondary electron emission is different, resulting in part of the image being abnormally bright and part becoming dark. If the charged surface potential of the sample repels more secondary electrons from the sample, the secondary electron yield will increase and the image brightness will be abnormally increased. In some cases, surface charging can even cause image contrast reversal.

Image an amorphosis and drift: Under the action of the local electrostatic field formed by the surface potential of the sample, if the incident electron beam deflects in a certain direction, the image drift will be formed; if the trajectory of the incident electron and the emitted secondary electron is irregularly deflected by the electrostatic field, it will also cause distortion and an amorphosis of the SEM image. Different incident electron beam energies will lead to different degrees of false images.

Image mirroring: As the negative charge on the dielectric surface increases, the surface potential causes the energy of the incident electron beam to increase and the landing site to shift. In the case of a very strong negative charge, the incident electron is deflected at a large angle. If it is bombarded to other positions of the sample chamber, such as the side wall, and the secondary electron is excited here, the sample surface acts as an "electrostatic mirror," and the secondary electron imaging signal collected by the secondary electron collector forms the shape image of the vacuum chamber. In some cases, the emitted secondary electrons are accelerated and bombarded to other parts of the vacuum chamber under the action of the strong electric field formed by the charged sample, and new secondary electron signals are generated, which are collected by the collecting electrode to form the shape image of the vacuum chamber, and it is superimposed on the surface image of the sample, which also produces the mirror effect.

Bright spots and bright lines: Irregular partial discharge often occurs on the surface of the charged insulator sample, causing irregular bright spots and bright lines in the image, resulting in discharge artifacts, that is, sporadic bright light signals that have no obvious relationship to the sample in time or space.

7.2.1.4 The Charged Contrast of the Dielectric

Although dielectric charging is a harmful interference factor in the normal imaging mode [3,4], on the other hand, people found that dielectric charging can produce an incentive mechanism of certain information and apply the charging phenomenon to obtain information that is not available in the absence of charging, so they began to explore how to effectively use the charging effect of dielectric samples to observe some information. The charged contrast is the contrast of the SEM image based on the charging effect of the dielectric sample. The surface potential of the charged sample can be used not only to form the contrast of the secondary electron image that reflects the surface information of the sample but also to form the contrast that reflects the internal microstructure information of the insulating sample. This voltage contrast is widely used in the SEM.

The potential change of the sample surface causes the difference of the secondary electron imaging signal current reaching the collecting electrode, so the contrast between light and dark will be formed on the secondary electron image, such as the P-N junction of the semiconductor and the bias integrated circuit. These local potential differences will affect the track and intensity of the secondary electron. The secondary electron seems to be pulled in the positive potential or high potential area and is not easy to escape, so the SEY in the positive potential area is less and is darker on the image; on the contrary, in the area of negative potential or low potential, the secondary electrons are easily pushed out, the yield is higher, and the image is brighter, which is the potential contrast.

The charged contrast of SEM has a wide range of applications in the field of micro-nano electronic devices. There are many related research reports. Such potential contrast can be used to study the process structure of materials and devices, and the defect location of chips. For example, open-circuit, short circuit, or other abnormalities can easily occur in some special positions of semiconductor chips, which are often not on the surface of the sample. In these cases, neither the surface topography contrast nor Z contrast can play a significant role. When scanning the sample surface with an electron beam for a certain period, if there is a defect in the chip, the potential of the surface position corresponding to the defect may be different from that around, and then the image shows obvious bright or dark spots. Through this method, the location information of chip defects can be obtained quickly.

7.2.1.5 Deep Layer Information Contrast of Charging Effect

Using the charged contrast can not only reflect the surface information of the sample but also observe the deep layer information of the sample covered by the insulating layer without damage[5], which meets the requirements of many new devices and IC manufacturing process development on the detection method.

In the process of modern IC manufacturing and processing, a layer of SiO_2 insulating layer is usually made between the two layers of circuits, and a layer of SiO_2 or Si_3N_4 insulating protection layer is needed on the surface of the fabricated integrated circuit. Semiconductor materials sometimes need to be covered with insulating layers such as SiO_2 during the process of manufacturing integrated circuits. In order to ensure the high reliability, yield, and qualification rate of integrated circuits, it is often necessary to detect the semiconductor materials covered with the insulating layer and the semi-finished and finished products of integrated circuits, especially the defects or micro-structures of the underlying semiconductor materials and integrated circuits without damaging the surface insulating layer. The thickness of the insulating film of the tested sample with an insulating protection layer can reach hundreds of nanometers to several microns. For the detection of this sample, the use of a high-energy electron beam will penetrate the insulating film, cause radiation damage to the device, and affect the inherent operating state of the device, so the low-energy electron beam is generally used for observation.

Researchers at Osaka University in Japan observed the topography of the internal wire covered by the insulating film for the first time in the charge contrast experiment of integrated circuits. The imaging of the deep layer structure information of the insulating sample was realized when no excitation signal was applied to the IC sample, and the incident electron with an energy of 1 keV could not penetrate the insulating film to reach the wire [6]. Toshiba Corporation in Japan carried out the alignment detection of the IC layout in the low-energy electron beam lithography system using the charged contrast and observed the alignment slot mark under the insulating film.

The charged contrast formed during secondary electron imaging is due to the different surface potential generated by the accumulated charge at the interface of the buried layer structure in the insulating film, which changes the local electric field on the surface, resulting in the difference between the secondary electron return to the surface and the secondary electron imaging current, which changes the image brightness. Chinese researchers used the MC method to simulate the electron and charge transport process when the electron beam irradiates the IC sample under the same conditions and theoretically confirmed the formation mechanism of this contrast [7]. Therefore, under the negative charge condition of the dielectric film, the slot mark can be observed through the charged contrast even if it cannot be penetrated by the incident electron beam.

7.2.2 Secondary Electron Multiplication of Dielectric Surface with Buried Layer Structure

For a long time, in order to reduce the charge of the sample when observing the surface of the dielectric sample in the SEM, it is usually necessary to carry out the surface gold-plating treatment on the sample or use the incident electron near the second critical energy. In general, the SEM can only be used to observe the topography of the material surface. However, in some experiments, it is often found that the charged dielectric sample has an ultra-penetrating internal detection ability for electrons. Therefore, SEM can also be used to observe the internal buried layer results of the dielectric samples in LSI.

Although this phenomenon of electron ultra-penetration has been discovered for a long time, its micro-physical mechanism and related characteristics are still unclear, and the corresponding quantitative theoretical research is also very scarce. Professor Ura analyzed the SEM image of the buried structure sample under the positive charge condition by considering the return of the emitted secondary electrons[8], but there is still no clear theoretical explanation for the abnormal brightness of the image under the negative charge condition. For this reason, Miyoshi put forward the theory of the return of the secondary electrons and the re-excitation of the tertiary electrons to explain the abnormal brightening phenomenon under the negative charge condition, but it still lacks strong quantitative theoretical support.

In view of this, in this section, the charging and imaging characteristics of SiO_2 dielectric samples with buried layer structure are investigated from the perspective of theoretical simulation. By establishing the corresponding model of dielectric charging and secondary electron emission under the irradiation of scanning electron beam, the charging characteristics and the multiplication mechanism of the electron emission of the buried layer structure sample during the electron irradiation process are analyzed, and the influence of the relevant incident electrons and sample structure on the secondary electron multiplication mechanism is studied, and the abnormal charging phenomenon of the buried layer structure in the experiment is explained.

7.2.2.1 Experimental Background and Calculation Model

This chapter mainly simulates the conditions of SiO_2 samples with metal-buried layer structure under a negatively charged irradiation environment based on

Miyoshi and Ura's study[8]. In the experiment, this kind of SiO_2 structure with metal-buried layer is widely used in large-scale integrated circuits. The irradiated electron energy used in the experiment was 5 keV.

The corresponding scattered electron range is less than the buried layer thickness of SiO_2 by 1 μm. The internal metal structure seen in the experimental SEM image reflects the ultra-penetrating effect of the dielectric sample irradiated by the electron due to charging.

In the SEM image, the detection and observation of the internal structure of the sample focus on the edge of the internal structure. By comparing the changes in the buried edge area of the image in the boxed area at different times can be found that the buried edge area is darker than the surrounding image at the initial stage of irradiation; and as the irradiation continues, the edge area gradually brightens and finally reaches saturation brightness.

In previous studies, it was believed that in the case of negative charge, the emitted secondary electrons would leave the sample surface because the emitted secondary electrons were repelled by the negative charge on the surface, which could not explain the phenomenon of darkening and then brightening at the edge of the buried layer [9–11]. Therefore, based on the possible mechanism of the return and re-emission of the emitted secondary electrons under the condition of negative surface charge proposed by Prof. Ura, this section will use the numerical simulation method to verify this kind of ultra-penetration phenomenon of the electron microscope to the dielectric with buried layer structure, and further analyze the relevant microscopic physical mechanism and the corresponding sensitivity analysis.

In this chapter, the influence of the secondary electrons emitted in the edge area of the buried layer is analyzed. Figure 7.20 shows the cross-section diagram of SiO_2 with metal buried layer structure simulated in this chapter, where the relevant dimension parameters, such as the total thickness of the sample H, the thickness of the buried layer $H_{B,}$ and the thickness of the metal layer are consistent with the experimental conditions, which are 3, 1, and 0.5 μm, respectively. Due to different growth conditions, the electronic mobility of SiO_2 materials can vary in a large range, while the electronic mobility of SiO_2 film as a passivation layer of integrated circuits is relatively low. Therefore, the electronic mobility of the SiO_2 layer in this chapter is $1 \times 10^{-5}\,cm^2/(V\,s)$, trapped density is $2 \times 10^{16}\,cm^{-3}$.

When the incident electron irradiates the surface of the SiO_2 sample, because the scattering range of the incident electron is smaller than the thickness of the buried layer, the

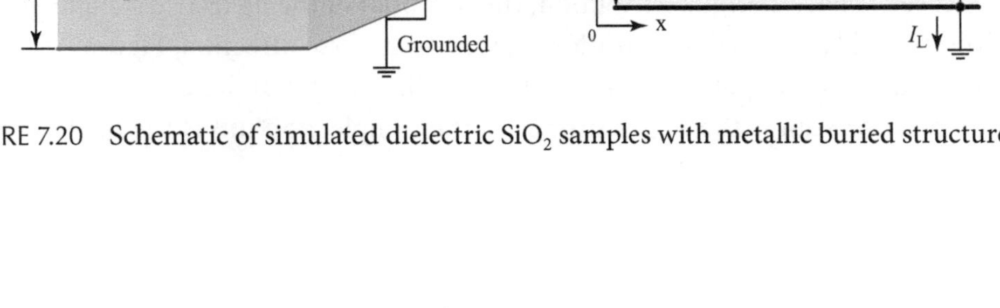

FIGURE 7.20 Schematic of simulated dielectric SiO_2 samples with metallic buried structures.

incident electron will only interact with the SiO_2 sample of the buried layer to generate secondary electrons and emit from the surface. The secondary electrons emitted with different energies and angles will move under the action of the space electric field generated by the surface charge. In this chapter, Rung-Kutta is used to track the trajectory of the emitted electrons, we have:

$$m\ddot{x}(t) = -eE_X(x,z,t),$$
$$m\ddot{z}(t) = -eE_Z(x,z,t). \tag{7.21}$$

Under the action of non-uniform space electric field, some of the secondary electrons emitted will directly emit, and some of the secondary electrons emitted from the surface will return to the sample surface here. The returned secondary electrons will react with the sample again to produce new secondary electrons. In this way, the emitted electrons eventually exit from the local field on the surface or stay inside the sample. After the simulation of all the incident electrons is completed, the final emitted electrons are counted. The ratio of the finally emitted secondary electrons r_{OE} is the ratio of the number of secondary electrons r_{OE} and the original number of primary and secondary electrons:

$$r_{OE} = \frac{N_{OE}}{N_{SE}} \tag{7.22}$$

7.2.2.2 Dielectric Charging Characteristics with Buried Layer Structure

Next, the charge and potential distribution characteristics of the dielectric sample with internal buried layer structure are analyzed. The material of the dielectric sample is SiO_2, and the relative dielectric constant ε_r is 2.2; the total thickness of the sample H and the thickness of the buried layer H_B are 3 and 1 μm, respectively. The conditions for scanning the electron are the same as in the experiment. The incident electron energy is 5 keV and the beam current is 20 pA. Considering the scanning area of the electron and the image magnification in the electron microscope, the equivalent incident current density is 5 nA/cm^2. Figure 7.21 shows the distribution of the internal charge near the edge of the buried layer in the sample (3 μm each on the left and right and edge at $x = 3\,\mu m$) when the scanning electron is continuously irradiated for 46 s. According to the cross-section structure of the sample, in Figure 7.21, the buried metal layer area with the grounding is in $x > 3\,\mu m, 1\,\mu m < z < 2\,\mu m$, and the charge in this area will be directly lost through the grounding. In the longitudinal z direction of the sample, the accumulated charge will be gradually transported downward; in the transverse direction, due to the different internal structure, the internal charge will also move to the side with the metal buried layer, so that the charge density in the buried layer is slightly higher than that in the non-buried layer.

In this case of space charge distribution, the potential of the dielectric sample near the edge of the buried layer is still non-uniform in the transverse direction. Figure 7.22 shows the internal and external potential distributions of the sample under corresponding conditions. Where $z = 3\,\mu m$ is the upper surface of the sample, $x > 3\,\mu m$ is the buried area of

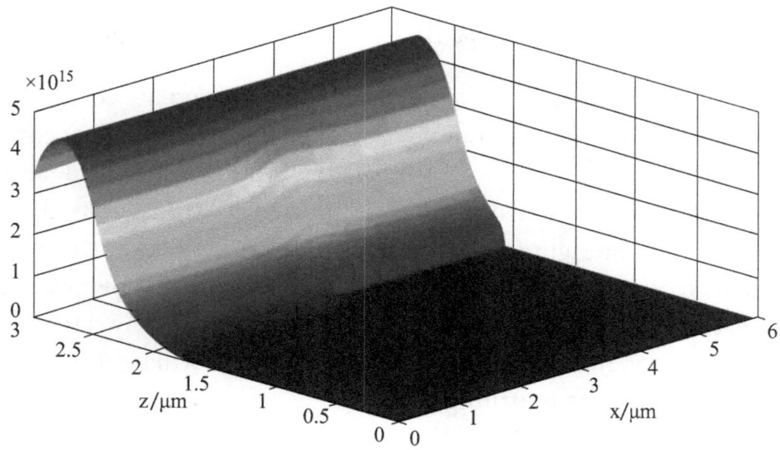

FIGURE 7.21 Charge distribution in the sample near the edge of the buried layer, the exposure time is 46 s.

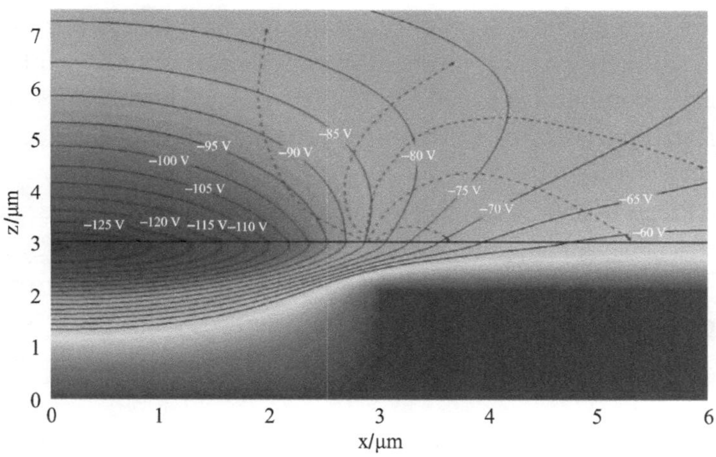

FIGURE 7.22 Internal and external potential distribution near the edge of the sample buried layer at the surface potential difference of 70.5 V.

the sample. Due to the difference in the internal structure of the sample and the difference in the internal charge distribution, the space potential inside and outside the sample is not consistent in the transverse x direction. As can be seen from the contour line in Figure 7.22, for the interior of the sample, because the lower plate of the sample and the buried metal are grounded, and the internal charge is distributed downward from the surface of the sample, according to the Poisson's equation for solving the potential, the surface potential is stronger at the left end farther away from the grounded part, and gradually decreases toward the bottom and right side of the sample, corresponding to the electric field line pointing from the lower right to the upper left. Here, the surface potential contrast V_{SC} is defined as the difference between the maximum value and the minimum value of the surface potential, that is, $V_{SC} = V_S|_{max} - V_S|_{min}$. In Figure 7.22, the surface potential contrast is $V_{SC} = 70.5$ V.

For the space electric field outside the sample, by considering the continuous condition of the electric vector of the surface boundary and the normal boundary condition of the electric field, the non-buried area on the left is a downward electric field, and the electric field force on the electron is upward, and the secondary electrons emitted from this area can be emitted out. On the right side of the edge of the buried layer, there will be a local upward electric field, in which the electrons will be subjected to a downward electric field force. Therefore, the secondary electrons emitted from this zone may return to the sample surface under certain conditions, and the returning electrons will be accelerated due to the lower negative potential of the returning position.

7.2.2.3 Multiplication Mechanism of Secondary Electrons

Under the action of the above space electric field, some of the emitted secondary electrons may return to the surface of the sample and be accelerated, so as to excite the secondary electrons again. The repeated return and re-excitation of the emitted secondary electrons will cause the cascade multiplication effect of the secondary electrons in the local space field. In order to study the physical mechanism and the corresponding influence law of the secondary electron multiplication caused by the space local field, we first study the trajectory and return characteristics of the secondary electron in the space electric field.

7.2.2.3.1 Return Properties of Secondary Electrons When the emitted secondary electrons emit at different angles and energies, their trajectories are also affected by different spatial electric fields. Figure 7.23 shows the trajectory of the secondary electrons under different emission conditions, and the surface potential contrast at this time is $V_{sc} = 70$ V, where

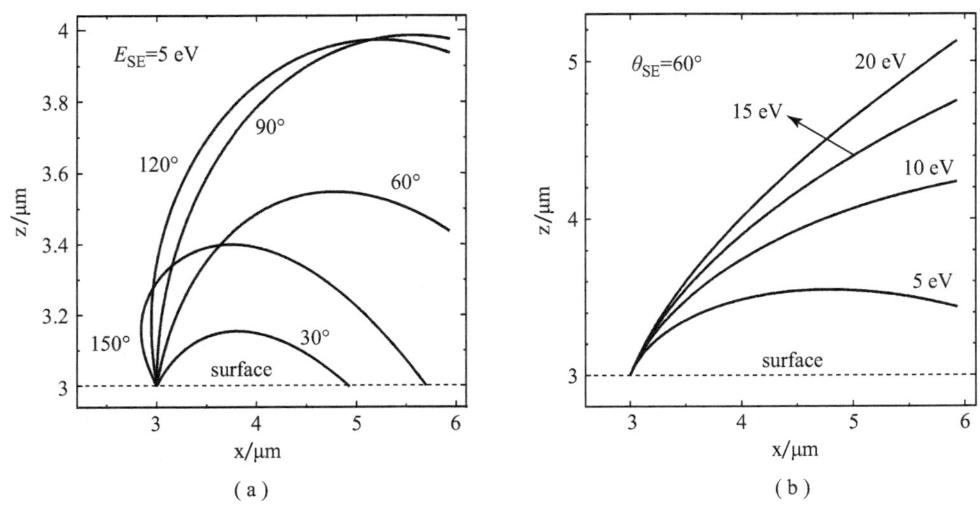

FIGURE 7.23 Secondary electron trajectories under different ejection conditions (a) different emission angles and (b) different emitted electron energies.

Figure 7.23a shows the cases at the energy of the emitted electrons of 5 eV and the emission angles of 30°, 60°, 90°, 120° and 150° (the emission angle here is the included angle between the emitted electrons and the positive x direction of the sample surface). The positions of the emitted secondary electrons are all at the edge of the surface buried layer structure $x=3\,\mu m$. Comparing the five different exit trajectories in the figure, it can be seen that the smaller the emission angle, the easier it is for the electrons to return to the sample surface under the action of the space electric field force. And considering the asymmetry of the space electric field, the electrons are subjected to the downward electric field force only in the right space, and the electrons in the whole space are subjected to the rightward electric field force. Therefore, the emitted electrons are deflected to the right, and the return position is also the right surface of the edge of the buried layer.

Similarly, the effect of the local electric field force on the emitted electrons of different energies is also different. Figure 7.23b shows the trajectories of the emitted secondary electrons with different energies (the emitted electron energies corresponding to the four trajectory curves are 5, 10, 15, and 20 eV, respectively), where the c angle is 60°, the emission position is still at the edge of the surface buried layer, and the surface potential contrast is $V_{sc}=70\,V$. From the comparison curve, it is easy to see that the electrons with lower emission energy are easier to return to the sample surface.

7.2.2.4 Cascade Multiplication Mechanism of Emitted Electrons

According to the above analysis, the part of the emitted electrons that meet certain conditions can escape from the local electric field on the sample surface to exit, while the other part will return to the sample surface under the effect of the local electric field force on the surface, and continue to incident into the sample interior and collide with the sample to generate the secondary electron N'_{SE}, and so on, until the electrons finally stay in the sample interior or escape from the emission. The final emitted electron N_{SE} includes the secondary electron N_{SE} that directly escapes and the sum of all secondary electrons $\sum_{i=1}^{n} N'_{SE}$.

Here, all secondary electrons are collectively referred to as tertiary electrons $N_{TE} = \sum_{i=1}^{n} N'_{SE}$, which meets

$$N_{OE} = N_{SE} + N_{TE} \tag{7.23}$$

Since the return and multiplication mechanism of the emitted electrons is caused by the surface charge, the surface charge intensity becomes the main factor affecting the total number of emitted electrons N_{OE}. Figure 6.36 shows the change curve of the ratio of electrons emitted from the edge of the surface buried layer r_{OE} with the surface potential contrast V_{sc}. Here, the ratio of emitted electrons r_{OE} is defined as the ratio of the total number of electrons finally escaped from the local electric field N_{OE} to the total number of secondary electrons originally emitted N_{SE0}, namely, $r_{OE} = \frac{N_{OE}}{N_{SE0}}$. The ratio of the emitted electrons includes the ratio of the emitted secondary electrons r_{SE} and the ratio of the emitted tertiary electrons r_{TE}; the ratio of the emitted secondary electrons r_{SE} is defined as the ratio of the directly escaped secondary electrons N_{SE} to the total number of the original emitted secondary electrons N_{SE0}, that is, $r_{SE} = \frac{N_{SE}}{N_{SE0}}$, and the ratio of the tertiary electrons emitted r_{TE} is defined as the ratio of the final number of tertiary electrons emitted N_{TE} to the original number of secondary electrons emitted N_{SE0}, that is, $r_{TE} = \frac{N_{TE}}{N_{SE0}}$. As shown in Figure 7.24, as the surface potential contrast increases, the ratio of secondary electrons emitted r_{SE} decreases due to the increase in surface charge, and the electrons are easier to return. In addition, the returned secondary electrons are accelerated stronger, and more tertiary electrons are generated under the cascade effect, and the corresponding tertiary electron ratio r_{TE} increases. Since the total emitted electrons include secondary and tertiary electrons, under the opposite effect of the two, the ratio of total emitted electrons r_{OE} decreases first and then increases with the increase of surface potential contrast V_{sc}. In the range of surface potential contrast $0 < V_{sc} < V_{sc1}$, the ratio of total emitted electrons $r_{OE} < 1$, that is, the emitted electrons are suppressed by the surface electric field, and it is the smallest when $V_{sc} = V_{sc0}$. In the range of $V_{sc} > V_{sc1}$, the ratio of total emitted electrons $r_{OE} > 1$, that is, the return and re-emission of the emitted electrons under the local electric field are enhanced, and it increases with the increase of surface potential contrast V_{sc} (Figure 7.24).

Due to the non-uniformity of the transverse electric field, the ratio of electrons emitted from different positions under the action of the surface electric field force is also different. Figure 7.25 shows the distribution of the emitted electron ratio r_{OE} (x) at different positions near the edge of the buried layer, whereas Figure 6.37a shows the electron microscope image of the buried layer structure [3]. The area in the box in the figure is the area near the edge of the buried layer, which corresponds to the analysis range in Figure 6.37b. The SEM

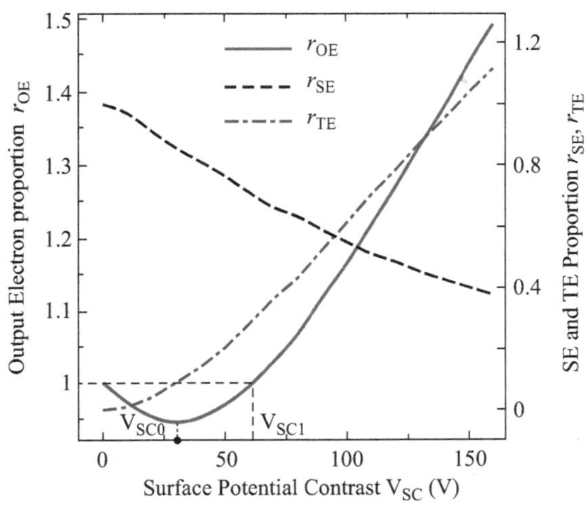

FIGURE 7.24 Curve of the ratio r_{OE} (including the ratio of ejected secondary electrons r_{SE} and the ratio of ejected tertiary electrons r_{TE}) of output electrons at the edge of the surface buried layer as affected by the surface potential contrast V_{SC}.

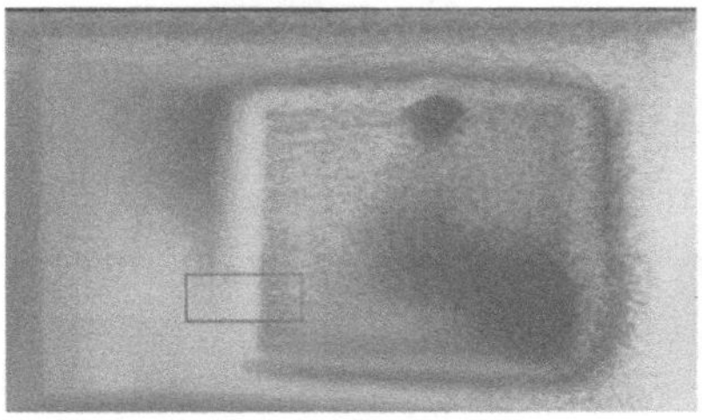

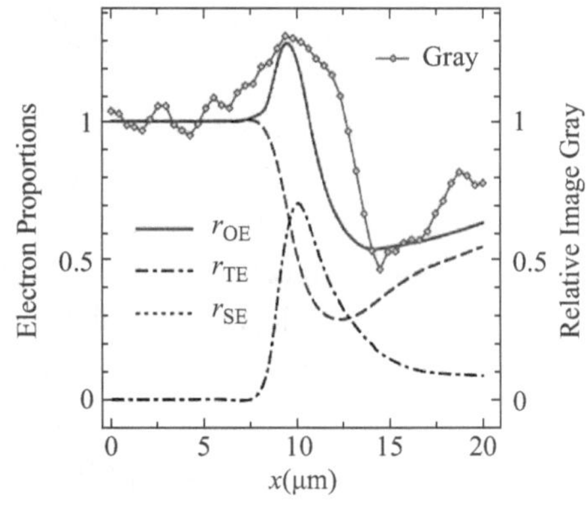

FIGURE 7.25 Distribution of ejected electron ratio at different positions in the transverse direction: (a) Scanning electron microscope images obtained in the experiment and (b) comparison of simulated ejected electron ratios with the grayscale of the electron microscope image under the same conditions.

environment and conditions are consistent with those described above, and the scanning time is 90 s. It can be seen that the left side of the box is gray, the middle part is bright, and the right side is relatively dark, which corresponds to the image gray curve in Figure 6.37b. In Figure 6.37b, r_{OE}, r_{SE}, and r_{TE} are the ratio curves of the emitted electrons at different positions simulated under the same conditions. The total emission electron curve is basically consistent with the image gray curve. For the electrons emitted from the left side, according to the previous discussion, because the space electric field is upward, the secondary electrons emitted from the original can basically be directly exit, so the ratio of emitted electrons is maintained at around 1, which corresponds to the medium brightness of the image. However, for the exit of the edge of the buried layer (i.e., $x = 10\,\mu m$ in the figure), since the returned electrons are greatly accelerated, the emitted electrons are cascaded and multiplied, the ratio of emitted electrons $r_{OE} > 1$, and the corresponding image is brighter. For the electrons emitted from the right area, although some of the electrons return under the action of the electric field, the number of electrons emitted has not increased due to the lack of larger electric field acceleration, so the ratio of emitted electrons $r_{OE} < 1$, and the corresponding image is also darker. This simulation is basically consistent with the experiment and also provides a more reasonable theoretical explanation for the abnormal brightening phenomenon at the edge of the buried layer (Figure 7.25).

The non-uniform charging characteristics of the sample and the electron emission multiplication mechanism are studied by establishing the charging and electron emission models of the dielectric sample with the buried layer structure irradiated by the electron. First, the corresponding physical structure model is established according to a typical buried layer structure in the actual large-scale integrated circuit. Then, the numerical physical model is used to calculate the charging characteristics of the sample under electron irradiation and the influence of surface non-uniform charging on the trajectory and quantity of emitted electrons. Finally, different samples and incident conditions are simulated, and the effects of relevant parameters on the multiplication mechanism of charged and emitted electrons are analyzed.

The main conclusions are as follows:

1. The electrons emitted from the non-uniformly charged surface include the directly emitted secondary electrons and the tertiary electrons excited by the returning electrons. The emitted electrons with smaller exit angle or energy are more likely to return under the action of the local electric field in the space caused by the charged sample.
2. With the increase of surface potential contrast, the emitted secondary electrons can more easily return and excite more tertiary electrons, while the total number of emitted electrons shows a tendency to first decrease and then increase. With the continuous scanning electron irradiation, the potential contrast of the sample surface gradually increases, and the corresponding number of emitted electrons also decreases first and then increases and the grayscale of the corresponding area of the electron microscope image also shows a change rule of dark first and then bright.

REFERENCES

1. Joy D C, Joy C S. Low voltage scanning electron microscopy. *Micron*, 1996, 27(3–4): 247–263.
2. Zhang H B, Feng R J, Ura K. Utilizing the charging effect in scanning electron microscopy. *Science Progress*, 2004, 87(4): 249–268.
3. Frank L, Zadražil M, Müllerová I. Scanning electron microscopy of nonconductive specimens at critical energies in a cathode lens system. *Scanning*, 2001, 23(1): 36–50.
4. Wong W K, Phang J C H, Thong J T L. Charging control using pulsed scanning electron microscopy. *Scanning*, 1995, 17(5): 312–315.
5. Nagase M, Kurihara K. Imaging of Si nano-patterns embedded in SiO2 using scanning electron microscopy. *Microelectronic Engineering*, 2000, 53(1–4): 257–260.
6. Ura K. Contrast mechanism of negatively charged insulators in scanning electron microscope. *Microscopy*, 1998, 47(2): 143–147.
7. Zhang H B, Li W Q, Wu D W. Contrast mechanism due to interface trapped charges for a buried SiO2 microstructure in scanning electron microscopy. *Journal of Electron Microscopy*, 2009, 58(1): 15–19.
8. Miyoshi M, Ura K. Negative charging-up contrast formation of multilayered structures with a nonpenetrating electron beam in scanning-electron microscope. *Journal of Vacuum Science & Technology B: Microelectronics and Nanometer Structures Processing, Measurement, and Phenomena*, 2005, 23(6): 2763–2768.
9. Wang F, Feng G, Zhang X, et al. Mechanism of electron multiplication due to charging for a SiO2 sample with a buried microstructure in SEM: A simulation analysis. *Micron*, 2016, 90: 64–70.
10. Feng R J, Zhang H B, Wang S Y, Ura K. Local field of surface and redistribution properties of secondary electrons emitted from insulating thin film with negative charge. *Chinese Journal of Semiconductors*, 2004, 25(1): 87–92.
11. Okyay G, Héripré E, Reiss T, et al. Soot aggregate complex morphology: 3D geometry reconstruction by SEM tomography applied on soot issued from propane combustion. *Journal of Aerosol Science*, 2016, 93: 63–79.

CHAPTER 8

Research on the Application of Secondary Electron Emission Phenomenon in Electron Multipliers

Baojun Yan and Shulin Liu

8.1 INTRODUCTION TO ELECTRON MULTIPLIERS

8.1.1 Principle of the Electron Multiplier

As a device for detecting weak signals, electron multiplier has become an invaluable tool in the fields of high-energy physics, nuclear detection, space exploration, and scientific instruments. Electron multiplier is a vacuum electronic device that can detect a wide range of particles, including electrons, photons, and ions. Its fundamental operating principle is based on the secondary electron multiplication effect.

As illustrated in Figure 8.1a, when incident particles with a specific energy (such as electrons) bombards the surface of the material, resulting in the emission of internal electrons from the surface of the material, this is known as the secondary electron emission phenomenon of the material. In general, the secondary electron spectral composition is comprised primarily of true secondary electrons (with an energy of less than 50 eV), elastically scattered electrons (with an energy that is close to that of the incident electron), and inelastically scattered electrons with an energy that lies between the two categories. The ratio of the number of emitted secondary electrons to the number of incident electrons is referred to as the secondary electron emission coefficient (δ), and the relationship between the δ value and the incident electron energy is illustrated in Figure 8.1b. The maximum value of δ ($\delta = \delta_m$) corresponds to an incident electron energy of E_m. When the incident electron energy is lower or higher, corresponding to $\delta < 1$, the surface of the sample is negatively charged, which indicates that the secondary electron suppression effect occurs.

DOI: 10.1201/9781003562863-8

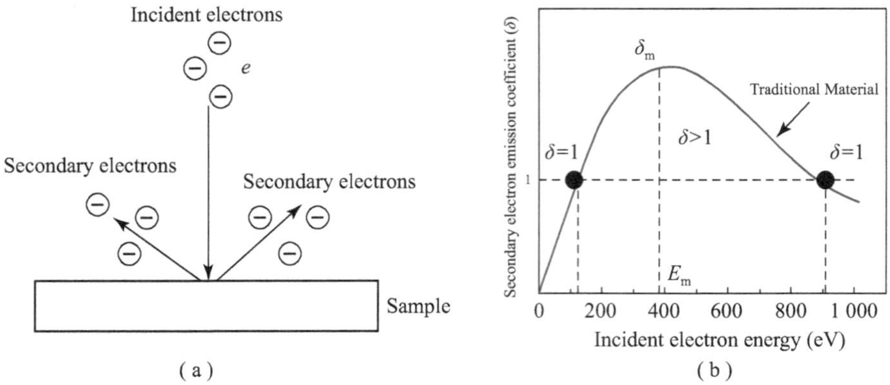

FIGURE 8.1 Secondary electron emission phenomena (a) and their typical characteristics (b).

This is a very important factor in the development of particle accelerators, spacecraft, and other fields. This phenomenon is of great interest in the field of particle physics, space exploration, and other related disciplines. When the incident electron energy is equal to $\delta = 1$, it indicates that the number of incident electrons and the number of outgoing secondary electrons are equal, and the surface of the sample is not electrically charged. When the number of outgoing secondary electrons is greater than the number of incident electrons, i.e., when $\delta > 1$, it is called the secondary electron multiplication effect, which is the basis of the work of the electron multiplier.

The performance of the electron multiplier (e.g., gain and lifetime) is dependent on the secondary electron emission characteristics of the thin film material, the applied operating voltage, and other factors. In general, thin film materials with high δ values are utilized to prepare electron multipliers. Furthermore, the secondary electron multiplication effect results in a positively charged surface of the material, which can impact the performance of the electron multiplier (e.g., dynamic range and count rate) if the electron replenishment is not promptly implemented to reduce the positive potential of the surface. Consequently, the electron multiplier necessitates not only a high δ-value of the material but also a robust electron replenishment capability of the material, which represents a significant consideration in the design process of the electron multiplier.

The group constructed a secondary electron emission characteristics test device based on the findings of related research conducted at home and abroad. Figure 8.2 illustrates the device. The incident energy of the electron beam generated by the pulsed electron gun is 100 eV to 10 keV. The sample stage can be moved up and down and rotated, with an up-and-down distance of 350 mm and a rotation angle of 85°. The maximum diameter of the test sample is 28 mm, and the working vacuum of the system is in the range of 10^{-6} Pa, while the limiting vacuum is in the range of 10^{-7} Pa. These values are sufficient to meet the conventional emission characteristics of the secondary electrons. The device is composed of a four-layer concentric nested spherical structure, with the outermost layer being the shielding layer, the collection pole, the outer grid, and the inner grid in that order. The spherical structure not only enhances the collection efficiency of secondary electrons but

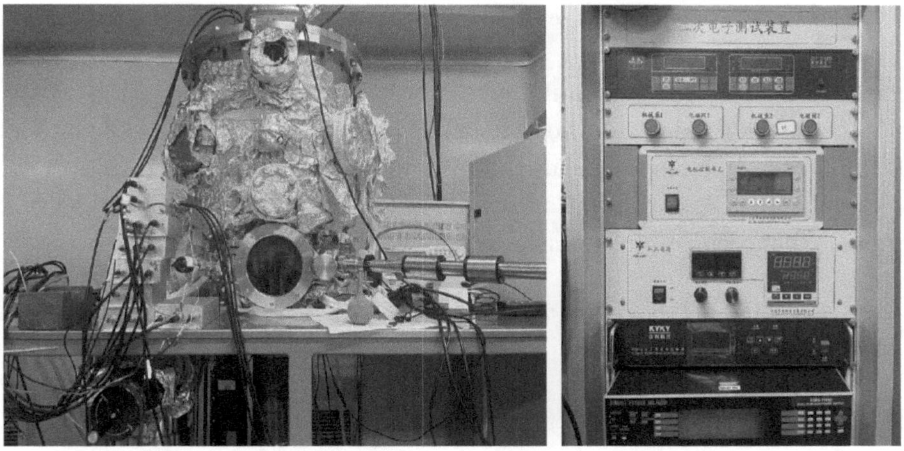

FIGURE 8.2 Constructed test setup for secondary electron emission characterization.

- Electron multiplier
  - Discrete dynode
    - Beryllium oxide material
    - Magnesium oxide material
    - Other materials
  - Continuous dynode
    - Channel electron multiplier (CEM)
    - Microchannel plate (MCP)
    - Microsphere plate (MSP)

FIGURE 8.3 Classification of electron multipliers.

also ensures that the secondary electrons generated by the sample can be fully collected after the sample stage is rotated, thus enabling the measurement of the δ-value at different incidence angles. The group has conducted experimental studies on the secondary electron emission characteristics of metallic, semiconductor, and insulating thin-film materials on this device. These studies provide a solid foundation for the enhancement of electron multiplier performance and the optimization of processing techniques.

8.1.2 Classification and Production Process of Electron Multipliers

For decades, electron multipliers have been utilized as detectors for weak signals, including electrons, ions, and photons. Currently, two types of electron multipliers are in common use: the discrete dynode electron multiplier (DDEM) and the continuous dynode electron multiplier, as illustrated in Figure 8.3.

Beryllium oxide and magnesium oxide are the most typical emission layer thin-film materials, which are made of copper-beryllium alloys or silver-magnesium alloys by high-temperature sensitization process, and were firstly applied in the discrete dynode electron photomultiplier, whose output current can be more than 100 μA. However, the disadvantage of these materials is that their performance is unstable in atmospheric environments. In recent years, new oxide materials have been employed as the discrete dynode emission layer thin film materials, exhibiting enhanced stability. However, the manufacturing cost is relatively high.

Continuous dynode electron multipliers are mainly made of lead silica glass, with a few made of ceramics. Continuous dynode electron multipliers include Channel Electron Multiplier (CEM), Microchannel Plate (MCP), and Micro Sphere Plate (MSP), of which the first two are the most used today. Continuous dynode electron multipliers can be used in environments that require constant exposure to the atmosphere.

8.1.2.1 Discrete Dynode Electron Multiplier (DDEM)

When using the DDEM, a voltage divider made of a resistor string is used to provide voltage to each of the discrete dynodes, and the total operating voltage is added to both ends of the resistor string. To detect electrons, for example, the incident electrons bombard the surface of the emission layer of the first discrete dynode to produce secondary electrons, secondary electrons bombard the second discrete dynode under the action of the electric field to produce new secondary electrons, and so on and so forth, and finally, the inductive signals are generated on the anode, and a gain of 10^3–10^8 times can be realized.

Discrete dynode emission layer thin film materials mainly include beryllium oxide (BeO), magnesium oxide (MgO), alkali–antimony, gallium phosphide (GaP), and phosphorus gallium arsenide (GaAsP), while the substrate material is used in the copper-beryllium alloy, silver-magnesium alloy, nickel metal, and stainless steel. The emission layer of traditional DDEM is made by high-temperature sensitization or magnetron sputtering process, etc. Domestic units such as Xi'an Jiaotong University, Lanzhou Institute of Physics, CAST, CNNC (Beijing) Nuclear Instrument Co., Ltd. (CNNC 261 factory), Nanjing Sanle Group and North Night Vision Technology Co., Ltd. have done a lot of research work in this area.

In the research of the new process, the Institute of High Energy Physics, Chinese Academy of Sciences (IHEP, CAS) has prepared alumina films with a thickness of about 10 nm on different metal substrates by using the Atomic Layer Deposition (ALD) technique, and investigated in detail the relationship between the secondary electron emission coefficient δ, the electron penetration depth, and the incident electron energy, and thus fabricated an 18-level box grid type DDEM, as shown in Figure 8.4. The DC gain can reach 10^6 and the pulse gain can reach 10^8 during the test, which fully verifies the feasibility of the new process in the preparation of DDEM. The biggest advantage of this preparation

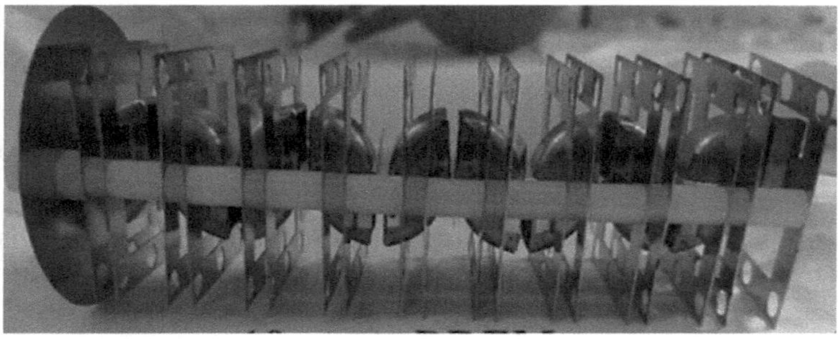

FIGURE 8.4 Physical photo of a new DDEM developed based on ALD technology.

method is that the performance of DDEM is no longer dependent on the type and stoichiometric ratio of the substrate material.

8.1.2.2 Channel Electron Multiplier (CEM)

The CEM can detect positive and negative ions, as well as electrons and photons. Its operational principle is based on the secondary electron multiplication effect of materials. A specific working voltage (2,000 V–3,000 V) is applied to both ends. When incident particles bombard the CEM flare position, secondary electrons are generated by excitation under the action of the electric field after successive impacts. This results in the phenomenon of secondary electron multiplication, which ultimately leads to the output of amplified electron signals via the channel. The signal is then collected and processed by the subsequent electronic device. The CEM has a gain of up to 10^8, with good single-electron detection capability, the signal peak and baseline position are easily distinguishable, and the discriminator threshold is relatively simple to set.

CEMs have been manufactured by several manufacturers in the past, necessitating an overview of their historical development. The concept of CEMs, comprising secondary electron emitting and resistive layers, was first proposed in 1930. The first successful CEM was manufactured by Goodrich and Wiley of Bendix Research Lab in the United States in 1958. Concurrently, Mullard Research Lab in the United Kingdom and L.E.P. Laboratory in France conducted extensive research. In 1973, the Electro-Optics division of the Bendix Company in the United States began mass production of the first CEM, which was subsequently trademarked as Channeltron. The division was subsequently acquired and renamed Gallieo Electro-Optics. Subsequently, Burle purchased Gallieo's Scientific Detectors division and formed a wholly owned subsidiary, Burle Electro-Optics, which was subsequently acquired by Photonis in 2005. Currently, CEMs are primarily supplied by Photonis (France), DeTech (USA) and Dr. Sjuts Optotechnik GmbH (Germany) also has mass production capabilities. The initial generation of CEMs operated in pulsed mode and were primarily utilized in space science research, where their compact size and minimal power consumption made them optimal detectors for planetary environment and solar system exploration research. Due to the stable performance of the glass surface of CEM under multiple vacuum atmospheric cycles, CEM has been applied in laboratory analytical instruments since 1970, especially in mass spectrometry instruments, receiving increasing attention. Currently, there are over dozens of CEM products in various shapes to choose from, but they are all provided by foreign manufacturers.

In the early days of the development of CEM in China, a great deal of research was conducted. In 1976, a multiplier development group was established by Shenyang Glass Test Plant and Lanzhou Institute of Physics, CAST. By optimizing the glass formula, melting, cleaning, reduction, and other process links, a high-lead glass with superior performance was successfully developed. The produced CEM has a body resistance of $10^9\ \Omega$ and a gain of up to 10^8. At that time, the product has been preliminarily tested in fields such as high-tech instruments, including ultra-high vacuum ionization gauges, Auger electron spectrometers, mass spectrometers, heavy ion time-of-flight spectrometers, and others. Unfortunately, in recent years, no commercial CEM products have been manufactured in China.

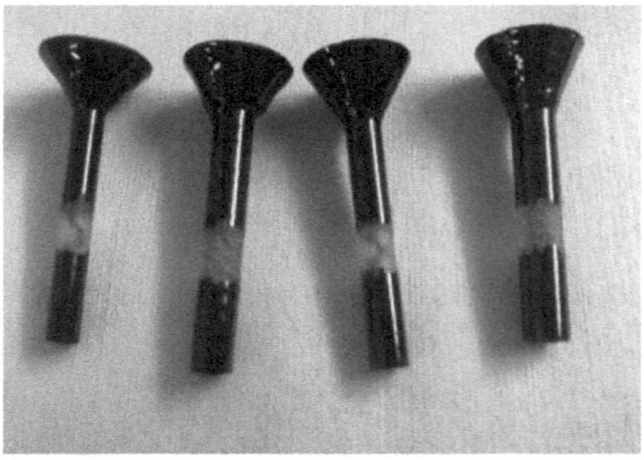

FIGURE 8.5 Physical photos of self-developed CEM.

The Institute of High Energy Physics of the Chinese Academy of Sciences has received substantial support from national policy in the field of key core components, and the demand for domestic electron multiplier products for high-end scientific instruments is increasing. To break the monopoly of this product, the institute has carried out extensive research in several areas, including lead-silica glass formulation, glass molding process, hydrogen burning reduction process, performance test methods, and so forth. This research has been conducted based on previous research and in conjunction with the latest technological developments in foreign countries. A substantial body of research has been conducted on the formulation of lead-silica glass, the glass molding process, the hydrogen-burning reduction process, performance test methods, and related topics. A low-resistance glass formula has been successfully developed and an innovative straight-tube CEM product with a spiral channel structure has been produced, with a resistance of less than 100 MΩ. The DC gain can reach 10^6, the pulse gain can reach 10^8, and the product has been trialed in mass spectrometers. Figure 8.5 shows the physical photo of the product.

8.1.2.3 Microchannel Plate (MCP)

A comprehensive examination of CEM data revealed that the gain is not contingent upon the length of the glass tube, L, or the aperture, d. Instead, the gain is determined by the ratio of L to d (length-to-diameter ratio), L/d. Consequently, as long as L/d remains constant, the glass tubes can be manufactured into smaller sizes with the same gain. The development of glass fiber technology has enabled the reduction of the glass tube aperture to the micrometer scale, providing a technical basis for the creation of MCP. The first MCP with certain performance was developed around 1960 at Bendix Research Lab in the United States. It contained several thousand glass channels with a channel aperture of 150 microns. In the present era, MCPs typically comprise 10^4–10^7 parallel straight tube-type channels, which are round or square in appearance. Each channel aperture has a diameter d of 6–50 μm, as illustrated in Figure 8.6, with an aspect ratio of 40:1–120:1. The channel axial and surface normal angle (bias angle) is approximately 6°–15°, the body resistance is 20–500 MΩ, and the open area ratio is 55%–80%.

The MCP is a two-dimensional electron multiplier with detection and imaging capabilities. Primary incident particles (e.g., electrons) bombard the surface of the channel, generating secondary electrons. Accelerated by an internal electric field, the resulting secondary

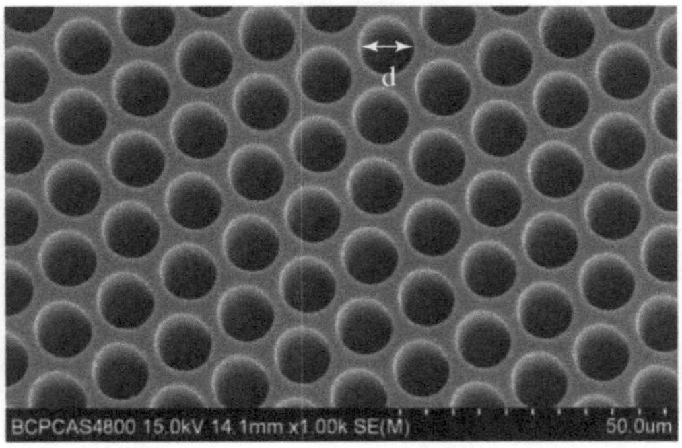

FIGURE 8.6 Physical photo of microchannel plate.

electrons bombard the channel surface and generate new secondary electrons, and after successive multiplication, many secondary electrons are output from the end of the channel. At the same operating voltage, the larger the value of δ, the larger the gain of MCP. However, at higher operating voltages, the gain of MCP will be saturated, which is due to the insufficient ability of the channel material to replenish electrons, i.e., the space charge saturation effect occurs. Typically, for an MCP with an aspect ratio of 40:1, the gain of the single MCP is in the range of 10^3–10^4, the gain of the dual MCPs is about 10^5–10^7, and the gain of the triple cascaded MCPs is up to 10^8. Therefore, to reach the level of detecting single photoelectrons, at least two MCPs are required.

In addition to traditional preparation processes, several novel MCP preparation processes have emerged in both domestic and international contexts. These include the semiconductor conductive glass process, the silicon corrosion array process, the anodic aluminum oxide process, and the 3D printing process. These nascent technologies still face significant technical challenges and have not yet been fully industrialized. In the traditional MCP manufacturing process, the resistive layer and the secondary electron emission layer are formed directly after the high-temperature hydrogen reduction process. This makes it challenging to adjust the physical and chemical properties of these two layers of materials individually. Furthermore, the hydrogen reduction treatment is used in this process, which makes the process complex and difficult to control. In terms of new process research, in 2009, Arradiance, an American company, f pioneered the use of ALD technology in borosilicate glass microporous arrays, preparing nanometer-thick resistive layer materials and secondary electron-emitting layer materials. The company was able to adjust the characteristics of these two layers of materials individually, thus overcoming the technical limitations inherent in the traditional process. The novel process eliminates the hydrogen reduction step in the conventional process. The resistive and secondary electron-emitting layers can be grown independently in the glass channel according to demand using ALD technology. The new MCP exhibits a DC gain of up to 4×10^4.

In China, the Institute of High Energy Physics of the Chinese Academy of Sciences (CAS) and the Xi'an Institute of Optics and Precision Mechanics of the CAS have carried out a great deal of research in the development of a new type of MCP using ALD technology. As shown in Figure 8.7a for MCP without hydrogen reduction, after growing the resistive and secondary electron-emitting layers using ALD technology, see Figure 8.7b.

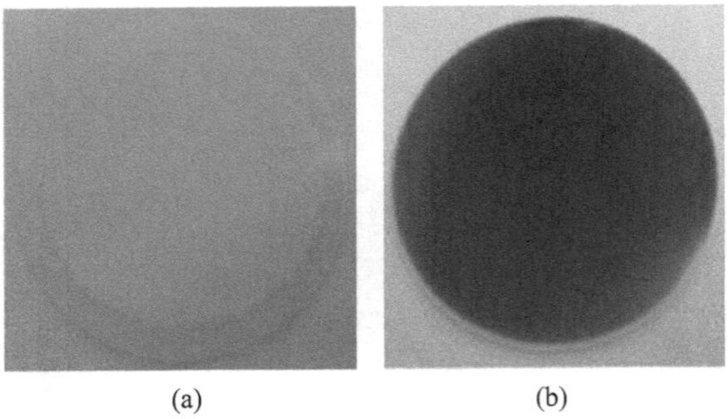

FIGURE 8.7 (a) MCP without hydrogen reduction, (b) after growth of resistive and secondary electron-emitting layers via ALD.

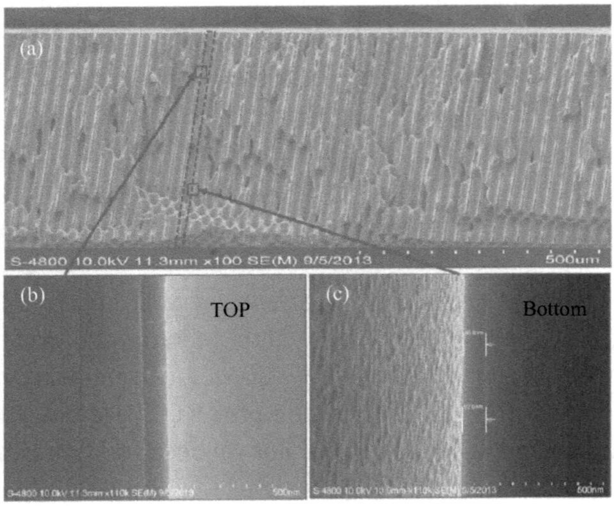

FIGURE 8.8 (a) Cross-section of ALD-MCP; single-channel magnification (b) upper and (c) lower.

The thickness uniformity of the thin film material inside the channel was observed using a scanning electron microscope, and the MCP cross-section is shown in Figure 8.8a, and Figure 8.8b and c show the upper and lower parts of the same channel, respectively. The measured non-uniformity of film thickness is better than 2.26%, indicating that ALD has excellent coating performance in large aspect ratio structures.

During the preparation of the MCP, impurities such as water vapor may remain on the surface of the glass channel. Additionally, residual gas molecules in the vacuum chamber may enter the MCP channel when the gain is tested. These factors can give rise to an ion feedback phenomenon in the MCP. As illustrated in Figure 8.9, the secondary electrons generated in the channel during the continuous multiplication process will continue to impact the channel surface and the residual gas molecules in the channel, resulting in the formation of free positive ions, particularly in the tail of the channel. The impact of the channel surface under the reverse acceleration of the electric field will produce "pseudo-secondary electrons," which will result in the MCP exhibiting pseudo-secondary

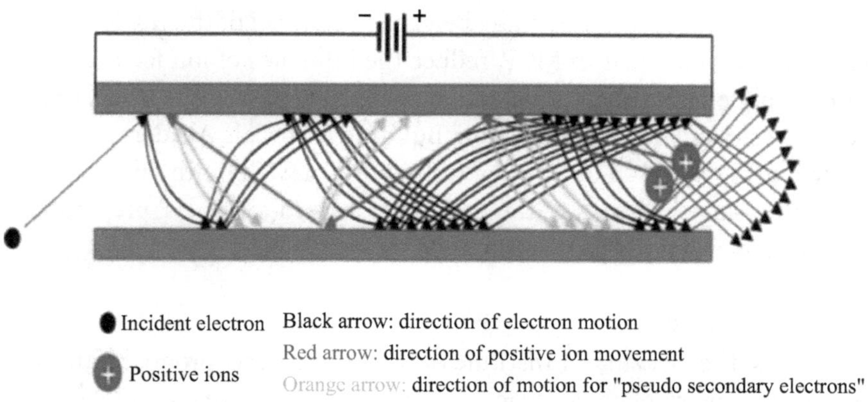

FIGURE 8.9 Schematic diagram of ion feedback in MCP channels.

electrons and a "pseudo-gain" of the MCP. For pulse testing, the ion feedback phenomenon generates a post-pulse signal. Consequently, to suppress the effects of ion feedback, MCPs are subjected to rigorous degassing procedures prior to utilization. In parallel, two or three MCPs are stacked to suppress the ion feedback phenomenon. Furthermore, it is also possible to suppress ion feedback by transforming the straight channel structure of MCP into a curved-channel structure.

In the development of curved-channel MCPs, a lot of research work has been carried out worldwide. In the 1970s, Boutot et al. in France developed a curved-channel MCP with an *L*/*D* ratio of 80:1 to improve the gain of the MCP and suppress the ion feedback phenomenon, and the gain of the single MCP could reach 10^6. In the 1980s, Timothy et al. in the US tested a curved-channel MCP with an *L*/*D* ratio of 144:1, and the gain could reach 2×10^6, and the resolution reached 30%. In the 1980s, Timothy et al. tested a curved-channel MCP with an *L*/*D* ratio of 144:1 with a gain of 2×10^6 and a resolution of 30%. NASA finally adopted the curved-channel MCP and applied it to early space UV detectors (e.g., Hubble Space Telescope). In the 1990s, Hassler et al. in the US analyzed the application of large aspect ratio curved-channel MCP in vacuum UV and extreme UV detection. They found that there was a mismatch in the position of the apertures at both ends of the curved-channel MCP, resulting in a detector position resolution error of one aperture size. The root cause of this phenomenon is that the manufacturing of curved-channel MCPs is extremely difficult. In the early days, only Galileo Electro-Optics (now Photoins Company) successfully developed a large aspect ratio of the curved-channel MCP products. Unfortunately, this product is no longer being developed abroad.

China initiated research into curved MCPs in the year 2000. The Xi'an Institute of Optics and Precision Mechanics of CAS developed curved MCPs with annular edges using high-temperature molding. Changchun Institute of Optics, Fine Mechanics and Physics of CAS employed high-temperature molding to develop curved MCPs with a radius of curvature of 150 mm and an aspect ratio of 80:1. These MCPs exhibited a single-chip of approximately 10^4 and a double-chip gain of approximately 10^6. The MCPs were constructed in a cascade of three MCPs, with a gain of 1.1×10^7 when 3,000 V was applied to both ends.

However, the above studies did not really bend each channel of the MCP. In order to realize the bending of all channels of MCP, reduce the influence of ion feedback, and achieve the purpose of single photoelectron detection using a single MCP, in 2020, the Institute of High Energy Physics of CAS and the China Building Materials Academy Co. Ltd. cooperated to carry out the research on the key technology of MCP with curved channels, and achieved stage-by-stage progress in the process of curved channels' preparation, and the method of performance testing and evaluation.

8.1.2.4 Microsphere Plate (MSP)

In the early 1990s, the Weizmann Institute of Science in Israel proposed and successfully developed a microsphere plate electron multiplier that can work in a low vacuum state, which has a broad application prospect. Microsphere plate is composed of 40–70 μm diameter glass beads irregularly stacked disk-like sintered body, thickness of about 1 mm, can be manufactured into a variety of shapes, its internal glass beads gap constitutes a microchannel. The surface of the treated microspheres has secondary electron emission properties, and the two end surfaces are coated with electrodes to form a microsphere plate electron multiplier. Its structure and working principle are shown in Figure 8.10. When a certain voltage is applied between the two electrodes, electrons are accelerated and multiplied in the electric field. The avalanche multiplication of electrons occurs in the channels between the microspheres, thus acting as a barrier to ion feedback.

To form a stable electric field and supply enough electrons, the microsphere plate material needs to have two conditions: (1) higher δ value. Because the microsphere plate gain depends on the secondary electron emission characteristics of the device material. If the δ value is low, the noise factor of the device is high and the gain is low; the higher the δ value, the fewer the number of collisions of electrons against the channel wall, and the lower the voltage applied to the microsphere plate electrode. (2) Has a sufficiently high body resistance. After applying high voltage, it can supply enough current to compensate for the released secondary electron flow. If the resistance is too high, the secondary electrons cannot be continuously supplied; if it is too low, thermal damage will occur. In addition, the material used to manufacture the microsphere plate should be electronically conductive rather than ionically conductive.

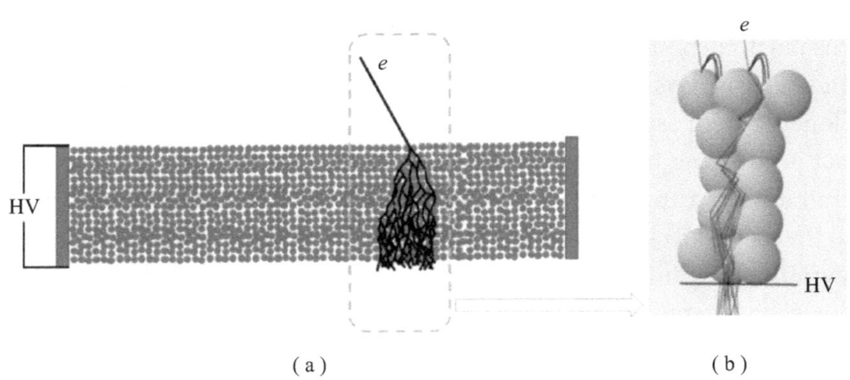

FIGURE 8.10 Structural view (a) and partial enlargement (b) of the microsphere plate.

The so-called ionic conductivity, that is, ionic charge carriers undergo extensive transfer within the material under the driving force of the applied electric field, while electronic conductivity is the movement of electrons from one atom to another atom.

8.1.3 The Operational Mode of the Electron Multiplier and Its Performance Indicators

The electron multiplier is employed to detect weak signals. Consequently, the characteristics of weak signals determine the performance indicators that electron multiplier should possess. Take the detection of photons as an example, under strong light irradiation, as shown in Figure 8.11 ①, the actual number of photoelectrons detected by the electron multiplier, as shown in ②, with an oscilloscope to observe the output signal pulses of the electron multiplier, as shown in ③, the actual measurement is a DC analog signal, as shown in ④. Under strong light, the output signal pulses of the electron multiplier exhibit narrow spacing and overlap, resulting in an analog waveform. When the input photon intensity is reduced, as shown in ⑤; the electron multiplier detects discrete photoelectrons, as shown in ⑥; and finally, a discrete pulsed signal is output, as shown in ⑦. The method of digitally counting such pulsed discrete signals is often called photon counting. When the number of incident photons is further reduced, the electron multiplier is required to have the ability to detect a single photoelectron, i.e., single photoelectron detection, which is more demanding on the performance of the electron multiplier. In the analog method test, the average value of the AC components in Figure ③ is used as the signal, while in the pulse method test, the number of each pulse in Figure ⑦ is used as the signal. The latter approach can be used by a discriminator to separate the signal from the noise pulses, thereby enhancing the signal-to-noise ratio.

Although there are numerous types of electron multipliers, as a detector, they almost have the same performance index. In the case of the analog method, the principal indices are DC gain, saturation effect, dark current, and lifetime; for the pulse method, the main indexes are pulse gain, peak-to-valley ratio, resolution, count rate, saturation effect, collection efficiency, detection efficiency, time characteristics, dynamic range, and post-pulse rate.

Here, taking MCP as an example, we briefly introduce its operating mode and some performance indexes.

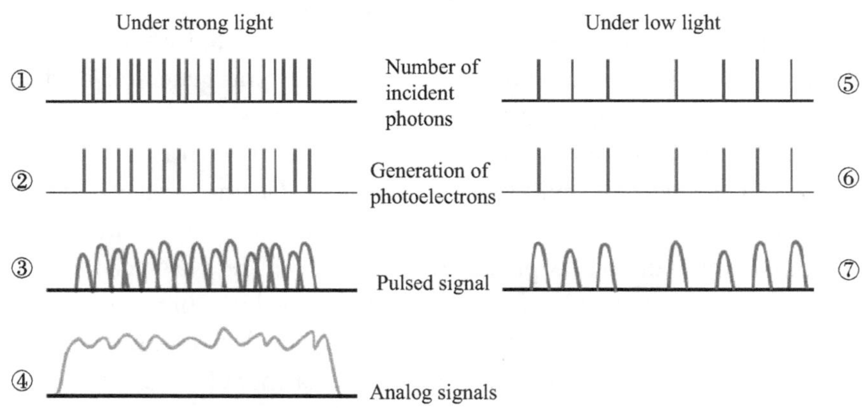

FIGURE 8.11 Output signal of electron multiplier under different incident photon intensities.

8.1.3.1 DC Performance

8.1.3.1.1 DC Gain The DC gain of the MCP is defined as the ratio of the anode output current to the input current. The gain of the MCP is primarily influenced by the potential difference between the input and output surfaces of the MCP and the input current density of the MCP. By increasing the potential difference applied between the end faces of the MCP and the impact energy of the electrons in the channel, it is possible to obtain a greater number of secondary electrons, resulting in a larger gain of the MCP. Typically, the gain of a single MCP is approximately 10^2–10^4 when the operating voltage is 800 V. To achieve a stable gain, the MCP must be treated by electron scrubbing and thermal baking to remove impurities adsorbed in the channel prior to use.

8.1.3.1.2 Saturation Effect When the input signal is sufficiently large and its output signal is no longer proportional to the input signal strength, this phenomenon is referred to as the saturation effect. The saturation effect of the MCP is, on the one hand, due to the space charge effect inside the channel, which results in the end potential of the channel no longer growing linearly along with the channel. Consequently, the end electric field strength decreases, and the δ-value decreases. Conversely, when the output current is large, the number of electrons escaping from the channel end to the channel wall increases, thus leading to saturation of the gain. This is because the number becomes too large for the conduction current to replenish, thus leading to saturation of the gain.

The saturation effect of the MCP operating in DC mode is affected by the conduction current. As the anode output current of the MCP increases to a range of 5%–10% of the conduction current, a decline in gain is observed. Consequently, it is possible to enhance the conduction current and increase the DC dynamic range by reducing the bulk resistance of the MCP. A reduction in the bulk resistance of the MCP is conducive to an improvement in the saturation characteristics. However, it is not possible to reduce the bulk resistance indefinitely, as if the conduction current is too large, its heating power increases in accordance with Joule's law, which leads to an unstable performance of the MCP due to the decrease in semiconductor resistance with increasing temperature.

8.1.3.1.3 Dark Current When there is no input signal to the MCP, the current output at the anode is called the dark current. Its sources are mainly hot electron emission currents, ionic currents due to ionization of residual gases, currents generated by field-induced emission from the inner wall of the channel, leakage currents such as those from the electrodes and their support materials, cosmic rays, and currents generated by radioactive isotopes contained in the electron multiplier material, etc. Cosmic rays are mainly muons, Cherenkov photons produced by Cherenkov radiation, and in addition, since the glass may contain a certain amount of ^{40}K, ^{40}K produces electrons by β-decay, which in turn produces dark currents through the multiplier system. Typically, the dark current of a single MCP is less than 0.5 pA/cm^2, so it can be described in terms of the dark count rate, and the value is typically less than 2 Hz/cm^2 for assemblies with two or three MCPs.

8.1.3.1.4 Lifetime MCP lifetime is the cumulative output charge per unit area (cumulative extracted charge) when its gain decreases to a certain degree, expressed in coulombs/cm^2. The lifetime of the MCP is also affected by the vacuum level of the operating environment, and the higher the vacuum level, the better it is for prolonging the lifetime.

8.1.3.2 Pulse Performance

8.1.3.2.1 Pulse Gain The incident pulse photons impact the MCP to produce photoelectrons, the photoelectrons impact the surface of the channel with a certain energy under the action of the electric field to produce secondary electrons, and the secondary electrons are multiplied under the action of the electric field in the channel, which are three mutually independent processes. It is generally believed that small probability events obey Poisson distribution, when the intensity of the light pulse is extremely weak, the number of photoelectrons excited by a single light pulse and the number of secondary electrons generated by photoelectrons impacting the MCP obey Poisson distribution. If the number of secondary electrons generated by photoelectrons impacting the MCP is >3, it is considered that the number of output electrons obeys a Gaussian distribution after continuous electron multiplication in the MCP channel.

Assumed that the photoelectrons generated by pulsed photon excitation are continuously multiplied, collected by the anode, and output as an electrical signal, the number n of photoelectrons output each time obeys a Poisson distribution, i.e., the probability is:

$$P(n,\mu)=\frac{\mu^n e^{-\mu}}{n!}, \tag{8.1}$$

where μ is the average value of light pulses collected by the anode and outputting a signal. If 90% of the light pulses do not produce a signal at the anode, the probability of $n=0$ is:

$$P(n=0,\mu)=\frac{\mu^0 e^{-\mu}}{0!}=e^{-\mu}=90\% \tag{8.2}$$

Then the probability that the anode collects a single photoelectron input ($n=1$) signal is:

$$P(n=1,\mu)=\frac{\mu^1 e^{-\mu}}{1!}=\mu^1 e^{-\mu}=9.486\% \tag{8.3}$$

Then the proportion of photoelectron peaks that show a single photoelectron input is:

$$\frac{P(n=1,\mu)}{1-P(n=0,\mu)}=94.86\% \tag{8.4}$$

That is, single photoelectron instances make up the majority of the photoelectron peaks and can be considered a single photoelectron spectrum.

Typically, the output charge spectrum contains noise and signal. Noise is the output of the anode when there is no input and is divided into two categories according to its source, one of which is caused by leakage currents, the so-called pedestal, which can be represented by a Gaussian function $S_0(x)$ of the number of photoelectrons by $n = 0$:

$$S_0(x) = \frac{1}{\sigma_0\sqrt{2\pi}} e^{-\frac{(x-Q_0)^2}{2\sigma_0^2}}, \tag{8.5}$$

where x is the charge amount of the output pulse, Q_0 is the average value of the output charge amount without photoelectron input, and σ_0 is the standard deviation of the pedestal.

Secondly, the noise is caused by thermal emissions from the inner walls of MCP channels, cosmic ray radiation, or radioactive background, and the distribution of such noise can be approximated as exponential decay $B_0(x)$:

$$B_0(x) = a\theta(x - Q_0)e^{-x}, \tag{8.6}$$

where a is a parameter related to the chance of occurrence of pedestal accompanying exponentially distributed noise, and $\theta(x - Q_0)$ is a step function, denoted as:

$$\theta(x - Q_0) = \begin{cases} 0, x < Q_0 \\ 1, x \geq Q_0 \end{cases} \tag{8.7}$$

The electron peaks of the output charge spectrum can be considered as the output charge distributions of the input photoelectron numbers n of 1, 2, …, respectively, superimposed according to the Poisson distribution. Since the proportion of multi-photoelectrons in the photoelectron peaks is small and negligible, the photoelectron charge distribution is mainly $S_1(x)$, i.e.:

$$S_1(x) = \frac{1}{\sigma_1\sqrt{2\pi}} e^{-\frac{(x-Q_1)^2}{2\sigma_1^2}}, \tag{8.8}$$

where Q_1 is the mean value of the output charge at single-electron input and σ_1 is the standard deviation of the charge distribution.

Therefore, the single-electron spectrum $S(x)$ obtained from the test can be expressed as:

$$S(x) = S_{noise} + S_1(x) = B_0(x) + S_0(x) + S_1(x), \tag{8.9}$$

where S_{noise} is the output charge distribution of the noise, i.e. $S_{noise}(x) = B_0(x) + S_0(x)$.

In the anode output charge of the electron multiplier, if most of the signal is a single photoelectron component, and the proportion of multi-photoelectron components is very small, it can be said that the response of the electron multiplier at this time is a single photoelectron response, and the pulse gain can be obtained from the following formula:

$$\text{Gain} = \frac{Q_1}{e}, \tag{8.10}$$

where e is the charge of a single electron, equal to 1.602×10^{-19}C. In the actual measured charge spectrum, if the noise is low, the shift of the signal peaks due to the noise is small, the peak position of the single photoelectron peak before deduction of the noise is Q_{SPE}, and the peak position of the pedestal is Q_0, and the gain can be expressed according to the following equation:

$$\text{Gain} = \frac{Q_{SPE} - Q_0}{e} \tag{8.11}$$

The experimental setup for testing the pulse performance of the MCP is shown in Figure 8.12. Ch1 of the pulse generator provides the driving voltage for the laser diode, so that the pulsed light generated by the laser diode is irradiated through the quartz window onto the MCP located in the vacuum chamber, and the photoelectric effect occurs on the surface of the MCP to generate photoelectrons, which are continuously multiplied under the action of the electric field in the channel to output a large number of electrons, which are collected by the anode and sent to the charge digit converter (QDC) plug-in. The synchronization pulse output from Ch2 is fed to the gate signal terminal of the QDC through a discriminator, and the charge data of the QDC measurement signal is transmitted to a computer for data analysis. By adjusting the driving voltage of the laser diode, the signal portion of the charge spectrum measured by the QDC accounts for about 10% of the signal, which is known by Poisson distribution that the main component of the signal is a single photoelectron signal, and the charge spectrum at this point is called a single photoelectron spectrum.

8.1.3.2.2 Peak-to-Valley Ratio and Resolution The single photoelectron peak resolution of the charge spectrum of the output signal can be defined as the ratio of the Full Width of

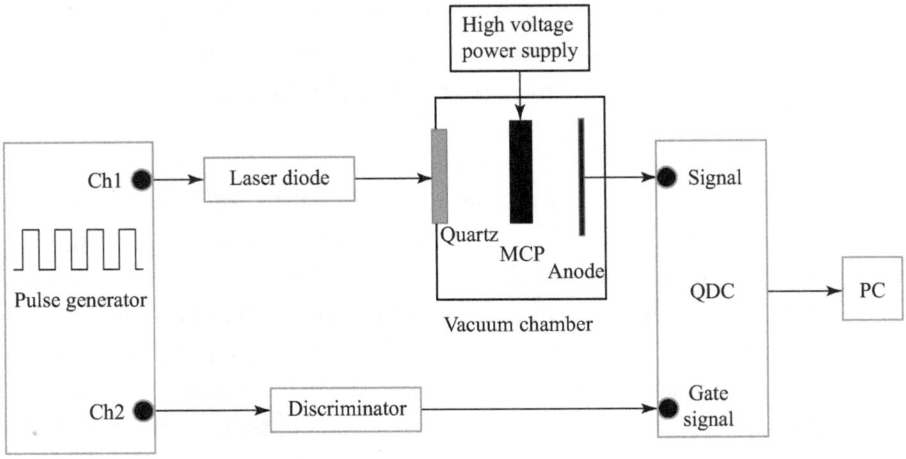

FIGURE 8.12 Schematic diagram of the single-photon test setup for the MCP.

Half Maximum (FWHM) to the gain. Ideally, the single photoelectron peak distribution is Gaussian, then:

$$\text{FWHM} = 2(2\ln 2)^{\frac{1}{2}}\sigma_1 = 2.355\sigma_1 \tag{8.12}$$

peak-to-valley ratio (*P*/*V*) is defined as the ratio of the single photoelectron peak to the valley between the pedestal and the single photoelectron peaks.

8.1.3.2.3 Count Rate and Saturation Effects A certain high voltage needs to be applied when the MCP operates, and the current flowing through the MCP at this time is the conduction current. In the pulse operation mode, the electrons escaping from the surface of each channel of the MCP are replenished by the conduction current, and the time required for replenishment is called the dead time. Consider the MCP as a flat capacitor with a plate spacing of 1 mm, a dielectric constant $\varepsilon \sim 8$, and a total capacitance of 100 pF. For an MCP with 10^6 channels, each channel occupies, on average, a capacitance $C_{\text{ch}} \sim 10^{-16}$ F. The total resistance of the MCP is about $10^6\ \Omega$, and the resistance of each channel is $R_{\text{ch}} \sim 10^{14}\ \Omega$. When an electron is incident on one of the channel walls at the input, producing a secondary electron cascade multiplication, the charge in the channel must be replenished. Since the number of electrons grows exponentially with the path, the vast majority of the charge near the output is not replenished in time and must be recharged from R_{ch} to C_{ch} with the time constant $\tau = R_{\text{ch}}C_{\text{ch}}$, so the dead time of each channel is about 10 ms, and because the work of each channel is basically independent of each other, the dead time of the entire MCP is about 10 ns, and therefore can withstand the highest count rate of up to 10^8 Hz.

The peak anode output current of an MCP assembly operating in pulse mode is affected by different factors under different frequency input conditions. When a low-frequency pulse is input, it is determined by the number of electrons that can be released from a single channel and the number of input channels due to the space charge effect; when a higher-frequency pulse is input, it is determined by the ratio of the sum of the output charges generated by all the input pulses per unit of time to the conduction current. That is, when the frequency of the input pulse is too high, because the dead time of a channel has not yet ended and multiple pulses come in for that channel, these pulses are not able to efficiently multiply the electrons within the channel, affecting the total output charge and leading to a decrease in gain.

8.1.3.2.4 Time Characteristics The time characteristics of the MCP output signal include rise time, fall time, transit time, and transit time spread. The rise time is defined as 10%–90% of the pulse front edge, and the fall time is 90%–10% of the pulse trailing edge. The transit time refers to the interval from the time when the incident particles hit the surface of the MCP to the time when the anode starts to output a signal. The fluctuation of the transit time is called transit time spread (TTS), and the TTS is generally characterized by the half-height and full width of the transit time spread; therefore, the TTS can better reflect the time resolution of the electron multiplier, and the time resolution of the MCP can reach several tens of ps.

TABLE 8.1 Conventional Performance Specifications for Electron Multipliers

Metrics	DDEM	CEM	Single MCP	Dual MCP	Triple MCP	Dual MSP
Gain	10^4–10^8	10^4–10^8	10^2–10^5	10^5–10^7	10^6–10^8	10^4–10^8
Dark current	~2 pA	~2 pA		~1 pA		~5 pA
Peak-to-valley ratio	>2	>3	–	>2.5	>2.5	>2
Rise time	~5 ns	~5 ns		~1 ns		~2 ns
TTS	~5 ns	~2 ns		~100 ps		~2 ns
Lifetime (C/cm^2)	>0.05	>0.02	>0.01	>0.05	>0.1	>0.05
Dark count rate (pcs/cm^{2}s)	~5	~5	–	~1	~2	~1
Spatial resolution	–	–	Good	Ordinary	Bad	Ordinary

8.1.3.2.5 Dynamic Range The dynamic range of an MCP is the detection range from the weakest to the strongest signal at constant gain. Generally, for DC testing, the dynamic range refers to the highest linear output current when the MCP gain is certain; for pulse testing, the dynamic range refers to the highest count rate of the MCP. The dynamic range of the MCP is limited to some extent by the ability of the resistive layer on the inner wall of the channel to electronically compensate for the secondary electron-emitting layer; therefore, appropriately lowering the resistance of the MCP can expand the dynamic range of the MCP.

8.1.3.3 General Performance Specifications of Electron Multipliers

The conventional performance specifications of various electron multipliers are shown in Table 8.1.

8.2 APPLICATION OF ELECTRON MULTIPLIER

An electron multiplier is a vacuum device that can detect various particles (such as electrons, photons, and ions) and has a wide range of applications, which can be roughly divided into two classes. One class is to use the electron multiplier to detect electrons, photons, or ions directly, and the electron multiplier needs to work in an open vacuum environment; the other is to make the electron multiplier in a closed vacuum environment, forming an independent device, according to the different detection particles, can be combined with the corresponding conversion materials, the detection of particle signals converted into electrical signals, such as photomultiplier tubes (PMT) and the image intensifier. In these two types of applications, according to the different anode structures behind the electron multiplier, in addition to detecting weak signals, it can also be position-resolved or imaging detection, such as multi-anode photomultiplier tubes, enhancement of charge-coupled devices and so on.

8.2.1 Application and Research Progress of Electron Multipliers in Particle Physics Experiments

8.2.1.1 Electron Multipliers Used in Domestic and Foreign Neutrino Experimental Devices

In 1914, the British physicist J. Chadwick discovered the continuous energy spectrum structure of the decay product electrons when he studied the beta decay of ^{210}Bi using a magnetic spectrometer. At that time, according to the laws of conservation of mass and conservation of charge, the products of β decay emitted by the nucleus of ^{210}Bi were ^{210}Po nuclei and electrons. This definite two-body problem is given by the laws of conservation of energy and conservation of momentum that the decay product electron should have a fixed energy. The continuous energy spectrum of the electron leads to the fact that the above model of beta decay does not satisfy the law of conservation of energy. Austrian physicist W. Pauli put forward the neutrino hypothesis in 1930 that in the β-decay end-state products, in addition to the new nucleus and electrons, there is also an uncharged and very small mass particles, which take away part of the energy, resulting in a continuous distribution of the electron energy spectrum. In 1934, the Italian-American physicist E. Fermi established the theory of beta decay, pointing out that the β-decay process belongs to the weak interaction, and formally proposed the naming of neutrinos.

The Standard Model is the current theoretical model in the field of particle physics, which describes three other interactions besides the gravitational interaction. The Standard Model contains six flavors of quarks, six flavors of leptons, four gauge bosons, and one Higgs boson, for a total of 17. In the Standard Model, the neutrinos belong to the leptons, which are massless, uncharged, with a spin of 1/2, and participate in the weak interactions only by exchanging $W^{\pm}$ or Z^0, corresponding to the current-carrying interactions and the neutral current interactions, respectively. Neutrinos are divided into three generations, including electron neutrinos, muon neutrinos and tau neutrinos. Direct detection of neutrinos is very difficult due to the extremely small cross-section of neutrinos interacting with matter. With the development of materials, machines, electronics, and detector technology, it was not until 1956 that the American physicists F. Reines and C. Cowan used a 1,400 L liquid scintillator detector to detect the electron antineutrinos emitted from the reactor of the Savannah River Nuclear Power Station of the U.S. Atomic Energy Commission, thus realizing the first direct detection of neutrinos. Since then, muon neutrinos and tau neutrinos were also experimentally discovered in 1962 and 2000. To study the nature of neutrinos in depth, several large-scale neutrino detection devices have been constructed at home and abroad, in which photomultiplier tubes with electron multipliers play an important role.

8.2.1.1.1 Overseas Neutrino Experimental Setup The Super Kamiokande Neutrino Detection Experiment (Super-K) is a large neutrino detector constructed by the University of Tokyo, Japan, in the abandoned arsenic mine of the Mozumi Mine in Kamioka-cho, Gifu Prefecture, Japan, which was completed and began data collection in 1996. This neutrino detector is located at a depth of 1,000 m underground and is a large Cherenkov detector. In the 1990s, its scale has reached 50,000 tons of ultrapure water, 11,000 twenty-inch photomultiplier tubes, and 1,885 eight-inch photomultiplier tubes, of which the photomultiplier tubes internal electron multiplier is a discrete dynode electron multiplier (DDEM), these photomultiplier tubes are all produced by Hamamatsu Corporation of Japan. In 1998, the Super-K detector detected and confirmed the transformation of muon neutrinos into tau neutrinos, a phenomenon of atmospheric neutrino oscillations. This was the first observation of neutrino oscillations.

The Sudbury Neutrino Observatory (SNO) is located in a nickel mine near Sudbury, Ontario, Canada. Officially opened in May 1999, the center of the detector is located

2,039 m below ground and is a large Cherenkov detector. SNO detector is a spherical Plexiglas structure with a diameter of 12 m. Inside the detector is 1,000 tons of heavy water as a detection medium, and 9,456 twenty-inch photomultiplier tubes (DDEM) manufactured by Hamamatsu, Japan, are used to detect the Cherenkov light signals generated. Cherenkov light signal. The SNO experiment can accurately detect the three interactions of solar neutrinos: band flux, neutral flux, and elastic scattering. If solar neutrino oscillations exist, the SNO experiment can accurately analyze the total number of neutrinos before and after the oscillations, as well as the fraction of each product. Furthermore, in 2001, the SNO experiment collaborative group confirmed the phenomenon of solar neutrino oscillations based on observations.

In addition to the above experiments, neutrino experiments that have been or are being run include the Double Chooz experiment in France, the RENO experiment in South Korea, and the IceCube experiment in Antarctica, etc. These experiments use a large number of photomultiplier products with electron multipliers as the core.

8.2.1.1.2 Neutrino Experimental Setup in China The Daya Bay Reactor Neutrino Experiment (DYB) is a large-scale neutrino experiment led by China with international cooperation, which is located near the reactors of the Daya Bay Nuclear Power Station and the Ling'ao Nuclear Power Station in the eastern part of Shenzhen City, Guangdong Province of China, and its main physical goal is to accurately determine the third oscillation angle θ_{13} of the neutrino through precise measurements of the neutrino spectrum of the reactor. The DYB experimental detectors include an anti-neutrino detector and a cosmic ray anticoincidence detector. The former is a three-layer coaxial cylindrical structure. The innermost layer is a Plexiglas canister 3 m in diameter and 3 m high, containing 20 tons of gadolinium-doped liquid scintillators to capture the neutrino signals produced by anti-beta decays; the middle layer is a steel canister 5 m in diameter and 5 m high. It is loaded with 37 tons of white oil for shielding the external radioactive background; the steel canister is equipped with 192 photomultiplier tubes with DDEM structure, all of which are imported. The anticoincidence detector, composed of a top resistive plate detector and two layers of water-based Cherenkov detectors outside the anti-neutrino detector, is used to measure and exclude cosmic ray instances. The DYB experiment began construction in 2007, and in 2012, the international collaboration group of the DYB experiment announced that it had discovered a new mode of neutrino oscillations and accurately measured the third oscillation angle. In 2016, the DYB experiment collaboration team was awarded the international "Breakthrough Prize in Fundamental Physics" for the discovery of new neutrino oscillation modes.

The Jiangmen Underground Neutrino Observatory (JUNO), a Class A Pilot Project of the Chinese Academy of Sciences (CAS), is a next-generation, large-scale neutrino detection experiment led by China with international participation, and the JUNO cooperative group currently consists of 77 universities and research institutes from 17 countries and regions. JUNO experiment is located at 700 m underground under the Dashi Mountain near Jinji Town, Jiangmen City, Guangdong Province, China, and is effectively shielded from cosmic rays. The main physical objectives of the experiment are to accurately measure the neutrino mass order and the oscillatory parameters θ_{12}, Δm_{21}^2, and Δm_{31}^2 using the reactor neutrino oscillations. The detector structure of the JUNO experiment consists of a central detector, a water-based Cherenkov detector, and a top tracker detector. The central detector is a Plexiglas sphere and stainless-steel grid structure filled with 20,000 tons of liquid scintillator and covered with 17,612 twenty-inch photomultiplier tubes and 25,600 three-inch photomultiplier tubes to detect anti-neutrino beta decays. The water-based Cherenkov detector is a central detector external pool containing 35,000 tons of ultrapure water and 2,400 twenty-inch photomultiplier tubes, which are used to provide cosmic ray muon events for the anti- coincidence test. The JUNO experiment is expected to be completed and put into operation in 2025, with an unprecedented energy resolution of $3\%/\sqrt{E}$, making it one of the most important large-scale neutrino experiments in the world.

8.2.1.2 Advances in Photomultiplier Tubes Based on Microchannel Plates as Core Devices

8.2.1.2.1 The Current State of Large-Size MCP-PMT Research in China In the JUNO experiment, the number of 20-inch photomultiplier tubes required was approximately 20,000 units. Such a vast demand necessitates a simultaneous consideration of both product performance and cost reduction when selecting photomultiplier tubes. At the outset of the JUNO project, there were no commercially available international products that fully met the requisite specifications. In order to mitigate the project risk, reduce the procurement price of photomultiplier tubes, and enhance the manufacturing capability of domestic photomultiplier tubes, the Institute of High Energy Physics of the Chinese Academy of Sciences spearheaded the establishment of the Large-Size Microchannel Plate Photomultiplier Tubes (MCP-PMT) Development Cooperative Group in 2012. The primary participating units included North Night Vision Technology Co., Xi'an Institute of Optics and Precision Mechanics of CAS, Nanjing University, CNNC (Beijing) Nuclear Instrument Co., Ltd. (CNNC 261 factory), etc. Among the aforementioned entities, Northern Night Vision and Xi'an Institute of Optics Precision Mechanics are responsible for the production of MCP-PMT, while the Institute of High Energy Physics is responsible for the overall design of the project, the organization of key technologies, the arrangement of the engineering plan, the sealing of the low background glass shell and its transition joints, the performance test of the MCP-PMT, and the debugging of the optimal working state of the MCP components, and so on.

In order to ensure that the new 20-inch MCP-PMT performance meets the design requirements of JUNO, it is necessary that its electron multiplier MCP performance be consistent with the following parameters: (1) high detection efficiency, that is, combined with the electro-optical design, so that photoelectrons are incident to the MCP effective area and effective multiplication. It is not only required to open the area ratio of the sufficiently high, and the photoelectrons incident to the channel between the openings are detected; (2) high gain, low noise, with good single-electron detection capability; (3) Wide dynamic range; (4) High time resolution; (5) long lifetime; (6) good uniformity, it is necessary that the cathode, collection efficiency, and uniformity of each channel of the MCP be of an acceptable standard; (7) ion feedback is not apparent, the after-pulse rate caused by ion feedback must be less than 10%.

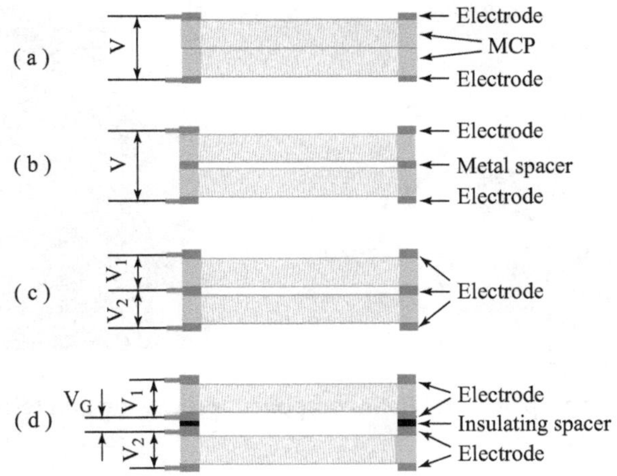

FIGURE 8.13 Four structures of the dual MCPs: (a) close proximity type, (b) gapped two-electrode type, (c) gapped three-electrode type, and (d) gap-voltage type.

The dual MCP assembly with a specific bias angle is capable of effectively suppressing ion feedback and achieving higher gain. Additionally, the manufacturing process and assembly are relatively simple, which has led to the adoption of this design scheme in the PMT. According to the differing requirements of the user, the dual MCPs assembly can be assembled into four distinct structures, as illustrated in Figure 8.13. These include the two MCPs in direct contact with each other or separated by an annular spacer, with or without voltage applied to the gaps. The advantages and disadvantages of each configuration are described in detail below.

As illustrated in Figure 8.13a, the initial configuration is relatively straightforward: the dual MCPs are in close proximity, and the voltage is supplied through the input and output electrodes. However, this approach presents two key limitations. First, the voltage of each MCP cannot be adjusted independently. Second, the channel of the dual MCPs is challenging to fully align, which affects the amplitude statistics of the output signal. This, in turn, results in a single electron peak resolution that is not optimal. The second structure is the gap two-electrode type, as shown in Figure 8.13b. The dual MCPs are separated by a metal ring to form a gap, thus resolving the issue of hole alignment accuracy. However, it is still not possible to adjust the MCP voltage individually. The third structure is the gap three-electrode type, as shown in Figure 8.13c. A voltage can be applied to the metal electrode ring between the MCPs, which addresses the issue of adjusting the MCP voltage individually. However, the gap voltage between the two MCPs cannot be altered, and the single-electron peak resolution of the output signal charge spectrum remains unoptimized. The fourth structure is the gap-voltage type, as depicted in Figure 8.13d. The inputs and outputs of the dual MCPs are connected with electrodes, and the two middle electrodes are separated by an insulating ring. This structure is the most complex, but the bias and gap voltages of the dual MCPs can be adjusted independently, and the gap size can be altered by using an insulating ring of an appropriate thickness as required, which results in a more precise resolution of the output signal charge spectrum. The new 20-inch MCP-PMT

FIGURE 8.14 Microchannel plates and their component processing and testing equipment.

development project employs a gap-voltage type MCP component structure, which exhibits high gain, good uniformity, good ion feedback suppression, and performance optimization that can be achieved by changing the dual MCPs and their gap voltages.

To conduct a comprehensive test and evaluation of the performance of the MCP, the Institute of High Energy Physics of the Chinese Academy of Sciences has independently designed a microchannel plate component processing and testing equipment, as illustrated in Figure 8.14. This equipment can carry out preliminary inspection, degassing treatment, comprehensive performance testing, and optimal working state tuning of MCPs and their components, among other functions. The equipment can perform the following functions: standard metrology for the routine testing of a single MCP; degassing treatment of MCP and its components, including vacuum baking and electron scrubbing; comprehensive performance testing and optimal working state tuning of MCP components, including gain, body resistance, dark current, dark count rate, pulse charge spectrum, dynamic range, and field-of-view quality tests, among others. The group has conducted extensive testing and research on this equipment, and some of the findings are presented below.

The effects of incident electron energy E_p, gap voltage V_G, and microchannel plate (MCP) voltage on the performance of the MCP assembly are investigated. The MCP has an outer diameter of 32.8 mm, an aperture of 10 μm, an aspect ratio of 48:1, and a bias angle of 12°. Figure 8.15 illustrates the gain and single-electron-peak resolution of the MCP assembly as a function of E_p. It can be observed that the gain increases with the increase of E_p. However, when $E_p > 400$ eV, the gain trend slows down and gradually enters a plateau, indicating that the MCP assembly has entered the saturation operation mode. The δ value of the channel material is a crucial factor in determining the resolution of the single-electron peak. Other conditions being held constant, a larger δ value results in a smaller half-height full width of the single-electron peak and thus a superior single-electron peak resolution. When the incident electron energy is low, the δ value of the first impact is relatively small, and the MCP assembly has not yet reached the saturation operating mode. Consequently, the single-electron peak resolution is poor. As the incident electron energy E_p increases, the single electron peak resolution initially improves and then deteriorates, which is correlated with the variation of the δ value of the channel material with E_p.

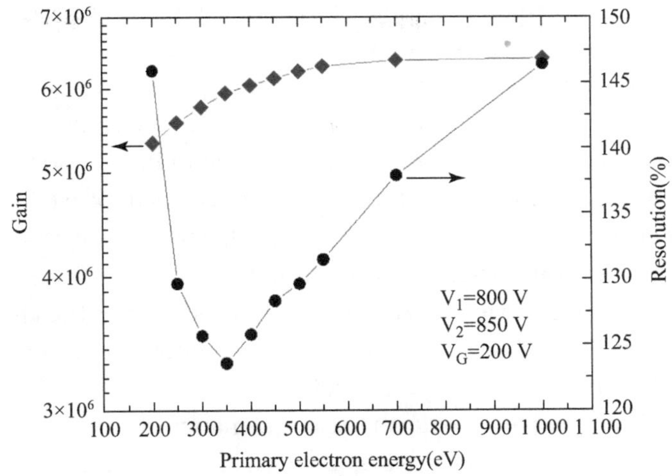

FIGURE 8.15 Gain and single-electron peak resolution of the output signal charge spectrum as a function of incident electron energy.

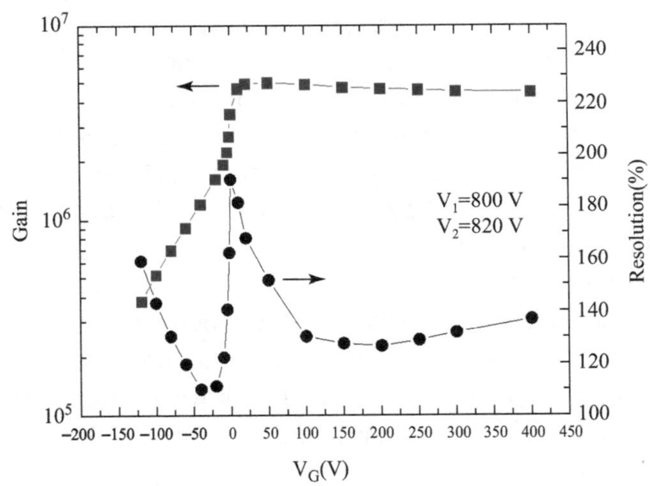

FIGURE 8.16 Gain and single-electron peak resolution of the charge spectrum of the output signal as a function of gap voltage V_G.

The impact of the gap voltage (V_G) on the gain and single-electron peak resolution of the MCP assembly was examined, as illustrated in Figure 8.16. When the bias voltages of MCP-1 and MCP-2 are $V_1 = 800\,V$ and $V_2 = 820\,V$, respectively, the gain is observed to be very low, approximately 3×10^5, when V_G is reversed and set to −120 V. The gain then grows rapidly with the increase of V_G until V_G reaches 0 V. The gain reaches a maximum of about 5×10^6 when $V_G = 20\,V$. The gain decreases slowly with a further increase of V_G. The relationship between the single-electron peak resolution with V_G is more complicated. The single-electron peak resolution starts to get better when V_G increases from −120 V. The single-electron peak resolution reaches a local optimum when V_G is about −50 to −30 V.

The gain of the single-electron peak resolution reaches a maximum value of about 5×10^6 with further increase in V_G. This is due to the reverse gap electric field prevents the lower kinetic energy electrons in the "electron cluster" coming out of the end of MCP-1 to reach the input surface of MCP-2, and only the higher density electrons in the middle region reach MCP-2, which participate in the saturation of the multiplication channel to a high degree, and then the resolution of the single-electron peak is improved. When V_G reaches 0 V, the single-electron peak resolution deteriorates sharply. In the absence of an electric field in the gap, the "electron clusters" emitted from the end of the MCP-1 channel are dispersed to the input surface of MCP-2. A significant proportion of the channel at the edge of the covered area exhibits a low electron density and is poorly saturated, which results in a deterioration of the single-electron peak resolution.

The effect of ALD coating technique on the performance of conventional MCP components was investigated. The bias voltages applied on the conventional MCP components were $V_G=-50$ V, $V_1=850$ V, and $V_2=850$ V. They were tested to have a gain of 1.8×10^6, a single-electron peak resolution of 90%, and a peak-to-valley ratio of 3.7. The MCPs of the group were left for one year in a laboratory environment, and then a 40-cycle alumina thin film was coated on the surface of the MCPs by using ALD technology. The performance of the coated MCP assembly was tested with operating bias voltages of $V_G=-30$ V, $V_1=850$ V, and $V_2=800$ V. It was tested to have a gain of 5.2×10^6 and a single-electron resolution of 80.6%. What's more, the peak-to-valley ratio can be increased to 11.2. The coating can significantly improve the single-electron peak resolution and peak-to-valley ratio of the MCP assembly, thus improving the single-electron detection capability.

The optimal operating state of the MCP assembly is obtained by studying the pulse performance of the dual MCPs assembly: that is to say, applying enough voltage to the dual MCPs or allowing the MCP assembly to work in saturation mode by adjusting the appropriate reverse gap voltage is important for improving the single-electron detection capability of detectors consisting of dual MCPs, and the above research work contributes significantly to the performance of MCP-PMTs. The above research work has made important contributions to the performance optimization design of MCP-PMT and the final finalization of the product.

8.2.1.2.2 The Current State of Research on Large-Sized MCP-PMT Abroad MCP-based photodetectors offer several advantages, including compact structure, lightweight, excellent time resolution and good spatial resolution, and two-dimensional imaging capability. Despite the rapid development of other solid-state detectors, vacuum-based MCP detectors continue to play an important role in the field of high-energy physics, where the ability to rapidly detect weak photon signals is crucial. In comparison to solid-state detectors, it has the advantages of rapid response speed, high gain, low noise, and high spatial resolution.

The Large Area Picosecond Photodetector (LAPPD) Collaboration was established under the auspices of the U.S. Department of Energy and includes Argonne National Laboratory, Incom Corporation, the University of Chicago, and the University of California, Berkeley. The group is working on a new MCP-based large area (203×203 mm^2) vacuum-encapsulated photodetector with imaging, high temporal and spatial resolution, vacuum encapsulation in borosilicate float glass, and a flat plate detector design for cascade splicing.

Key design features of the LAPPD detector include:

1. The detector employs two-chip new MCPs stacked in a Chevron structure. The new MCPs are made by growing resistive and secondary electron-emitting layers in borosilicate glass capillary arrays using ALD technology.
2. The modular all-glass detector package comprises a conductive RF microstrip line, sealed through the glass, which bonds the sidewalls to the bottom anode plate in a gas-tight manner. This eliminates the need for metal electrode pins to penetrate the detector's vacuum package while simultaneously making electrical connections to the inside of the device.
3. The ALD technique is utilized to prepare a resistive layer material on the glass support to have a certain resistance, which acts as a voltage divider circuit through which a voltage is applied to the MCP assembly, avoiding the need to individually electrode the two-chip MCPs inside the detector.
4. An RF microstrip line anode is used with an analog bandwidth of 1.5 GHz and good temporal and spatial resolution.

Currently, the large-size MCP-PMT developed by the LAPPD has an effective area of 373 cm^2, single photoelectron detection capability, gain of 10^7, dark count rate of approximately 1 kHz/cm^2, time resolution of approximately 65 ps, and quantum efficiency of greater than 30% at 365 nm. Incom has initiated the production of MCP detectors with apertures of 10 and 20 μm, which have been utilized in a multitude of fields, including high-energy physics experiments.

8.2.2 Application of Electron Multipliers in Other Fields

8.2.2.1 Application in Mass Spectrometry

A mass spectrometer is an instrument used to determine the mass of a substance. Its basic principle is to ionize the sample to be tested into charged ions, which are separated under an electromagnetic field and then detected by an ion detector to obtain the mass-to-charge ratio and the relative intensity of the mass spectra. This information is then used to deduce the mass of the molecules in the analyte. The fundamental components of a mass spectrometer include an injection system, an ion source, a mass analyzer, a detector, a data analysis system, and a vacuum system. Ion detectors are typically distinguished by a number of key characteristics, including high gain, low noise, a wide dynamic range, a long operational life, high sensitivity, and a fast response time. Detectors are primarily classified into two categories: gainless and gain-based.

The gainless structure is comprised primarily of a Faraday cup and a signal amplification circuit. Assuming that each ion carries only one charge, its charge is 1.6×10^{-19} C. If the galvanometer measures a charge of 1×10^{-9} A, this indicates that 6.25×10^{9} ions are detected per second, and thus a minimum of 10^4 ions can be detected for a galvanometer with a range of 1×10^{-15} A. The signal strength of the galvanometer is weak when the number of ions is small, which results in a low signal-to-noise ratio.

Ion detectors with gain structure include: DDEM, CEM, and MCP. The DDEM detector exhibits a wide dynamic range, particularly when utilized in conjunction with an analog/pulse dual mode switching circuit, which enables a dynamic range of more than nine orders of magnitude. The CEM detector's advantage lies in its compact size, which renders it well-suited for portable mass spectrometers. In pulse mode, the signal-to-noise ratio is particularly impressive. To enhance the sensitivity of these two detectors to heavy ions, a converter dynode can be incorporated in front of the detector. This converter dynode can be optimized for ion/electron conversion efficiency by applying a higher bias voltage. These two detectors are frequently utilized as detectors for ion trap analyzers. The MCP detector has several advantages, including a large detection area, high gain, and fast response. It is particularly well-suited for imaging detection. Its time half-width of the current signal is typically less than 1 ns, and it is the most used detector on time-of-flight mass spectrometers.

8.2.2.2 Applications in Space Exploration

The extreme ultraviolet light detector is employed to study the distribution of space He+ ions. This is achieved by detecting the 30.4 nm extreme ultraviolet light scattered by the resonance of He^+ ions in the Earth's plasma layer. This is a crucial aspect of the study of the solar wind and the physical problems of the Earth's plasma. The detection of extreme ultraviolet light in deep space is dominated by MCP-based position-sensitive anode detectors, which are composed of an incident window, a photocathode, a cascade MCP, a positon-sensitive anode, and readout electronics. These detectors exhibit several advantages, including a high signal-to-noise ratio, high sensitivity, a wide dynamic range, and good drift resistance. They have been widely used in foreign space detectors.

In recent years, a substantial amount of research has been conducted in China on the development of MCP position-sensitive anode detectors. Given the relative maturity of photocathode technology, the core components of MCP photon-counting imaging detectors in China are the MCP, position-sensitive anode, and readout electronics. MCP typically employs a two- or three-chip cascade configuration to detect single photon signals. A variety of position-sensitive anodes have been developed, with multi-anode microchannel arrays, wedge-stripe anodes, delay lines, cross-stripe anodes, and vernier anodes representing the most mature. Among these, wedge-stripe anode technology has been the subject of more extensive study in China, due to its relatively simple fabrication process and excellent performance. In terms of readout electronics, the design of readout electronics for synchronized measurement of photon arrival time and position based on MCP position-sensitive anode detectors includes the design of hardware circuits such as photon arrival timing circuit, multi-channel pulse peak acquisition, start signal generation, etc., as well as the control of peak acquisition, measurement of photon arrival time, data caching and transmission, and so on.

By enhancing the functionality of the MCP, optimizing the structure of the position-sensitive anode and readout electronics, and improving their performance, we can achieve a higher spatial resolution. However, the ultimate spatial resolution is limited by the aperture of the MCP channel. The advancement of the MCP preparation process has led to the emergence of small aperture MCP as a key area of focus within the industry. Examples of this include the development of MCP with a channel aperture of 2 μm, which has been introduced to the international market, and MCP with channel apertures of 5 and 4 μm, which are currently available in China. The smaller channel aperture can further enhance the performance of MCP-based position-sensitive anode detectors, particularly in terms of spatial resolution, which is of significant value for space exploration.

8.2.2.3 Applications in Atom Frequency Standards

The atom frequency standard, also known as the atomic clock, is a precision measuring instrument that has been developed through cross-disciplinary research in quantum mechanics, electromagnetism, and modern electronics. Atomic clocks are used extensively in a variety of fields, including communication network synchronization, time and frequency standards, geodetic instruments, precision navigation and positioning, and astronomical observation. China has been developing cesium beam atomic clocks since the 1960s. The small magnetically selected state cesium atomic clock is comprised of two main components: the frequency standard circuit and the cesium beam tube. The cesium beam tube is the core component of the cesium atomic clock, and its technical index and service life are largely determined by it.

The operational lifespan of a cesium beam tube is significantly influenced by the performance of the electron multiplier. It has been observed that the gain of the electron multiplier declines with the passage of time when using cesium beam tubes. Internationally, the aging mechanism of active thin-film electron multipliers was studied through accelerated aging tests. The test revealed that the carbon contamination layer deposited on the electrode surface is the primary cause of reduced secondary electron emission and subsequent decline in gain. To identify the root cause and optimize the electron multiplier performance, we utilized an electron multiplier performance test system and conducted a comprehensive analysis of the factors affecting its lifetime. The analysis indicates that the operational lifespan of the electron multiplier is contingent upon the longevity of the first-stage multiplier electrode. To optimize the operational lifetime of the electron multiplier, it is essential to set a reasonable number of multiplier electrodes and working voltage. In the Chinese market, the method of raising the voltage was used to regulate the gain of the electron multiplier. This method allows for the qualitative assessment of the fast and slow degree of gain attenuation of the electron multiplier based on the enhancement of operating voltage. In general, the faster the operating voltage is enhanced, the faster the gain attenuation is, while the slower the operating voltage is enhanced, the slower the gain attenuation is. Furthermore, the attenuation of the gain of the electron multiplier is also related to the size of the incident ion current. In this regard, a small incident ion current can slow down the gain attenuation. To enhance the long-term reliability of cesium atomic clocks, it is essential to elevate the δ-value of the dynode material in the future. This will optimize the performance of the electron multiplier, reduce the starting operating voltage of the multiplier, and lower the intensity of the cesium ions incident on the first dynode of the multiplier. This will slow down the attenuation of the multiplier gain and prolong the service life of the atomic clocks.

8.2.2.4 Applications in Neutron Detection

Conventional neutron imaging detectors include film photography, scintillation crystal-coupled photomultiplier tubes, and gas detector-coupled CCDs. All these neutron imaging detectors have inherent limitations, such as radiation-induced noise that affects the performance of the CCD. Integrated detectors are unable to perform time-of-flight measurements at high count rates, and are unlikely to be available in large format sizes.

The exceptional temporal and spatial resolution of MCP detectors enabled Nova Corporation of the United States to become the inaugural enterprise to employ MCP for the imaging detection of thermal neutrons in 2000. They doped ^{10}B in the traditional MCP glass and realized for the first time the spatial resolution of thermal neutron imaging is better than 30 μm, and the detection efficiency is more than 20%. For the MCP doped with ^{10}B in the channel wall, the incident thermal neutrons and their interactions produce α particles and ^7Li, and the α-particles and ^7Li particles pass through the glass and enter into several neighboring channels, generating some free electrons in the process. In addition, other charged particles such as protons, OH^- ions, and alkali ions may also be released by the α-particles and ^7Li particles into the channels. The free electrons generated on the surface of the channel move under the action of the electric field and hit the channel wall to generate secondary electrons, which are multiplied in the channel, and finally, a large number of electrons escape from the end of the channel and are collected by the position-sensitive anode, realizing thermal neutron imaging detection.

The recoil method for detecting neutrons is to measure the ionization caused by the recoil nucleus after elastic scattering between neutrons and atomic nuclei in the detection medium, in order to infer the properties of the original neutrons. This process results in the inversion of the original neutrons' nature. However, this method is only suitable for the detection of fast neutrons. Typically, materials with a high hydrogen content are employed for the detection of fast neutrons, including plexiglass (polymethylmethacrylate), polyethylene, and other polymers. In approximately 2010, Arradiance (USA) was the inaugural organization to employ atomic layer deposition (ALD) technology. It was then that they realized the detection of fast neutrons by growing resistive layers and secondary electron-emitting layers in microporous arrays made of plexiglass. The interaction between fast neutrons and hydrogen produces recoiling protons, which give rise to free electrons in neighboring channels. These electrons produce secondary electrons in the channel, which, after successive multiplication, result in the detection of fast neutrons. In 2013, Nova's patent introduced the use of ALD technology to realize thermal neutron detection by sequentially growing a neutron-sensitive layer, a resistive layer, and a secondary electron-emitting layer in the channel of a conventional MCP. The materials utilized for the neutron-sensitive layer include B_2O_3, BN, and Gd_2O_3. The resistive layer is AZO, while the secondary electron-emitting layer is MgO. At present, the company has commercialized neutron-sensitive MCP detector products.

A substantial body of research has been conducted in China on the development of neutron-sensitive MCP detectors. Northern Night Vision Technology Co., Ltd., Tsinghua University, and Changchun University of Science and Technology doped neutron-sensitive nuclides (^{10}B or $^{155,157}Gd$) into MCP glass and removed the natural radioisotope components in the glass to achieve the detection of thermal neutrons by MCP. This resulted in the development of a large array, low-noise, neutron-sensitive MCP with a diameter of 106 mm. The team employed ALD technology to grow neutron-sensitive materials on the channel surface of a conventional MCP and conducted tests to assess the detector performance. The ALD technique was employed to fabricate neutron-sensitive materials on the channel surface of a conventional MCP, after which the detector performance was evaluated. The current study focus in the field of neutron-sensitive MCP detectors is on the improvement of several key parameters, including detection efficiency, neutron/gamma suppression ratio, time and space resolution, preparation process optimization and stabilization.

8.2.2.5 Application in the Field of National Defense

The low-light-level night vision instrument represents a crucial piece of night combat equipment, with the image intensifier serving as its fundamental component. The function of the image intensifier is to magnify images that cannot be observed by the human eye in low light to a level that can be observed. One of the core components in the image intensifier is the MCP.

In the second generation of image intensifiers, the multiplication and amplification of photoelectrons generated by the photocathode was achieved for the first time using the MCP. This resulted in a notable enhancement of the image intensifier's performance, leading to a reduction in the size of the image intensifier and an increase in brightness gain. The third-generation image intensifier is largely similar to the second generation, with the primary distinction being the photocathode material. The second-generation intensifier employs a multialkali photocathode, whereas the third-generation intensifier utilizes a GaAs/GaAsP photocathode, which exhibits enhanced sensitivity and a broader spectral response range.

The Gen 3 anti-ion feedback film product represents another performance improvement over the Gen 3 image intensifier, which is often referred to as the Gen 3+ image intensifier. The Gen 3+ product was developed by ITT Night Vision in the United States as a response to competition from another U.S. manufacturer, Litton, which offered a Gen 3 product without an anti-ion feedback film. ITT discovered that the performance and reliability requirements of the Army-mandated Gen 4 product could be met by thinning rather than removing the anti-ion feedback film. Furthermore, the retention of the anti-ion feedback film serves to protect the structure of the GaAs photocathode. Consequently, Gen 3+ products utilize an anti-ion feedback film with a thickness of approximately 1–3 nm. Nevertheless, the presence of the anti-ion feedback film still results in the loss of approximately 25% of the photoelectrons emitted from the photocathode. However, these losses are considerably lower than those observed in typical Gen 3 tubes. Furthermore, the signal-to-noise ratio of Gen 3+ products with anti-ion feedback film is highly satisfactory, approaching that of the Gen 3 products without anti-ion feedback film. Concurrently, Gen 3+ products with anti-ion feedback films exhibit enhanced reliability. Consequently, Gen 3+ with anti-ion feedback film is the image intensifier employed by the U.S. military with greater frequency.

8.3 DISCUSSION ON THE DEVELOPMENT TREND OF ELECTRON MULTIPLIER

In recent years, semiconductor detectors based on Si, Ge, CdZnTe, InGaAs, and other materials have been rapidly developed for particle detection of electrons and photons. These detectors and their electronics can meet the technological needs in fields such as high-energy physics, nuclear medicine, space exploration, and quantum communication. Currently, the three traditional vacuum electron multipliers, DDEM, CEM, and MCP, are still widely used and play an important role in particle detection.

The performance of a vacuum electron multiplier is contingent upon the characteristics of the surface secondary electron-emitting thin film material. One of the current development trends of electron multipliers is to study new secondary electron-emitting thin film materials with the objective of producing high δ-values under low-energy incident electron bombardment. This approach ensures that the electron multiplier has sufficient gain, while also extending its service life, reducing system noise and the development cost by reducing the working voltage, and solving some of the problems that exist in the practical application of the electron multiplier.

The relationship between the δ value and the incident electron energy E is illustrated in Figure 8.17. The black line represents conventional materials, and when the incident electron energy E satisfies $0<E\leq E_1$, the number of internal secondary electrons excited by incident electrons in the object is also increasing. Consequently, the δ-value gradually increases from 0 to 1 without the secondary electron multiplication effect. For $E_1<E\leq E_{m1}$, the δ-value gradually reaches the maximum value δ_{m1}, with E_{m1} typically being 500 eV or greater. As $E\geq E_{m1}$, the depth of incident electrons into the sample increases, resulting

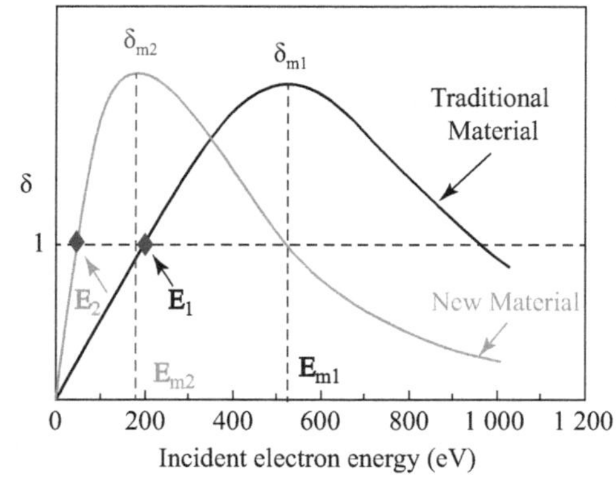

FIGURE 8.17 Relationship between the secondary electron emission coefficient δ of a material and the incident electron energy E_p.

in the generation of excited internal secondary electrons primarily occurring within the depth of the sample. Additionally, most secondary electrons are transported to the surface with increased energy loss, leading to a reduction in the number of escaped secondary electrons. The red line represents a typical feature of new secondary electron-emitting thin film materials. In these materials, the peak E_{m2} of δ_{m2} is shifted to the low-energy end, or the energy of the incident electrons corresponding to the energy E_2 is lowered when the production of electron multiplication is allowed. In general, technical methods for developing new secondary electron-emitting thin films can be explored based on insulating thin-film materials with good secondary electron-emitting properties through methods such as elemental doping and surface modification.

8.3.1 The Doping Elements Enhance the Secondary Electron Emission Properties of the Materials

As for the theoretical analysis of doping elements to improve the secondary electron emission performance of materials under low-energy incident electrons, according to the solid energy band theory, the escape work is equal to the difference between the Fermi energy level and the vacuum energy level. The escape work of insulating materials is often relatively large. In theory, the Fermi level position of insulating materials can be changed by doping elements, thereby reducing the escape work of the material to achieve the purpose of regulating the secondary electron emission performance of the material. Therefore, insulating materials doped with other elements can obtain high δ values. In addition to proper doping, try to make the material with the least impurities and structural defects, i.e., as few electron traps and composite centers as possible in the film, so that the low-energy incident electrons produce the maximum migration distance of secondary electrons and the maximum depth of escape of secondary electrons, which can lead to high δ values. Figure 8.18 shows the energy band structure of undoped and doped insulating materials; E_f is the Fermi energy level, Φ is the escape work, E_c is the conduction band edge, E_v is the valence band edge, and E_0 is the vacuum energy level. Therefore, the escape work Φ of the material can be reduced by appropriate amount of doping, which improves the secondary electron emission ability of the material.

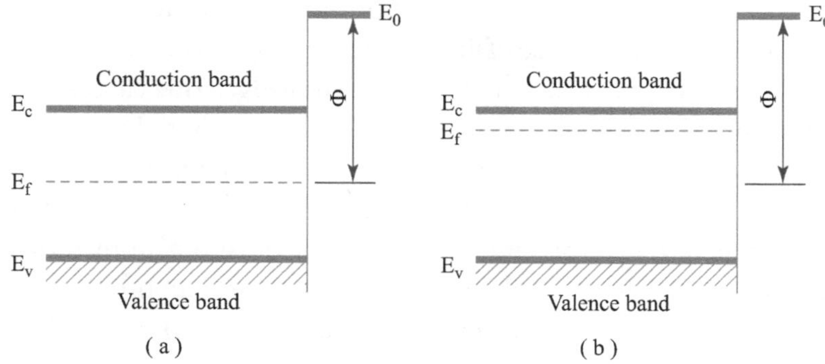

FIGURE 8.18 Band structure diagram of (a) undoped and (b) doped insulating materials.

It is not always the case that more doping of an element is better. Consequently, the greater the doping amount, the more defect energy levels, the more easily inner secondary electrons are excited in the material and lose their energy, resulting in the excitation of true secondary electrons only under high-energy incident electrons. Therefore, it is crucial to determine the optimal doping amount in insulating materials to regulate the Fermi energy level and achieve high secondary electron emission under low-energy incident electrons.

8.3.2 Surface Modification Improves Secondary Electron Emission Properties of the Materials

Advanced ALD coating techniques are employed to grow a layer of N-type doped semiconductor material on the surface of an insulating material. This layer has a thickness on the order of nanometers and reduces the electron affinity potential of the material, thereby improving the secondary electron emission capability of the material. In order to obtain a suitable low electron affinity material, it is usually necessary to undergo a certain surface treatment. This is generally achieved by growing a layer of N-type doped material on the surface of the material that is easy to form positive ions in order to produce a strong energy band bending. In order for the reduction of the effective electron affinity potential to be effective, it is necessary that the excited electrons are able to pass through the energy band bending region. Therefore, the width of the energy band bending region should be narrow, with a maximum width that does not exceed the escape depth of the material. As illustrated in Figure 8.19, a layer of N-type doped semiconductor material with a thickness on the order of nanometers is grown on the surface of an insulating material to form a surface energy level, as shown in (a). The Fermi energy level of the surface state, the E_{SF}, is a special energy level in the semiconductor surface energy level. The semiconductor surface is neutral when the electrons occupy the position of the E_{SF}. For insulating materials, the surface state of the Fermi energy level E_{SF} is situated at a considerable height above the E_f, as illustrated in (b). Consequently, the surface charge will flow with the insulating material until the equilibrium state is reached, whereby the E_{SF} aligns with the E_f. This results in a surface energy level very empty and positively charged, the formation of a hole barrier, and an energy band that bends downward. This ultimately leads to a significant reduction in the effective electron affinity potential χ_{eff}. This phenomenon is depicted in (c). The surface barrier height, Φ_B, and the work function, W, can be determined by XPS and UPS, respectively, allowing for the calculation of the effective electron affinity potential, $\chi_{eff} = W - \Phi_B$. By deep-level doping of insulating materials, such as P-type doping, the Fermi energy level is brought close to the valence band edge. This causes the energy band bending brought about by the surface state energy level to result in the vacuum energy level E_0 being located below the conduction band edge. This in turn results in a negative electron affinity potential surface, as shown in Figure 8.19d. This approach is more conducive to enhancing the secondary electron emission capability of thin film materials. According to the solid energy band theory, the bandgap widths of insulation materials and surface modification layers are different and belong to heterojunction structures. The energy band

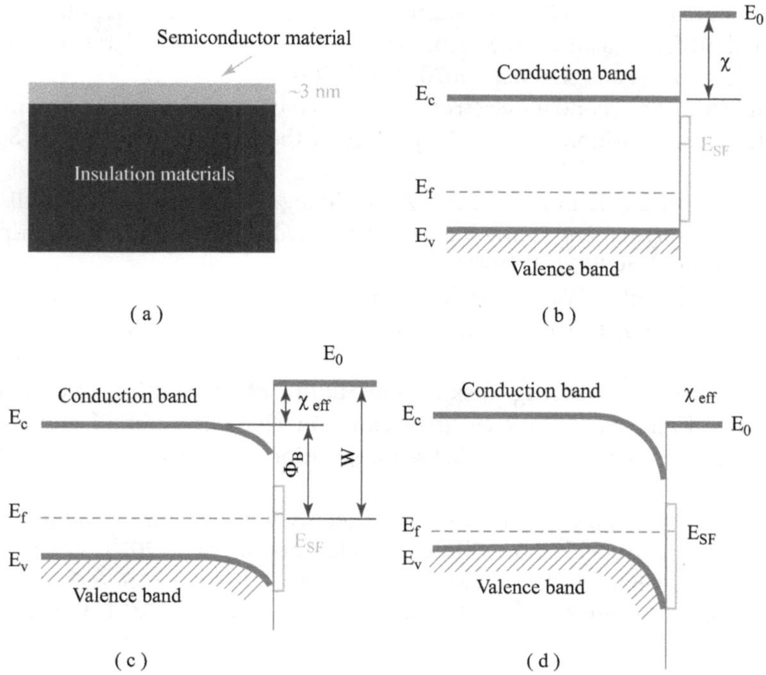

FIGURE 8.19 Effect of surface energy levels on the effective electron affinity potential of insulating materials. Heterojunction material structure (a) and its band structure diagrams of the surface state in the initial state (b), in the equilibrium state (c), and in the equilibrium state after deep doping (d).

structure of the heterojunction has an important influence on the generation of secondary electrons, such as valence and conduction band mismatch. This will promote or impede the emission of secondary electrons. Consequently, testing the valence band and conduction band mismatches of heterojunction materials can assist in comprehending the impact of heterojunction energy band structures on secondary electron emission.

The development of electron multipliers is based on the understanding of the mechanism of secondary electron emission. These two areas are mutually reinforcing. With the advancement of new secondary electron-emitting thin-film material preparation technology, test and evaluation technology, and theoretical research, it is anticipated that the performance of traditional electron multipliers will be enhanced and their applications expanded.

BIBLIOGRAPHY

1. Zhou Rongmei. *Photoemission, Secondary Electron Emission, and Photomultiplier Tube.* Chengdu: University of Electronic Science and Technology Press, 2015.
2. Yan Baojun, Liu Shulin, Wen Kaile, Wang Yuman, Zhang Binting, Xu Meihang, Wei Wenlu, Peng Huaxing. Research on the application of new thin film materials in electron multipliers. *Space Electronics Technology*, 2022, 19 (4): 85–92.

3. Wen Kaile, Liu Shulin, Yan Baojun, Wang Yuman, Zhang Binting, Cai Zhiyan. New neutralization method for measuring the secondary electron yield of insulative material. *Radiation Detection Technology and Methods*, 2020, 4: 319–326.
4. Wen Kaile. Study on secondary electron emission characteristics of thin film materials based on atomic layer deposition. Beijing: University of the Chinese Academy of Sciences, 2020: 28–90.
5. Li Jie, Yang Jishi, Liu Hulin, Liu Biye, Zhao Yuan, Hu Wenbo, Wu Shengli. Research on performance improvement technology for separated dynode electron multiplier. *Vacuum Electronics Technology*, 2021, 1: 18–24.
6. Li Chen, Luo Chongtai, Wang Duoshu, Long Jianfei. Preparation method of MgO secondary electron emission functional thin films. *Vacuum and Low Temperature*, 2009, 15 (4): 187–192.
7. Wei Wenlu, Yan Baojun, Heng Yuekun, Liu Shulin, Zhang Binting, Peng Huaxing, Wang Yuman, Wen Kaile. Secondary electron emission characteristics of alumina coating on metallic substrate prepared by atomic layer deposition. *Journal of Instrumentation*, 2023, 18: P02002.
8. Wei Wenlu. Research on a new type of separated dynode electron multiplier based on ALD coating. Beijing: University of the Chinese Academy of Sciences, 2023: 53–110.
9. Zhang Duoming. Channel electron multiplier. *Physics*, 1986, 10: 635–637.
10. Zhang Binting, Li Cuiping, Liu Shulin, Yan Baojun, Li Baoqiang, Wei Wenlu, Peng Huaxing. Experimental study of channel electron multiplier. *Radiation Detection Technology and Methods*, 2023, 7: 117–123.
11. Zhang Binting. Research on spiral channel electron multipliers. Beijing: University of the Chinese Academy of Sciences, 2023: 33–90.
12. Wiza Joseph Ladislas. Microchannel plate detectors. *Nuclear Instruments and Methods*, 1979, 162: 587–601.
13. Pan Jingsheng. Microchannel boards and their main characteristic properties. *Applied Optics*, 2004, 25 (5): 25–29.
14. Liu Shulin, Li Xiang, Deng Guangxu, Pan Jingsheng, Su Detan, Yan Cheng, Li Junguo. Development of low noise, high gain microchannel boards. *Applied Optics*, 2006, 27 (6): 552–557.
15. Huang Yonggang, Huang Ying, Zhang Yang, Liu Hui, Li Guoen. Research progress on silicon microchannel plates. *Applied Optics*, 2011, 32 (5): 960–966.
16. Yi Jialiang, Niu Lihong, Kuao Xiaomei, Zhou Junlan. Development of semiconductor glass microchannel plates. *Applied Optics*, 2007, 28 (2): 121–124.
17. Beaulieu David R, Gorelikov Dmitry, de Rouffignac R, Saadatmand Kourosh, Stenton Kelley, Sullivan Neal, Tremsin Anton S. Nano-engineered ultra-high-gain microchannel plates. *Nuclear Instruments and Methods in Physics Research A*, 2009, 607: 81–84.
18. Yan Baojun, Liu Shulin, Heng Yuekun. Nano-oxide thin films deposited via atomic layer deposition on microchannel plates. *Nanoscale Research Letters*, 2015, 10: 162.
19. Yan Baojun, Liu Shulin, Wen Kaile, Lin Yanjian, Wang Yuman. Thickness optimization of alumina thin film for microchannel plate detector. *19th International Conference on Atomic Layer Deposition (ALD 2019)*, July 21–24, 2019, Bellevue, WA, USA.
20. Wang Yuman, Liu Shulin, Yan Baojun, Qi Ming, Wen Kaile, Zhang Binting, Gu Jianyu and Yao Wenjing. The design of the AZO conductive layer on microchannel plate. *Nanoscale Research Letters*, 2021, 16: 55.
21. Timothy J. Gethyn, Bybee Rita. Preliminary-results with microchannel array plates employing curved microchannels to inhibit ion feedback. *Review of Scientific Instruments*, 1977, 48(3): 292–299.
22. Timothy J. Gethyn. Curved-channel microchannel array plates. *Review of Scientific Instruments*, 1981, 52: 1131–1142.

23. Hassler Donald, Rottman Gary, Lawrence Gregory. Position offsets in curved-channel microchannel plate detectors. *Applied Optics*, 1991, 30 (25): 3575–3581.
24. Han Suli, Ni Qiliang, Zhang Hongji, He Lingping, Chen Bo. The effect of microchannel board gain on the imaging performance of photon counting detectors. *Applied Optics*, 2014, 35 (3): 484–488.
25. Naaman Ron and Vager Zeev. An electron multiplier capable of working at low vacuum: The microsphere plate. *Review of Scientific Instruments*, 1996, 67 (9): 3332–3333.
26. Yao Wenjing, Liu Shulin, Yan Baojun, Zhao Gaofeng, Dong Yongwei, Wang Zhigang, Zhu Kejun, Zhang Binting, Wen Kaile, Wang Yuman, Gu Jianyu. Linear dynamic range of low impedance MCP in DC mode. *Infrared Technology*, 2022, 44 (3): 310–314.
27. Peng Huaxing, Yan Baojun, Liu Shulin, Zhang Binting, Wei Wenlu. Electronic multiplier pulse signal graphical offline analysis software. *Applied Optics*, 2022, 43 (6): 1107–1116.
28. Yao Wenjing, Liu Shulin, Yan Baojun, Zhang Binting, Wei Wenlu, Peng Huaxing, Yang Yuzhen. Development of a cold cathode electron source based on gold cathode MCP. *Journal of Mass Spectrometry*, 2023, 44 (1): 96–104.
29. Zhang Binting, Yan Baojun, Liu Shulin, Wen Kaile, Wang Yuman, Gu Jianyu, Yao Wenjing. Design of charge sensitive amplifiers for electronic multiplier devices. *Infrared Technology*, 2022, 44 (8): 792–797.
30. Zhang Xuantong. Measurement of neutrino oscillation parameters and development of photomultiplier tubes. Beijing: University of the Chinese Academy of Sciences, 2018: 1–22.
31. Summer Classic Armor. Jiangmen neutrino experiment photomultiplier tube and μ research on sub reconstruction. Beijing: University of the Chinese Academy of Sciences, 2016: 69–103.
32. Wang Yifang, Qian Sen, Zhao Tianchi, Tian Jinshou, Li Huirui, Cao Jun, Xu Xiangyan, Wang Xiaoyun, Liu Shudong, Liu Hulin, Liu Shulin, Liu Delin, Heng Yuekun, Cao Xibin, Shentu Jun. A new design of large area MCP-PMT for the next generation neutrino experiment. *Nuclear Instruments and Methods in Physics Research A*, 2012, 695: 113–117.
33. JUNO Collaboration. Mass testing and characterization of 20-inch PMTs for JUNO. *The European Physical Journal C*, 2022, 82: 1168.
34. Yang Yuzhen, Liu Shulin, Zhao Tianchi, Yan Baojun, Wang Peiliang, Yu Yang, Lei Xiangcui, Yang Luping, Wen Kaile, Qi Ming, Heng Yuekun. Single electron counting using a dual MCP assembly. *Nuclear Instruments and Methods in Physics Research A*, 2016, 830: 438–443.
35. Yang Yuzhen, Yan Baojun, Liu Shulin, Zhao Tianchi, Yu Yang, Wen Kaile, Li Yumei, Qi Ming. MCP performance improvement using alumina thin film. *Nuclear Instruments and Methods in Physics Research A*, 2017, 868: 43–47.
36. Yang Yuzhen. Research and performance optimization of MCP-PMT microchannel board components based on JUNO. Nanjing: Nanjing University, 2017: 23–77.
37. Minot Michael, Bennis Daniel, Bond Justin, Craven Christopher, O'Mahony, Aileen, Renaud Joseph, Stochaj Michael, Elam Jeffrey, Mane Anil, Demarteau Marcellinus, Wagner Robert, McPhate Jason, Siegmund Oswald Helmut, Elagin Andrey, Frisch Henry, Northrop Richard, Wetstein Matthew. Pilot production & commercialization of LAPPD™. *Nuclear Instruments and Methods in Physics Research A*, 2015, 787: 78–84.
38. Kaptanoglu Tanner, Callaghan Edward, Minfang Yeh, Gabriel Orebi D. Gann. Cherenkov and scintillation separation in water-based liquid scintillator using an LAPPD™. *European Physical Journal C*, 2022, 82: 169.
39. Taiwan Mass Spectrometry Society. *Principles and Applications of Mass Spectrometry Analysis Technology*. Beijing: Science Press, 2019.
40. Ni Qiliang. Research on soft X-ray extreme ultraviolet photon counting imaging detectors using curved microchannel plates and inductive charge position sensitive anodes. *China Optics*, 2015, 8 (5): 847–871.

41. Yan Qiurong. Research on time dependent single photon counting technology based on MCP position sensitive anode detectors. Beijing: Graduate School of the Chinese Academy of Sciences, 2012: 1–110.
42. Cheng Dapeng, Ma Yinguang, Wang Ji, Chen Jiang, Yang Jun. Analysis of factors affecting the lifespan of cesium beam tubes. *Vacuum and Low Temperature*, 2015, 21 (1): 48–50.
43. Pan Jingsheng, Sun Jianning, Han Xiaoming, Zhang Rong, Sun Sailin, Yang Yigang, Tian Yang, Jin Ge, Zhang Zhengjun, Shao Aifei, Su Detan. A large array thermal neutron sensitive microchannel board with high detection efficiency and low noise. *Acta Photonica Sinica*, 2014, 43 (12): 1204003.
44. Chrzanowski K. Review of night vision technology. *Opto–Electronics Review*, 2013, 21 (2): 153–181.
45. Wang Jinshu, Zhou Meiling. *Electronic Emission Materials*. Beijing: Beijing University of Technology Press, 2008.

CHAPTER 9

Metal's Photoemission Based on Secondary Electron Emission

Yifan Liu, Aigen Xie, and Hongjie Dong

9.1 PHOTOEMISSION PROPERTIES OF METAL SURFACES NEAR THE THRESHOLD FREQUENCY AND ITS RESEARCH METHODS

Before 1930, investigations into photoemission were primarily focused on metals. However, after 1930, practical photocathodes like Ag-O-Cs, $Na_2KSb(Cs)$, and negative electron affinity GaAs were developed. Metal photocathodes are known for their high response speed and suitability for environments with removable dynamic vacuum systems. Consequently, photoemission from metal photocathodes has gained importance in fields such as synthetic metals and high-energy physics, with photoelectric sensitivity (PS) and quantum efficiency (QE) being crucial topics. Numerous authors have researched PS and QE of metals, and the widely used classical models for PS and QE of metals are the three-step model and the Fowler-DuBridge model.

Electron-induced secondary electron emission SEE and photoemission have common mechanisms for the transport and escape of excited electrons. Thus, the theories of transport and escape of electron-induced secondary electrons can be used as those of transport and escape of photon-induced electrons. Therefore, from Sommerfeld's theory, experimental PS curves, theory of SEE, and the characteristics of transport of electron-induced secondary electrons with $E_{vac} < 1.0$ eV and the three steps of photoemission, the electron-induced secondary electron model ESEM for PS and QE of metal surfaces near threshold frequency γ_0 is developed. The ESEM includes the effects of frequency of incident light γ, Kelvin temperature T, properties of metals (Fermi energy and work function φ including Schottky effect), and applied field ξ. The PS curves show that for γ near γ_0 there is a marked increase in photoemission with T, for γ farther away there is no change of photoemission with T,

DOI: 10.1201/9781003562863-9

and for γ still father away there is a slight decrease in photoemission with T; these characteristics of PS curves were explained, respectively. Two methods of determining φ by using formulas deduced here for PS or QE curves of metal surfaces near γ_0 to fit corresponding experimental data were presented, respectively. The ESEM is compared with existing widely used classical models, i.e., the three-step model and the Fowler-DuBridge model [1–4]. Even though with different assumptions and settings, it is found that PS and QE curves of metal surfaces near γ_0 in ESEM agree well with corresponding curves in other models, and that the ESEM has following three advantages over other models. First, PS and QE curves of metal surfaces near γ_0 are investigated in ESEM. But only PS curves of metal surfaces near γ_0 were investigated in Fowler-DuBridge model, and only QE curves of metal surfaces near γ_0 were investigated in the three-step model. Second, the formulas for PS and QE curves of metal surfaces near γ_0 deduced in ESEM are simpler than corresponding formulas deduced in other models. Thus, it is more convenient to use the formulas for PS and QE curves of metal surfaces near γ_0 deduced in ESEM to calculate corresponding parameters. Third, the influences of T on photoemission were taken into account in three steps (excitation, transport, and escape) of photoemission in ESEM. But the influences of T on photoemission were mainly taken into account in excitation step of photoemission in other models. Thus, the influences of T on photoemission were explained more clearly in ESEM.

9.1.1 Energy Distribution of Free Electrons

According to Sommerfeld's theory of metals, the quantum states of free electrons at conduction band of metals can be expressed in terms of [5]:

$$q(E)=CE^{0.5} \tag{9.1}$$

where E is the energy measured from the bottom of conduction band of metal, C of a given metal is a constant.

The free electrons at conduction band of metals obey Fermi–Dirac statistics, and Fermi–Dirac statistics can be written as [4]:

$$f(E)=[e^{(E-E_F)/(K_BT)}+1]^{-1} \tag{9.2}$$

where K_B is the Boltzmann constant, and Fermi energy E_F can be expressed as [5]:

$$E_F=0.3646n^{(2/3)} \tag{9.3}$$

The unit of E_F is eV, n is the number density of free electrons at conduction band, the unit of n is number of free electrons at conduction band per nm^3, and n can be expressed as [5]:

$$n=(\rho N)/A_\alpha \tag{9.4}$$

where ρ is the material density, A_α is the atomic weight, and N is the number of free electrons in one atom.

The E_F of W, K_m (potassium), Ag, and Au were calculated with Eqs. (9.3) and (9.4) and corresponding parameters equal 13.0, 2.12, 5.0, and 5.52 eV, respectively. The E_F of Ta, Pd,

TABLE 9.1 The Calculated E_F of Ta, Pd, Hg, -Sn, β-Sn, and γ-Sn

M Metal	α-Sn	β-Sn	γ-Sn	Hg	Ta	Pd
N	4	4	4	2	2	1
ρ (g·cm^{-3})	5.75	6.54	7.28	13.59	16.654	12.023
A_α (μ_A)	118.71	118.71	118.71	200.59	180.95	106.42
E_F (eV)	8.7	9.5	10.2	6.86	8.43	6.09

Hg, α-Sn, β-Sn, and γ-Sn were calculated with Eqs. (9.3) and (9.4), and parameters are shown in Table 9.1.

The energy distribution of free electrons at the conduction band of metal can be obtained by combining Eqs. (9.1) and (9.2):

$$F(E)=q(E)f(E)=CE^{0.5}[e^{(E-E_F)/(K_BT)}+1]^{-1} \tag{9.5}$$

9.1.2 Mean Escape Depth of Photon-Induced Electrons

According to the electron-phonon interaction in insulators and the transport of electron-induced secondary electrons investigated by J. Llacer and E. L. Garwin [6], it is known that the mean escape depth of electron-induced secondary electrons with E in insulators can be approximately expressed as:

$$\lambda(E,\text{in})=C_1E_{\text{vac}}{}^m=C_1(E-\chi)^m \tag{9.6}$$

where C_1 of the same kind of insulator is a constant, χ is the electron affinity and the surface barrier of insulator, m is positive, E_{vac} is the energy measured from vacuum. According to Eq. (9.6), it is known that electron-induced secondary electrons in insulators with more energy have larger $\lambda(E, \text{in})$.

Experimental data [7–10] shows that the electron-induced secondary electron yield Y of insulators decreases with increasing temperature (T), whereas Y of metals remains almost constant. Previous research has shown that the number of photons increases with increasing T, leading to a decrease in the mean escape depth λ of electron-induced secondary electrons in insulators. This decrease in λ is responsible for the decrease in Y of insulators with increasing T. Thus, electron-phonon interactions have significant effects on λ and Y of insulators. Conversely, Y of metals does not decrease with increasing T, indicating that the mean escape depth λ of metals remains largely unchanged. Therefore, it can be concluded that electron-phonon interactions have minimal effects on λ and Y of metals. Additionally, it is observed that increasing T does not affect the surface states, and the processes of SEE also support the conclusion that electron-phonon interactions have little effect on λ and Y of metals.

The mean energy of electron-induced secondary electrons emitted from metals E_{AM} measured from vacuum is about a few eV, most of electron-induced secondary electrons in metal have $E_{vac}>1.0\,eV$. The electron-induced secondary electrons having $E_{vac}>1.0\,eV$ mainly lose energy by single electron-electron scattering in metal, single electron-electron scattering loses larger energy every scattering. Thus, the electron-induced secondary electrons with larger energy in metal have less λ. The electron-induced secondary electrons having $E_{vac}<1.0\,eV$ mainly lose energy by multiple electron-electron and electron-phonon scattering in metal; multiple electron-electron and electron-phonon scattering loses less energy every scattering. For example, electron-phonon scattering often loses about 0.03 eV every scattering. Thus, the electron-induced secondary electrons with less energy in metal have larger λ. Therefore, the mean escape depth of electron-induced secondary electrons with E in metal can be expressed as:

$$\lambda(E,\text{metal})=\begin{cases}\dfrac{200}{(E-E_F)^{1.3}}\left(\dfrac{E_F}{5.57}\right)^{0.45} & [0\le(E-\varphi-E_F)\le1.5E_F]\\[2ex] \dfrac{200}{(E-E_F)^{1.3}}\left(\dfrac{E_F}{5.57}\right)^{0.45}\left(\dfrac{1.5E_F}{E-\varphi-E_F}\right) & [1.5E_F\le(E-\varphi-E_F)\le2.0E_F]\\[2ex] \dfrac{150}{(2.0E_F+\varphi)^{1.3}}\left(\dfrac{E_F}{5.57}\right)^{0.45} & [2.0E_F\le(E-\varphi-E_F)\le50\,\text{eV}]\end{cases} \tag{9.7}$$

From Eq. (9.7), it is known that electron-induced secondary electrons in metals with less energy have larger $\lambda(E, \text{metal})$.

Electron-induced secondary electrons in metals mainly lose energy through single electron-electron scattering, and the energy loss through multiple electron-electron and electron-phonon scattering can be ignored in SEE from metals. The mean energy of electron-induced secondary electrons emitted from insulators measured from vacuum is only about 1.0 eV, and as these electrons mainly lose energy through multiple electron-electron and electron-phonon scattering in metals, the energy loss of electron-induced secondary electrons by multiple electron-electron and electron-phonon scattering cannot be ignored in electron-induced SEE from insulators. The mean energy of photon-induced electrons undergoing photoemission from metal near γ_0 is less than 1.0 eV, and the energy loss of these electrons by multiple electron-electron and electron-phonon scattering cannot be ignored in photoemission from metal near γ_0. Therefore, the mechanism of transport for photon-induced electrons in this scenario is similar to that of electron-induced secondary electrons in insulators, and the mean escape depth of hγ photon-induced electrons at initial E, $\lambda(E, \text{h}\gamma)$, has a similar expression as $\lambda(E, \text{in})$, where h is Planck constant, and E_{Have} of hγ photon-induced electrons at initial E is equal to $(E+\text{h}\gamma)$, with E_{Have} also measured from the bottom of the conduction band. Then, from Eq. (9.6), the fact that the surface barrier of insulator is χ and the fact that the surface barrier of metal is $(E_F+\varphi)$, we assume that the $\lambda(E, \text{h}\gamma)$ can be approximately expressed as:

$$\begin{aligned}\lambda(E,\text{h}\gamma)&=C_2{E_{vac}}^S\\ &=C_2(E_{Have}-E_F-\varphi)^S=C_2(E+\text{h}\gamma-E_F-\varphi)^S\end{aligned} \tag{9.8}$$

where C_2 of given metal and T is a constant, S is positive, and S of given metal and T is a constant.

The energy loss of photon-induced electrons due to electron-phonon interactions increases with increasing T. Thus, $\lambda(E, \mathrm{h}\gamma)$ decreases with increasing T. Therefore, from the fact that the $(E+\mathrm{h}\gamma - E_\mathrm{F} - \varphi)$ of Eq (9.8) is less than 1 eV, it can be concluded that S increases with increasing T. For the same energy distribution of electron-induced internal secondary electrons, the mean energy of electron-induced secondary electrons increases with increasing $(E_\mathrm{F}+\varphi)$. Thus, from the fact that SEE and photoemission have common mechanisms for the transport and escape of excited electrons, we assume that the mean energy of photon-induced electrons at E which undergo photoemission from metal near γ_0 also increases with increasing $(E_\mathrm{F}+\varphi)$. The electron-induced secondary electrons in insulators with more energy have larger $\lambda(E, \mathrm{in})$. Therefore, from the conclusion that photon-induced electrons undergoing photoemission from metal near γ_0 have a similar mechanism of transport as electron-induced secondary electrons in insulators do, it can be assumed that $\lambda(E, \mathrm{h}\gamma)$ increases with increasing $(E_\mathrm{F}+\varphi)$. Thus, from the fact that the $(E+\mathrm{h}\gamma - E_\mathrm{F} - \varphi)$ of Eq. (9.8) is less than 1 eV, it can be assumed that S decreases with increasing $(E_\mathrm{F}+\varphi)$. Therefore, from the conclusion that that S increases with increasing T, we assume that the S can be approximately expressed as:

$$S = C_3 T^p (E_\mathrm{F} + \varphi)^{-q} \tag{9.9}$$

where C_3, p, and q are approximately positive constants, respectively.

According to the theory of SEE, the probability that an electron-induced secondary electron having $E \geq (\varphi + E_\mathrm{F})$ and reaching the surface of the metal can pass over the surface barrier into a vacuum can be written as follows [11]:

$$P(E) = 1 - [(E_\mathrm{F} + \varphi)/E]^{0.5} \tag{9.10}$$

SEE and photoemission have common mechanisms for the transport and escape of secondary electrons. Thus, according to Eq. (9.10) and the fact that the electrons at initial E absorbing one hγ photon have $(E + \mathrm{h}\gamma)$, the probability that an electron at initial E absorbing one hγ photon, having $(E + \mathrm{h}\gamma) \geq (\varphi + E_\mathrm{F})$ and reaching the surface of the metal can pass over the surface barrier into a vacuum can be written as follows:

$$P(E,\gamma) = 1 - [(E_\mathrm{F} + \varphi)/(E + \mathrm{h}\gamma)]^{0.5} \tag{9.11}$$

Of course, the probability that an electron at initial E absorbing one hγ photon, having $(E + \mathrm{h}\gamma)$ $E \leq (\varphi + E_\mathrm{F})$ and reaching the surface of metal can pass over the surface barrier into a vacuum is zero.

9.1.3 Universal Formula for PS Curves

For a monochromatic incident light with given photon flux I_0 hitting on the metal having φ, the number of photons at γ absorbed by the metal can be expressed as:

$$n_n = [C_4(1 - R(\gamma))\gamma_0]/\gamma \tag{9.12}$$

where C_4 of a given I_0 is a constant, $R(\gamma)$ is optical reflectivity at γ, and the relation between φ and γ_0 can be written as [12–18]:

$$\varphi = \mathrm{h}\gamma_0 \tag{9.13}$$

The incident photon flux entering metals decreases gradually and becomes zero at R which is on the order of 1,000 Å, where R is the distance from the incident surface to the position at which incident photon flux entering metal becomes zero. The λ of metals are in the range of 5–15 Å, and the maximum escape depth of electron-induced secondary electrons emitted from metals T_λ which approximately equals 5λ are in the range of 25–75 Å. Thus, from the fact that R is much larger than T_λ, it is known that the incident photon flux inside T_λ can be approximately looked on as a constant $I_0(1-R(\gamma))$. According to the characteristics of electron-photon interaction and Fowler's theory of PS curves of metals near γ_0, the probability that the electron absorbs one photon having γ which do not differ greatly from γ_0 can be approximately considered as a constant C_5. Therefore, according to Eqs. (9.5) and (9.12), the number of electrons at initial E which absorb one photon having γ which does not differ greatly from γ_0 per unit path length of incident light inside T_λ can be expressed as [19]:

$$N(E) = CC_4C_5(1 - R(\gamma))\gamma_0\gamma^{-1}[e^{(E-E_F)/K_BT} + 1]^{-1}E^{0.5} \tag{9.14}$$

T_λ approximately equals 5λ. Thus, from the fact that electron-induced SEE and photoemission have common mechanisms for the transport of excited electrons, it is known that the maximum escape depth of photon-induced electrons at initial E which undergo photoemission from metal near γ_0 also approximately equals $5\lambda(E, \mathrm{h}\gamma)$. Therefore, from the three processes of SEE [20,21], energy band of metal, the fact that SEE and photoemission have common mechanisms for the transport and escape of secondary electrons and Eqs. (9.8), (9.11), and (9.14), the number of photon-induced electrons emitted from metals near γ_0 as a function of γ and T can be written as:

$$\begin{aligned} F(E,\gamma) &= \int_{E_F+\mathrm{h}\gamma_0-\mathrm{h}\gamma}^{E_F+\varphi} \left\{ \int_0^{5/\lambda(E,\mathrm{h}\gamma)} [N(E)e^{-x/\lambda(E,\mathrm{h}\gamma)}P(E,\gamma)]\,\mathrm{d}x \right\} \mathrm{d}E \\ &= \int_{E_F+\mathrm{h}\gamma_0-\mathrm{h}\gamma}^{E_F+\varphi} [N(E)\lambda(E,\mathrm{h}\gamma)P(E,\gamma)(1-e^{-5})]\,\mathrm{d}E \\ &= \frac{C_6\gamma_0}{\gamma} \int_{E_F+\mathrm{h}\gamma_0-\mathrm{h}\gamma}^{E_F+\varphi} \left\{ \begin{matrix} E^{0.5}(E+\mathrm{h}\gamma - E_F - \mathrm{h}\gamma_0)^S[e^{(E-E_F)/K_BT}+1]^{-1} \\ [1-[(\varphi+E_F)/(E+\mathrm{h}\gamma)]^{0.5}] \end{matrix} \right\} \mathrm{d}E \end{aligned} \tag{9.15}$$

where x is the distance from the incident surface to the position at which the electrons are induced by photon; for photoemission from metal near γ_0, $R(\gamma)$ can be approximately considered as a constant [3,22]; for given metal and I_0, the C, C_4, and C_5 are constants, respectively; for given metal and T, C_2 is a constant. Thus, for given metal, T and I_0, the C_6

equaling $(1-e^{-5})CC_2C_4\,C_5(1-R(\gamma))$ is also a constant. Therefore, it can be concluded that the C_2 of a given metal changes with T at the same speed as C_6 does.

From the courses of obtaining Eq. (9.12), it is known that the C_4 is proportional to I_0. Thus, from the relation of $C_6=(1-e^{-5})CC_2C_4\,C_5(1-R(\gamma))$, it is known that C_6 is also proportional to I_0. Therefore, from the fact that the photo-current is proportional to the value of Eq. (9.15), it is known that Eq. (9.15) can only give the general form of the whole family of PS curves of metals near γ_0, and that Eq. (9.15) can be used to express the universal formula for PS curves of metals near γ_0.

Based on Eqs. (9.9) and (9.15), E_F, φ obtained by some authors and experimental PS curves at various T and γ of metals near γ_0, the values of p, q, and C_3 are estimated to be 0.61, 0.94, and $3.13446\times10^{-3}(\text{eV})^{0.61}\ \text{K}^{-0.94}$. Thus, based on the estimated values of p, q, and C_3 and Eq. (9.9), the S of Eq. (9.15) can be expressed as:

$$S=3.13446\times10^{-3}(\text{eV})^{0.61}\text{K}^{-0.94}T^{0.94}(E_F+\varphi)^{-0.61} \tag{9.16}$$

On the basis of Eqs. (9.8) and (9.16), the $\lambda(E,\text{h}\gamma)$ can be expressed as:

$$\begin{aligned}\lambda(E,\text{h}\gamma)&=C_2(E+\text{h}\gamma-E_F-\varphi)^S\\&=C_2(E+\text{h}\gamma-E_F-\varphi)^{3.13446\times10^{-3}(\text{eV})^{0.61}\text{K}^{-0.94}T^{0.94}(E_F+\varphi)^{-0.61}}\end{aligned} \tag{9.17}$$

The p, q, and C_3 are the parameters of S, and S is the parameter of $\lambda(E,\text{h}\gamma)$. Note that $\lambda(E,\text{h}\gamma)$ is the mean escape depth of hγ photon-induced electrons at initial E which undergo photoemission from metal near γ_0, field emission from metal with $\xi<10^6$ eV/cm, and thermal emission from metal at $T\leq1{,}100$ K can be ignored. Thus, the method presented here of estimating the p, q, and C_3 by Eqs. (9.9) and (9.15), E_F, φ, and experimental PS curves at various T and γ of metals near γ_0 is practicable.

9.1.4 Determination of φ and Special Formula for PS Curves

The relation between the wavelength of incident light λ_{light} and γ can be expressed as:

$$C_{\text{light}}=\lambda_{\text{light}}\gamma \tag{9.18}$$

where C_{light} is the speed of light in vacuum and is equal to 3.0×10^8 m/s.

We attempt to use Eqs. (9.13), (9.15), and (9.16), h, K_B, E_F, T shown in Table 9.2, tentative γ_0 and tentative C_6 to fit the experimental relative PS of PS curves at $T=293$ K of K_m near γ_0 sometimes, and we can obtain $\gamma_0=5.435\times10^{14}$ Hz and relative $C_6=6{,}100$. Based on Eq. (9.13) and obtained $\gamma_0=5.435\times10^{14}$ Hz, the $\varphi=2.25$ eV of K_m is obtained. The obtained γ_0, φ, and relative C_6 of K_m are shown in Table 9.2 [19].

Some authors reported relative PS from clean metals at various temperatures as a function of mean escape depth λ_{light}. Using these experimental data and Eq. (9.18), relative PS curves as a function of γ near γ_0 were obtained. Eqs. (9.13), (9.15), and (9.16) with values from Tables 9.2 and 9.3 were used to fit these curves with tentative values of γ_0 and C_6 to obtain corresponding γ_0 and relative C_6. From Eq. (9.13) and obtained γ_0, φ of corresponding

TABLE 9.2 The Obtained γ_0, φ, and Relative C_6 of K_m, Ag, Au, W, and Hg

Metals	K_m	Ag		Au		W	Hg
E_F (eV)	2.12	5.0		5.52		13.0	6.86
φ (eV) determined by other methods	2.21	4.74;4.78		4.90;4.81		4.79	4.52
Obtained average φ (eV)	2.25	4.74		4.9		4.8	4.41
Obtained average γ_0 (10^{15}Hz)	0.5338	1.146		1.1815		1.16	1.064
T (K)	293	293	873	293	1013	1100	293
Obtained relative C_6	6100	800	700	1300	1510	1710	117
Obtained γ_0 at various T (10^{15}Hz)	0.5434	1.159	1.133	1.183	1.18	1.16	1.064
Obtained φ at various T (eV)	2.25	4.79	4.69	4.90	4.89	4.80	4.41

TABLE 9.3 The Obtained γ_0, φ and Relative C_6 of α-Sn, Ta, Pd, β-Sn, and γ-Sn

Metals	α-Sn	Ta		Pd			β-Sn	γ-Sn
E_F (eV)	8.7	8.43		6.09			9.5	10.2
φ (eV) determined by other methods	4.17	4.12		4.97;4.96; 5.2			4.28	4.39
Obtained average φ (eV)	4.2	4.11		4.97			4.22	4.37
Obtained average γ_0 (10^{15}Hz)	1.015	0.993		1.2			1.02	1.055
T (K)	673	293	973	305	730	1078	483	358
Obtained relative C_6	55	3900	3900	670	700	715	37	46
Obtained γ_0 at various T (10^{15}Hz)	1.015	0.9935	0.9925	1.2055	1.2	1.1955	1.02	1.055
Obtained φ at various T (eV)	4.2	4.11	4.11	4.99	4.97	4.95	4.22	4.37

metals were obtained. The obtained γ_0, C_6, and φ of clean metals at various temperatures are shown in Tables 9.2 and 9.3. For small changes in temperature (<1,100 K), the work function of a given clean metal can be considered constant. Hence, the average work function of a given clean metal is the average of the obtained work function at T<1,100 K; for example, the average value of φ of Ag shown in Table 9.2 equals [(φ = 4.79 eV of Ag at 293 K) + (φ = 4.69 eV of Ag at 873 K)]/2.

From the above courses of obtaining γ_0, φ, relative C_6 corresponding to relative values of PS, it is known that the special formula for PS curves of given metal near γ_0 as a function of γ and T can be obtained by Eqs. (9.13), (9.15), and (9.16) and relative PS of PS curves of metals near γ_0, and that absolute C_6 and special formula for PS of given metal near γ_0 as a function of γ and T can also be obtained by Eqs. (9.13), (9.15), and (9.16) and absolute PS of metals near γ_0.

Comparison between relative PS of PS curves of several metals near γ_0 calculated with Eqs. (9.13), (9.15), and (9.16) and parameters shown in Tables 9.2 and 9.3 and experimental ones are shown in Figures 9.1 and 9.2 [19].

9.1.5 QE and QE Curves

Based on Eqs. (9.12) and (9.15) and the definitions of QE and PS, the universal formula for QE curves at various T of metals near γ_0 can be written as [19]:

$$Y(E,\gamma)=C_7\int_{E_F+h\gamma_0-h\gamma}^{E_F+\varphi}\left\{E^{0.5}(E+h\gamma-E_F-h\gamma_0)^S[e^{(E-E_F)/K_BT}+1]^{-1}\left[1-\left(\frac{\varphi+E_F}{E+h\gamma}\right)^{0.5}\right]\right\}dE \quad (9.19)$$

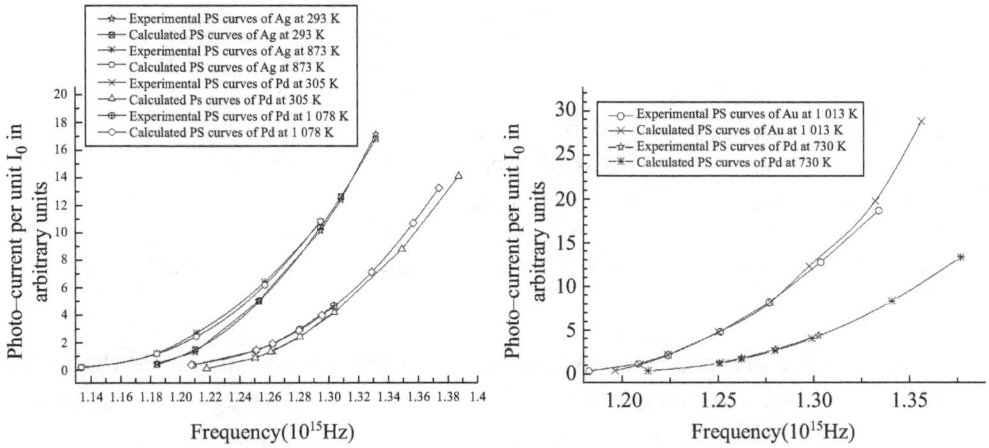

FIGURE 9.1 Comparison of calculation results and experimental results of PS curves of Ag and Pd near γ_0.

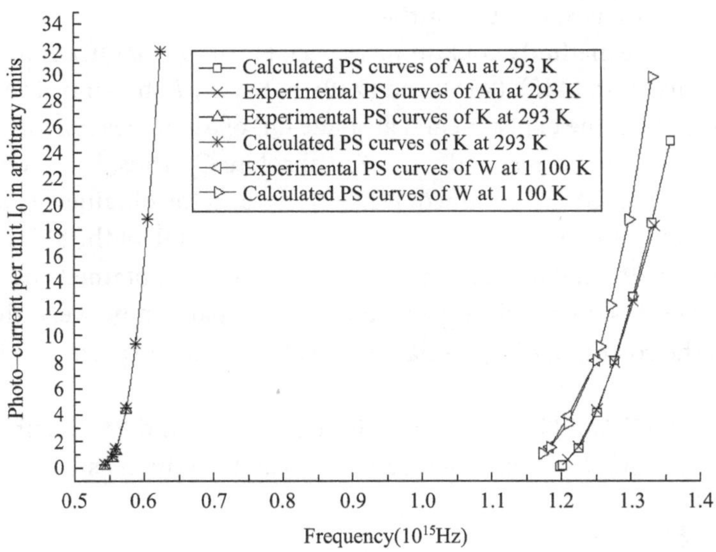

FIGURE 9.2 Comparison of calculation results and experimental results of PS curves of Au, K and W near γ_0.

For given metal and T, C_7 is a constant.

Similar to the method of obtaining a special formula for PS curves of metals near γ_0, we can attempt to use Eqs. (9.13), (9.16), and (9.19), h, K_B, E_F, T, tentative γ_0 and tentative C_7 to fit experimental relative QE of QE curves at T of given metal near γ_0 sometimes, and we can obtain γ_0 and relative C_7. Thus, based on Eq. (9.13) and obtained γ_0, the φ of the given metal is obtained. Therefore, the special formula for QE curves of the given metal near γ_0 which is composed of Eqs. (9.13), (9.16), and (9.19), h, K_B, E_F, T, obtained relative C_7, γ_0, and φ can be obtained.

TABLE 9.4 The Obtained C_6, γ_0, and φ of S1, S2, S3, S4, S5, and S6

Metals	S1	S2	S3	S4	S5	S6
E_F (eV)	7.0	7.0	7.0	7.0	7.0	7.0
φ (eV) obtained by other papers	5.55	5.0	4.39	4.35	4.31	4.49
φ (eV) obtained here	5.583	5.066	4.673	4.26	4.2	4.384
γ_0 (10^{15} Hz) obtained here	1.35	1.225	1.13	1.030055	1.0155475	1.06
T (K)	300	300	300	300	300	300
Absolute C_7 obtained here	0.0025	0.0007	0.0048	0.0038	0.0038	0.0035

D. H. Dowell have prepared six samples of oxygen-free Cu and measured their QE, the first sample S1 is the initial sample with 31% carbon surface coverage, the second sample S2 is the sample after exposure to 0.63 mC of H-ion beam has 11% carbon surface coverage, the third sample S3 is the sample after exposure to 2.1 mC of H-ion beam has 12% carbon surface coverage, the fourth sample S4 is the sample after exposure to 3.0 mC of H-ion beam has 10% carbon surface coverage, the fifth sample S5 is the sample after exposure to 10.23 mC of H-ion beam has 7% carbon surface coverage, the sixth sample S6 is the sample after heating to 503 K has 8% carbon surface coverage.

Similar to the above method of obtaining a special formula for QE curves of metals near γ_0, we attempt to use Eqs. (9.13), (9.16), and (9.19), h, K_B, E_F, T shown in Table 9.4, tentative γ_0 and tentative C_7 to fit the experimental absolute QE of S1–S6 near γ_0 sometimes, respectively; and we can obtain corresponding γ_0 and absolute C_7. Thus, based on Eq. (9.13) and obtained γ_0, the φ of S1–S6 are obtained, respectively, and the obtained γ_0, φ, and absolute C_7 are shown in Table 9.4. Therefore, the special formula for QE of the S1–S6 near γ_0, which is composed of Eqs. (9.13), (9.16), and (9.19), h, K_B, E_F, and T, obtained absolute C_7, γ_0, and φ shown in Table 9.4 is obtained, respectively. The comparison between the QE of S1–S6 calculated with the corresponding special formula for QE and experimental ones is shown in Figure 9.3.

Based on Eqs. (9.12) and (9.21) and the definitions of QE and PS, the universal formula for QE curves at various T of metals with ξ near γ_0 can be written as:

$$Y(E,\gamma,\xi)=$$

$$C_8\int_{E_F+h\gamma_0-3.79\times10^{-4}\sqrt{\xi}-h\gamma}^{E_F+\varphi-3.79\times10^{-4}\sqrt{\xi}}\left\{\frac{\dfrac{E^{0.5}(E+h\gamma-E_F-h\gamma_0+3.79\times10^{-4}\sqrt{\xi})^{S(\xi)}}{\left[e^{\left(\frac{E-E_F}{K_BT}\right)}+1\right]}}{\left[1-\left(\dfrac{\varphi+E_F-3.79\times10^{-4}\sqrt{\xi}}{E+h\gamma}\right)^{0.5}\right]}\right\}dE \qquad (9.20)$$

Similar to the above method of obtaining φ, γ_0, absolute or relative C_7, φ, γ_0, absolute or relative C_8 of Eq. (9.20) can be obtained by Eqs. (9.13), (9.22), and (9.20) and absolute or relative QE of metals with ξ near γ_0 first, next the special formula for QE or QE curves at

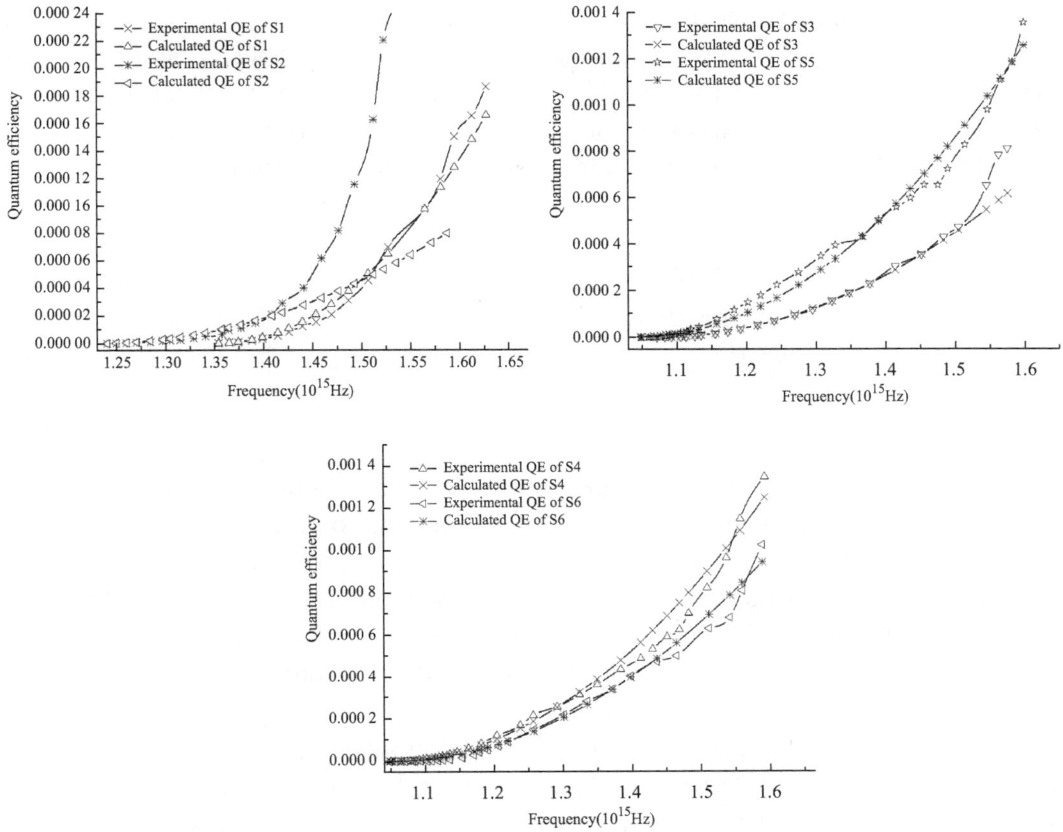

FIGURE 9.3 Comparison between experimental QE of S1–S6 and calculated ones.

various T of metals with ξ near γ_0 can be obtained by Eqs. (9.13), (9.22), and (9.20) and relative or absolute QE of metals with ξ near γ_0.

9.2 PHOTOEMISSION CHARACTERISTICS OF METAL SURFACES WITH FREQUENCIES MUCH LARGER THAN THRESHOLD AND THEIR RESEARCH METHODS

The photoelectric effect has important applications in various fields including astrophysics, material analysis, photon science, and X-ray sources. The energy distribution curves of photo-emitted electrons can provide valuable information about the photoelectric effect. Two formulas for these curves have been deduced for metals in the vacuum ultraviolet range and have been experimentally verified for Au, Ni, and Cu. Several parameters, such as absolute quantum efficiency, mean escape depth of photo-emitted electrons, and mean energy of photo-emitted electrons, are important in characterizing photoemission ability. Formulas for these parameters have also been deduced for metals in the vacuum ultraviolet range. Photoelectric cross-section is an important topic, but research on it at low photon energies is scarce. A formula for calculating photoelectric cross-section PCS based on density of states, molar mass of atom, work function, distance from the bottom of conduction band to Fermi level, number of electrons of conduction band, and other parameters has been deduced and has been used to calculate the photoelectric cross-section for electrons in the conduction band of Au and Ni. A universal formula for absolute quantum efficiency has also been deduced, which can be used to calculate photoelectric cross-section for metals and negative electron affinity semiconductors. The accuracy of these methods has been experimentally verified.

9.2.1 Processes of Photoemission

When N_0 photons at $\gamma < (E_F + E_g)/h$ enter perpendicularly into metals, the number of incident photons at x can be written as [22–24]:

$$N(x) = N_0 e^{-\alpha_\gamma x} = N_0 e^{-x/\lambda_\gamma} \tag{9.21}$$

where x is the distance from the incident surface to the position at which the photons reach, N_0 is the number of incident photons at $x = 0$, α_γ is the optical absorption coefficient at γ. From the energy band structures of metal shown in Figure 9.4 (E_g is the distance from the top of the valence band to the bottom of the conduction band of metals), the theory of quantum and law of energy conservation, it is known that in the case that N_0 photons at $\gamma < (E_F + E_g)/h$ enters perpendicularly into metals only the electrons in the conduction band of metal can be excited by the photons.

According to the characteristics of electron-photon interaction, it is known that the probability that all of the electrons in the same energy band of a given metal absorb one photon at a given γ can be approximately looked on as a constant C_γ. That is, the PCS at a given γ of the electrons in the conduction band of a given metal can be approximately looked on as a constant C_γ. Based on the energy band structures of metal shown in Figure 9.4, calculated number of electrons per atom per eV $g(E)$, the values of E_F, law of energy conservation and the fact that γ is $< (E_F + E_g)/h$, we can calculate the relative number of electrons in the conduction band of metal which may absorb one photon and become photo-excited electrons. For example, when the photons at $h\gamma = 7.4$ eV enter into Au with $E_F = 11.6$ eV [25], the electrons with $E \geqq (E_F - 7.4\text{ eV})$ (i.e. 4.2 eV) may absorb one photon and become photo-excited electrons. Thus, based on the $g(E)$ of Au [25], the fact that the electrons with $E \geqq 4.2$ eV may become photo-excited electrons and the fact that E_F of Au

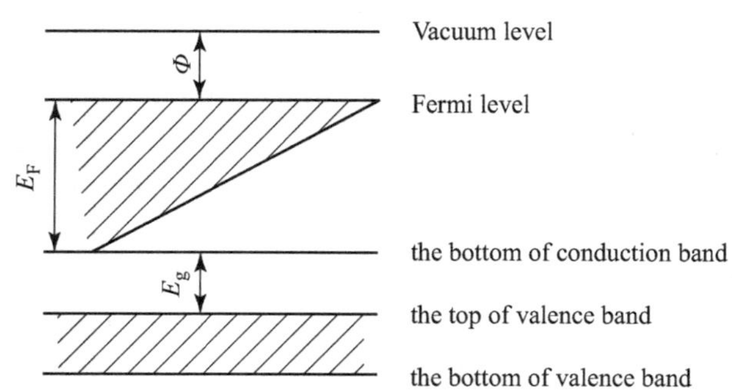

FIGURE 9.4 Schematic energy band structures of metals.

equals 11.6 eV, we can obtain the $n\%$ electrons in the conduction band of Au which may absorb one photon and become photo-excited electrons by calculating, and the obtained $n\%$ at $h\gamma = 7.4$ eV of Au equals 0.808 (by the same method, we obtain $n\%$ at different $h\gamma$ of Au, Ni and Cu, and the corresponding results are shown in Tables 9.1 and 9.2 respectively). Therefore, the α_γ of Eq. (9.21) can be written as:

$$\alpha_\gamma = (\lambda_\gamma)^{-1} = (s\rho C_\gamma n\% N_A)/A_\alpha \tag{9.22}$$

The unit of ρ is g/m³, the unit of A_α is g/mol, N_A denotes Avogadro constant.

From the fact that E_{ph} is the energy of $h\gamma$ photon-induced electrons with E measured from the bottom of the conduction band of metal, the relation of $E = E_{ph} - h\gamma$ can be obtained. Some authors have calculated $g(E)$ [25–27] and the energy distribution of electrons in the conduction band of metal with $V = 1.0\,m^3$ $G(E)$ (i.e. density of states of conduction band of metal [28,29]). According to the definitions of $G(E)$ and $g(E)$ and the relation of $E = E_{ph} - h\gamma$, the relation of $G(E)$ and $g(E)$ (i.e. the relation of $G(E_{ph} - h\gamma)$ and $g(E_{ph} - h\gamma)$) can be written as [28,29]:

$$G(E) = (\rho N_A g(E))/A_\alpha = G(E_{ph} - h\gamma) = (\rho N_A g(E_{ph} - h\gamma))/A_\alpha \tag{9.23}$$

Processes of photoemission may be considered three-step processes [26]: first, electrons are excited to become internal photo-emitted electrons; second, a portion of the internal photo-emitted electrons propagate to emission surface; and third, a portion of the internal photo-emitted electrons reaching the emission surface escape into the vacuum and become photo-emitted electrons. The three-step processes of photoemission from metal, which are investigated in detail in this work, are as follows.

From the fact that the PCS at a given γ of the electrons in the conduction band of a given metal can be looked on as a constant C_γ, definitions of $\alpha_{ph\gamma}$, $G(E_{ph} - h\gamma)$, and $g(E_{ph} - h\gamma)$ and Eqs. (9.21) and (9.23), in the case, that N_0 photons at $\gamma < (E_F + E_g)/h$ enter perpendicularly into metals, the number of internal photo-emitted electrons at x per unit path length of incident photons can be written as:

$$\begin{aligned} n(x) &= \alpha_{ph\gamma} N(x) = C_\gamma G(E_{ph} - h\gamma) N(x) = C_\gamma G(E_{ph} - h\gamma) N_0 e^{-\alpha_\gamma x} \\ &= \frac{C_\gamma \rho N_A g(E_{ph} - h\gamma) N_0 e^{-x/\lambda_\gamma}}{A_\alpha} [(E_{ph} - \Phi - E_F) \geq 0] \end{aligned} \tag{9.24}$$

where $\alpha_{ph\gamma}$ is the photoelectric absorption coefficient at γ.

Most secondary electrons have energy $E_{vac} > 1.0$ eV [30,31], and the mean energy of secondary electrons emitted from metal E_{am} is much larger than 1.0 eV; both E_{vac} and E_{am} are measured from vacuum. In the case that E_{am} is much larger than 1.0 eV, the mean escape depth of secondary electrons with $E_0(0 \leqq (E_0 - \Phi - E_F \leqq 1.5E_F))$ in metals can be expressed as [32,33]:

$$\lambda(E_0, \text{metal}) = \frac{2.0 \times 10^{-8}}{(E_0 - E_F)^{1.3}} \left(\frac{E_F}{5.57} \right)^{0.45} \quad [0 \leq (E_0 - \Phi - E_F) \leq 1.5E_F] \tag{9.25}$$

where E_0 is measured from the bottom of the conduction band, E_0 is the sum of E and the energy obtained from primary electrons by scattering.

According to the energy band structures of metals, the fact that the electrons with E absorbing one hγ photon have $(E+\mathrm{h}\gamma)$ (i.e. E_{ph}) and the theories of photoemission, we estimate that in the case that γ is farther away γ_0 {i.e. γ is$>[(3.0\ \mathrm{eV/h})+\gamma_0]$}, most of h$\gamma$ photon-induced electrons in metals have $E_{\mathrm{vac}}>1.0$ eV, and E_{mean} is much larger than 1.0 eV, where γ_0 is the threshold frequency. Thus, from the fact that SEE and photoemission have common mechanisms of escape and transport and the facts that most of secondary electrons in metals also have $E_{\mathrm{vac}}>1.0$ eV and E_{am} is also much larger than 1.0 eV, it is known that the hγ photon-induced electrons with E_{ph} which undergo photoemission from metal father away γ_0 have the same transport mechanisms as secondary electrons in metals do and that the mean escape depth of hγ photon-induced electrons with E_{ph} which undergo photoemission from metal father away γ_0 $\lambda(E_{\mathrm{ph}}, \mathrm{h}\gamma)$ has a similar expression as Eq. (9.25) does. Therefore, according to Eq. (9.25) and the fact that the electrons with E absorbing one hγ photon have E_{ph}, which corresponds to E_0, in the case that E_{mean} is much larger than 1.0 eV the $\lambda(E_{\mathrm{ph}}, \mathrm{h}\gamma)$ can be written as:

$$\lambda(E_{\mathrm{ph}},\mathrm{h}\gamma)=\frac{2.0\times10^{-8}}{(E+\mathrm{h}\gamma-E_{\mathrm{F}})^{1.3}}\left(\frac{E_{\mathrm{F}}}{5.57}\right)^{0.45}=\frac{2.0\times10^{-8}}{(E_{\mathrm{ph}}-E_{\mathrm{F}})^{1.3}}\left(\frac{E_{\mathrm{F}}}{5.57}\right)^{0.45}\quad[0\le(E_{\mathrm{ph}}-\Phi-E_{\mathrm{F}})\le1.5E_{\mathrm{F}}] \tag{9.26}$$

The probability that an internal secondary electron having $E_0\ge(\Phi+E_{\mathrm{F}})$ and reaching emission surface pass over surface barrier of metal into a vacuum [32–34] is expressed as:

$$P(E_0)=1-[(E_{\mathrm{F}}+\Phi)/E_0]^{0.5} \tag{9.27}$$

From Eq. (9.27), the fact that photoemission and SEE have common mechanisms of escape and the fact that E_0 corresponds to $E_{\mathrm{ph}}=E+\mathrm{h}\gamma$, the probability that an internal photo-emitted electron with E absorbing one hγ photon, having $E_{\mathrm{ph}}\ge(\Phi+E_{\mathrm{F}})$ and reaching emission surface pass over the surface barrier of metal can be written as [32,35]:

$$P(E_{\mathrm{ph}},\mathrm{h}\gamma)=1-[(E_{\mathrm{F}}+\Phi)/(E+\mathrm{h}\gamma)]^{0.5}=1-[(E_{\mathrm{F}}+\Phi)/E_{\mathrm{ph}}]^{0.5} \tag{9.28}$$

Based on Eqs. (9.26) and (9.28), the probability that an internal photo-emitted electron with E, which absorbs one hγ photon and is excited at x, can reach emission surface and pass over the surface can be written as:

$$P(E_{\mathrm{ph}},\mathrm{h}\gamma,x)=\{1-[(E_{\mathrm{F}}+\Phi)/E_{\mathrm{ph}}]^{0.5}\}e^{-\frac{x}{\lambda(E_{\mathrm{ph}},\mathrm{h}\gamma)}} \tag{9.29}$$

9.2.2 First SEE Model for Photoemission

Based on the fact that SEE has maximum escape depth five times the mean escape depth [32,33,35], it is easy to know that this one is also applied for photo-emitted electrons since both of them have common mechanisms of escape and transport. Thus, $f(E_{\mathrm{ph}}, \mathrm{h}\gamma)$ can be given as:

$$f(E_{\mathrm{ph}},\mathrm{h}\gamma)=\int_{0}^{\infty}(n(x)P(E_{\mathrm{ph}},\mathrm{h}\gamma,x))\mathrm{d}x=\int_{0}^{5\lambda(E_{\mathrm{ph}},\mathrm{h}\gamma)}\left\{n(x)e^{-\frac{x}{\lambda(E_{\mathrm{ph}},\mathrm{h}\gamma)}}\left[1-\left(\frac{E_{\mathrm{F}}+\Phi}{E_{\mathrm{ph}}}\right)^{0.5}\right]\right\}\mathrm{d}x$$

$$=\frac{C_{\gamma}\rho N_{\mathrm{A}}g(E_{\mathrm{ph}}-\mathrm{h}\gamma)N_{0}\lambda_{\mathrm{real}}}{A_{\alpha}}\left(1-e^{-\frac{5\lambda(E_{\mathrm{ph}},\mathrm{h}\gamma)}{\lambda_{\mathrm{real}}}}\right)\left[1-\left(\frac{E_{\mathrm{F}}+\Phi}{E_{\mathrm{ph}}}\right)^{0.5}\right] \tag{9.30}$$

$$[0\le(E_{\mathrm{ph}}-\Phi-E_{\mathrm{F}})\le1.5E_{\mathrm{F}}]$$

where λ_{real} of Eq. (9.30) can be written as:

$$\lambda_{\mathrm{real}}=\lambda(E_{\mathrm{ph}},\mathrm{h}\gamma)\lambda_{\gamma}/(\lambda(E_{\mathrm{ph}},\mathrm{h}\gamma)+\lambda_{\gamma}) \tag{9.31}$$

Based on the energy band structures of metals shown in Figure 9.4, Eq. (9.28) and the fact that the electrons with E absorbing one hγ photon have $E_{\mathrm{ph}}=(E+\mathrm{h}\gamma)$, it is known that in the case that N_0 photons at $\gamma<(E_{\mathrm{F}}+E_{\mathrm{g}})/\mathrm{h}$ enter into metals, the electrons with E in the range of $[(E_{\mathrm{F}}+\Phi-\mathrm{h}\gamma), (E_{\mathrm{F}}+\Phi)]$ absorbing one hγ photon have E_{ph} in the range of $[(E_{\mathrm{F}}+\Phi), (E_{\mathrm{F}}+\Phi+\mathrm{h}\gamma)]$, and that the photo-emitted electrons with E_{ph} in the range of $[(E_{\mathrm{F}}+\Phi), (E_{\mathrm{F}}+\Phi+\mathrm{h}\gamma)]$ may escape into the vacuum. Thus, from Eq. (9.30), it is known that in the case that N_0 photons at $\gamma<(E_{\mathrm{F}}+E_{\mathrm{g}})/\mathrm{h}$ enter perpendicularly into metals, the number of photo-emitted electrons can be written as:

$$N_{\mathrm{electrons}}=\frac{C_{\gamma}\rho N_{\mathrm{A}}}{A_{\alpha}}\int_{(E_{\mathrm{F}}+\Phi)}^{(E_{\mathrm{F}}+\Phi+\mathrm{h}\gamma)}\left\{N_{0}g(E_{\mathrm{ph}}-\mathrm{h}\gamma)\lambda_{\mathrm{real}}\left(1-e^{-\frac{5\lambda(E_{\mathrm{ph}},\mathrm{h}\gamma)}{\lambda_{\mathrm{real}}}}\right)\left[1-\left(\frac{E_{\mathrm{F}}+\Phi}{E_{\mathrm{ph}}}\right)^{0.5}\right]\right\}\mathrm{d}E_{\mathrm{ph}} \tag{9.32}$$

$AQE(\mathrm{h}\gamma)$ is defined as the number of photo-emitted electrons per absorbed photon [36,37]. Thus, from the definition of $AQE(\mathrm{h}\gamma)$, the fact that N_0 photons at $\gamma<(E_{\mathrm{F}}+E_{\mathrm{g}})/\mathrm{h}$ are absorbed by metals and Eq. (9.32), the $AQE(\mathrm{h}\gamma)$ from metals induced by photons at $\gamma<(E_{\mathrm{F}}+E_{\mathrm{g}})/\mathrm{h}$ can be written as:

$$AQE(\mathrm{h}\gamma)=\frac{C_{\gamma}\rho N_{\mathrm{A}}}{A_{\alpha}}\int_{(E_{\mathrm{F}}+\Phi)}^{(E_{\mathrm{F}}+\Phi+\mathrm{h}\gamma)}\left\{g(E_{\mathrm{ph}}-\mathrm{h}\gamma)\lambda_{\mathrm{real}}\left(1-e^{-\frac{5\lambda(E_{\mathrm{ph}},\mathrm{h}\gamma)}{\lambda_{\mathrm{real}}}}\right)\left[1-\left(\frac{E_{\mathrm{F}}+\Phi}{E_{\mathrm{ph}}}\right)^{0.5}\right]\right\}\mathrm{d}E_{\mathrm{ph}} \tag{9.33}$$

The C_{γ} of Au calculated with parameters (hγ, N_{A}, s, $\rho=1.93\times10^{7}$ g/m^3, $A_{\alpha}=197$ g/mol, $g(E_{\mathrm{ph}}-\mathrm{h}\gamma)$, E_{F}, Φ, n% and experimental $AQE(\mathrm{h}\gamma)$ shown in Tables 9.5) and Eqs. (9.22), (9.26), (9.31) and (9.33) are still shown in Tables 9.5. The λ_{γ} of Au calculated with Eq. (9.22) and parameters are also shown in Tables 9.5. The $f(E_{\mathrm{ph}}, \mathrm{h}\gamma)$ (in arbitrary units) of Au are calculated with parameters (E_{ph}, hγ, $g(E_{\mathrm{ph}}-\mathrm{h}\gamma)$, E_{F}, Φ, λ_{γ} and C_{γ} shown in Table 9.5) and Eqs. (9.26), (9.30) and (9.31). The comparison between these calculated $f(E_{\mathrm{ph}}, \mathrm{h}\gamma)$ of Au and experimental ones is shown in Figure 9.5.

TABLE 9.5 The Parameters of Au Calculated by the First SEE Model

M hγ(eV)	Experimental AQE(hγ)	Calculated AQE(hγ)	m%	n%	C_γ (10^{-23}m²)	λ_γ (10^{-9}m)	λ (10^{-9}m)	B	E_{mean} (eV)
7.4	0.000800	0.000696	0.0734	0.808	34.3	5.56	2.590	0.0241	1.38
7.7	0.00110	0.000975	0.106	0.845	35.2	5.18	2.56	0.0235	1.46
7.9	0.00140	0.00125	0.135	0.867	37.0	4.81	2.56	0.0231	1.49
8.1	0.00175	0.00158	0.167	0.887	37.8	4.60	2.55	0.0235	1.52
8.7	0.00320	0.00297	0.266	0.938	40.2	4.09	2.50	0.0276	1.82
9.0	0.00420	0.00392	0.320	0.957	41.9	3.84	2.45	0.0301	1.99
9.2	0.00510	0.00479	0.355	0.968	45.2	3.52	2.42	0.0303	2.07
10.4	0.01000	0.00952	0.506	0.995	45.4	3.41	2.16	0.0483	2.55
10.7	0.01200	0.0114	0.548	0.996	49.8	3.11	2.12	0.0509	2.70
11.0	0.01400	0.0133	0.597	0.998	53.1	2.96	2.04	0.0538	2.85
11.2	0.01520	0.0144	0.631	0.999	53.9	2.86	2.00	0.0553	2.93
11.5	0.01700	0.0161	0.682	0.999	54.5	2.83	1.96	0.0575	3.06

The C_γ at h$\gamma \leqq$ 11.0 eV of Ni calculated with parameters (hγ, N_A, s, $\rho = 8.9\times10^6$ g/m³, $A_\alpha = 58.69$ g/mol, $g(E_{ph} - \text{h}\gamma)$, E_F, Φ, n% and AQE(hγ) shown in Table 9.6) and Eqs. (9.22), (9.26), (9.31) and (9.33) are still shown in Table 9.6. The sum of E_F and Φ of Ni is < 11.6 eV. Thus, from Figure 9.4, when we use parameters (hγ, N_A, s, ρ, A_α, $g(E_{ph} - \text{h}\gamma)$, E_F, Φ, n% and experimental AQE(hγ) shown in Table 9.6) and Eqs. (9.22), (9.26), (9.31), and (9.33) to calculate the C_γ at hγ = 11.6 eV of Ni, the lower limit of integral [i.e. $(E_F + \Phi)$] in Eq. (9.33) should be replaced with "11.6 eV." The λ_γ of Ni calculated with Eq. (9.22) and parameters (s, N_A, ρ, A_α, n%, and C_γ shown in Table 9.6) are also shown in Table 9.9. The $f(E_{ph}$, h$\gamma)$ (in arbitrary units) of Ni calculated with corresponding parameters (E_{ph}, hγ, $g(E_{ph} - \text{h}\gamma)$, E_F, Φ, λ_γ, and C_γ shown in Table 9.6) and Eqs. (9.26), (9.30), and (9.31) are shown in Figure 9.6.

Based on Eq. (9.32) and three-step processes of photoemission, it is known that in the case that N_0 photons at $\gamma < (E_F + E_g)/\text{h}$ enter perpendicularly into metals, the number of internal photo-emitted electrons reaching the surface can be written as:

$$N_{\text{reach}} = \frac{C_\gamma \rho N_A}{A_\alpha} \int_{(E_F+\Phi)}^{(E_F+\Phi+\text{h}\gamma)} \left[N_0 g(E_{ph} - \text{h}\gamma)\lambda_{\text{real}} \left(1 - e^{-\frac{5\lambda(E_{ph},\text{h}\gamma)}{\lambda_{\text{real}}}} \right) \right] dE_{ph} \tag{9.34}$$

Based on Eqs. (9.26) and (9.32), it is known that in the case that N_0 photons at $\gamma < (E_F + E_g)/\text{h}$ enter perpendicularly into metals, the total escape depth of photo-emitted electrons can be written as:

$$N_{\text{depth}} = \frac{C_\gamma \rho N_A}{A_\alpha} \int_{(E_F+\Phi)}^{(E_F+\Phi+\text{h}\gamma)} \left\{ N_0 g(E_{ph} - \text{h}\gamma)\lambda(E_{ph},\text{h}\gamma)\lambda_{\text{real}} \left(1 - e^{-\frac{5\lambda(E_{ph},\text{h}\gamma)}{\lambda_{\text{real}}}} \right) \left[1 - \left(\frac{E_F+\Phi}{E_{ph}} \right)^{0.5} \right] \right\} dE_{ph} \tag{9.35}$$

According to Eq. (9.32) and the fact that the photo-emitted electrons with E absorbing one hγ photon have $E_{vac} = (E_{ph} - E_F - \Phi)$, it is known that in the case that N_0 photons at $\gamma < (E_F + E_g)/\text{h}$ enter perpendicularly into metals the total E_{vac} of photo-emitted electrons can be written as:

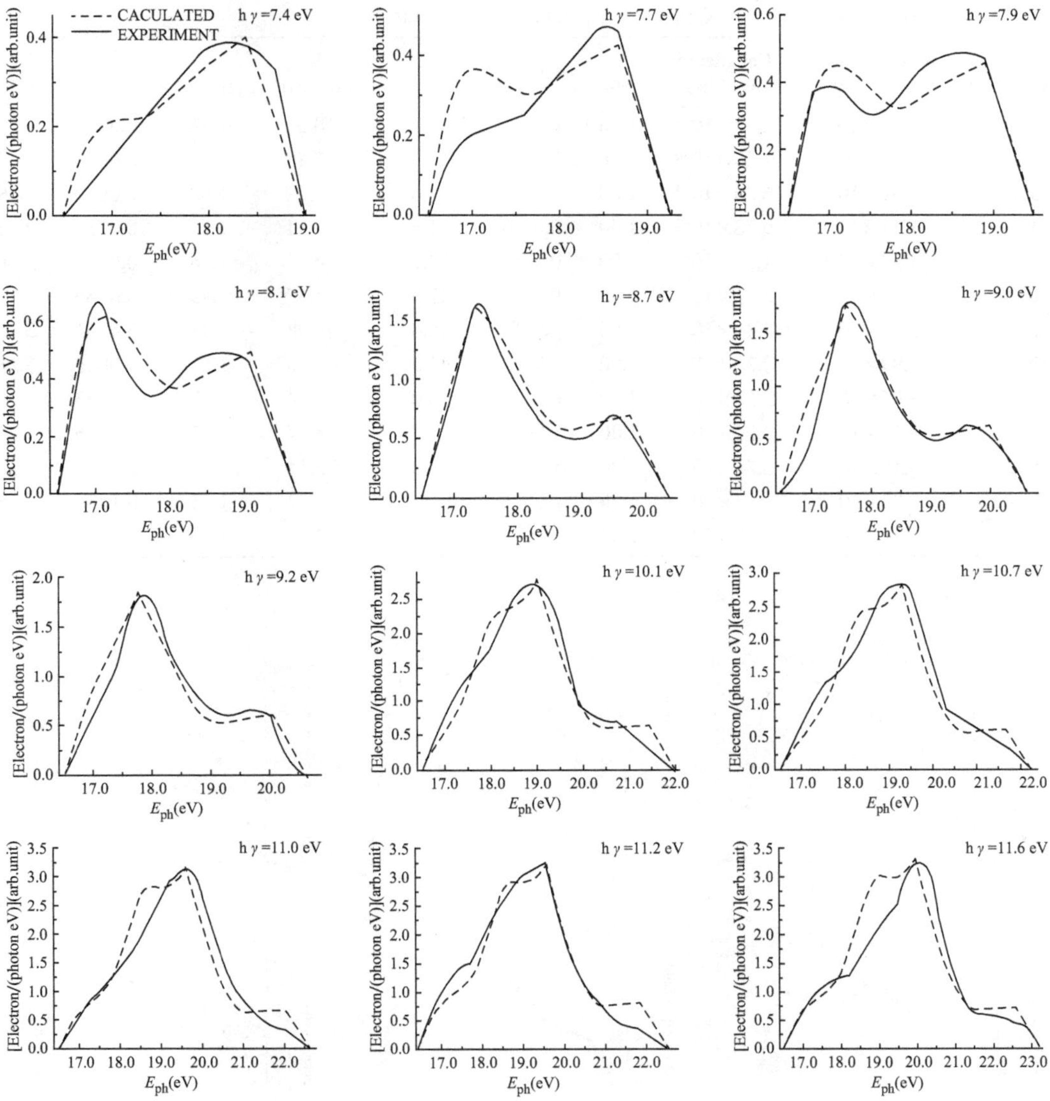

FIGURE 9.5 The comparison between the $f(E_{ph}, h\gamma)$ of Au calculated with Eq. (9.30) and experimental ones.

$$N_{\text{energy}} = \frac{C_\gamma \rho N_A}{A_\alpha} \int_{(E_F+\Phi)}^{(E_F+\Phi+h\gamma)} \left\{ \begin{array}{l} N_0 g(E_{ph} - h\gamma)(E_{ph} - E_F - \Phi)\lambda_{\text{real}} \left(1 - e^{-\frac{5\lambda(E_{ph},h\gamma)}{\lambda_{\text{real}}}} \right) \\ \left[1 - \left(\frac{E_F+\Phi}{E_{ph}} \right)^{0.5} \right] \end{array} \right\} dE_{ph} \tag{9.36}$$

Based on the definition of B and Eqs. (9.32) and (9.34), B can be expressed as:

$$B = N_{\text{electrons}} / N_{\text{reach}} \tag{9.37}$$

TABLE 9.6 The Parameters of Ni Calculated by the First SEE Model

M hγ(eV)	Experimental $AQE(h\gamma)$	Calculated $AQE(h\gamma)$	n%	m%	C_γ (10^{-23}m^2)	λ_γ (10^{-9}m)	λ (10^{-9}m)	B	E_{mean} (eV)
7.6	3.00×10^{-3}	2.67×10^{-3}	1.00	0.362	10.4	10.5	1.74	0.0520	1.83
8.0	5.17×10^{-3}	4.45×10^{-3}	1.00	0.415	14.8	7.40	1.67	0.0583	2.05
8.6	7.82×10^{-3}	6.70×10^{-3}	1.00	0.496	17.6	6.22	1.58	0.0667	2.39
9.2	1.15×10^{-2}	9.78×10^{-2}	1.00	0.605	21.3	5.14	1.51	0.0712	2.71
9.8	1.52×10^{-2}	1.27×10^{-2}	1.00	0.766	22.7	4.83	1.46	0.0712	2.96
10.4	2.49×10^{-2}	2.11×10^{-2}	1.00	0.921	32.5	3.37	1.43	0.0769	3.18
10.5	2.81×10^{-2}	2.40×10^{-2}	1.00	0.938	36.8	2.98	1.42	0.0791	3.23
10.6	2.99×10^{-2}	2.70×10^{-2}	1.00	0.953	41.3	2.65	1.41	0.0815	3.28
10.8	3.05×10^{-2}	2.62×10^{-2}	1.00	0.978	36.0	3.04	1.39	0.0853	3.35
11.0	3.14×10^{-2}	2.71×10^{-2}	1.00	1.00	34.6	3.17	1.37	0.0896	3.45
11.2	3.18×10^{-2}	2.78×10^{-2}	1.00	1.00	32.8	3.34	1.35	0.0965	3.56
11.4	3.20×10^{-2}	2.82×10^{-2}	1.00	1.00	31.1	3.52	1.33	0.103	3.68
11.6	3.22×10^{-2}	2.87×10^{-2}	1.00	1.00	29.7	3.69	1.3	0.11	3.81

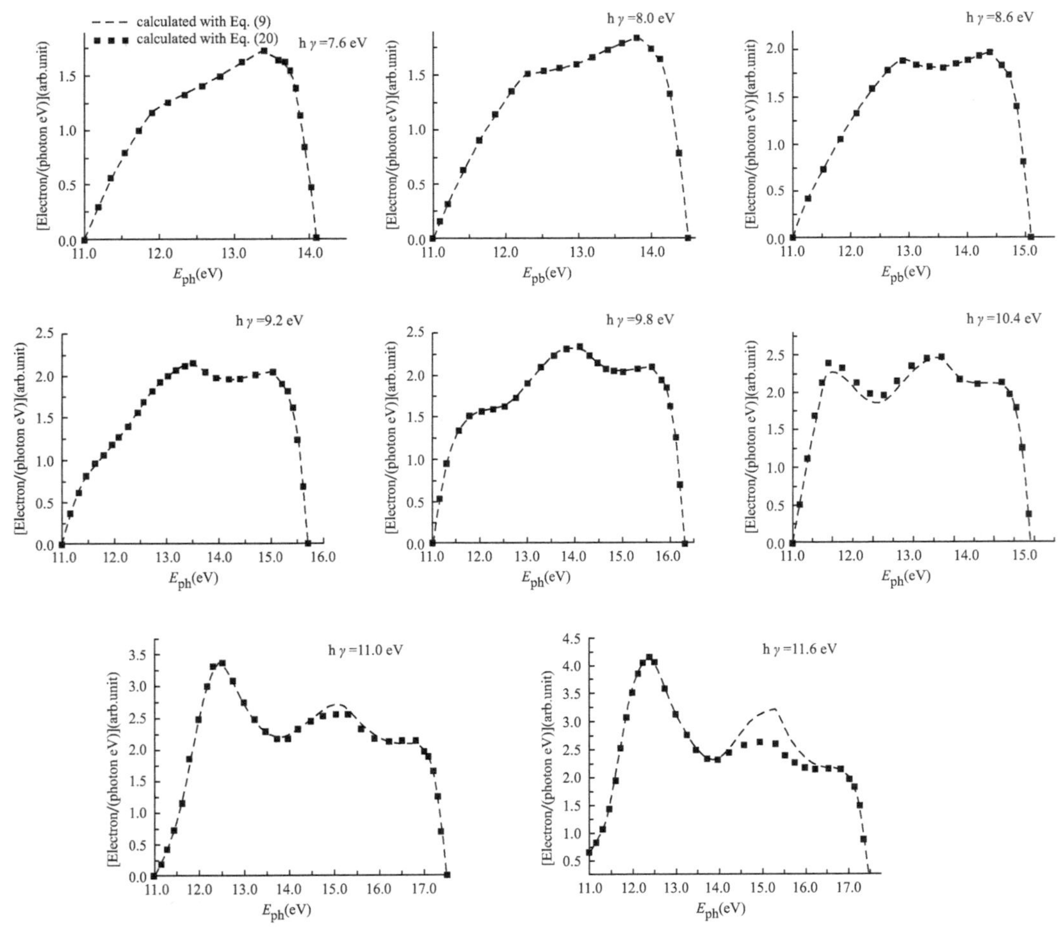

FIGURE 9.6 The $f(E_{ph}, h\gamma)$ calculated with Eq. (9.30) and Eq. (9.41) of Ni.

From the definition of λ and Eqs. (9.32) and (9.35), λ can be written as:

$$\lambda = N_{\text{depth}} / N_{\text{electrons}} \tag{9.38}$$

Based on the definition of λ and Eqs. (9.32) and (9.36), E_{mean} can be expressed as:

$$E_{\text{mean}} = N_{\text{energy}} / N_{\text{electrons}} \tag{9.39}$$

The B of Au and Ni are calculated with Eqs. (9.26), (9.31), (9.32), (9.34), and (9.37) and corresponding parameters (hγ, $g(E_{ph}-h\gamma)$, E_F, Φ, λ_γ shown in Tables 9.5 and 9.6), respectively; The λ of Au and Ni are calculated with Eqs. (9.26), (9.31), (9.32), (9.35), and (9.38) and corresponding parameters (hγ, $g(E_{ph}-h\gamma)$, E_F, Φ, λ_γ shown in Tables 9.5 and 9.6), respectively; The E_{mean} of Au and Ni are calculated with Eqs. (9.26), (9.31), (9.32), (9.36), and (9.39) and corresponding parameters (hγ, $g(E_{ph}-h\gamma)$, E_F, Φ, λ_γ shown in Tables 9.5 and 9.6), respectively. These calculated B, λ, and E_{mean} of Au and Ni are still shown in Tables 9.5 and 9.6, respectively.

The sum of E_F and Φ of Ni is < 11.6 eV. Thus, in the case that photons at hγ = 11.6 eV enter Ni the electrons in conduction band absorbing one hγ photon at least have E_{ph} = 11.6 eV. Therefore, when we use Eqs. (9.32) and (9.34)–(9.39) to calculate the corresponding parameters (B, λ, and E_{mean}) at hγ = 11.6 eV of Ni, the lower limit of integral in Eqs. (9.32) and (9.34)–(9.36) should be replaced with "11.6 eV," respectively. These calculated B, λ, and E_{mean} of Ni are still shown in Table 9.6.

From the energy band structures of metal shown in Figure 9.4, the law of energy conservation, the relation of $E_{ph} = E + h\gamma$, and the courses of deducing Eqs. (9.32)–(9.36), it is known that in the case that hγ is in the range of $((E_F + \Phi), (E_F + E_g))$ the lower limit of integral in Eqs. (9.32)–(9.36) should be replaced with "hγ," respectively.

9.2.3 Second SEE Model for Photoemission

From the above courses of calculating C_γ with Eqs. (9.22), (9.26), (9.31), and (9.33), it is known that in the cases where the AQE(hγ) or absolute $g(E_{ph}-h\gamma)$ is not known, the C_γ cannot be calculated with Eqs. (9.22), (9.26), (9.31) and (9.33). From the above courses of calculating λ_γ, $f(E_{ph}$, hγ), B, λ, and E_{mean}, it is also known that the λ_γ, $f(E_{ph}$, hγ), B, λ, and E_{mean} can be calculated by the first SEE model for photoemission on the basis of known C_γ. Thus, in the case that the AQE(hγ) or absolute $g(E_{ph}-h\gamma)$ is not known, λ_γ, $f(E_{ph}$, hγ), B, λ, and E_{mean} cannot be calculated by the first SEE model, either. So we present the following second SEE model for $f(E_{ph}$, hγ), C_γ, λ_γ, B, λ, and E_{mean} of metals; in the case that both AQE(hγ) and absolute $g(E_{ph}-h\gamma)$ are not known, the $f(E_{ph}$, hγ), B, λ, and E_{mean} still can be calculated by the second SEE model; in the case that absolute $g(E_{ph}-h\gamma)$ is not known, C_γ and λ_γ still can be calculated by the second SEE model.

Suppose that in the case that N_0 photons at $\gamma < (E_F + E_g)/h$ enter perpendicularly into metals, the number of photons at x does not decrease with increasing x and equals N_0. Thus, based on Eq. (9.23), the above assumption and the fact that the PCS at a given γ of the electron in the conduction band of a given metal can be looked on as a constant C_γ, in the case that N_0 photons at $\gamma < (E_F + E_g)/h$ enter into metals number of internal photo-emitted electrons per unit path length of incident photons can be written as:

$$n(x)=C_\gamma G(E_{ph}-h\gamma)N_0=\frac{C_\gamma \rho N_A g(E_{ph}-h\gamma)N_0}{A_\alpha}[0\leq(E_{ph}-\Phi-E_F)] \tag{9.40}$$

Therefore, from the three-step processes of photoemission, the conclusion that maximum escape depth of internal photo-emitted electrons with E_{ph} equals 5.0 $\lambda(E_{ph}, h\gamma)$ and Eqs. (9.29) and (9.40), $f(E_{ph}, h\gamma)$ can be expressed as:

$$f(E_{ph},h\gamma)=\int_0^{5\lambda(E_{ph},h\gamma)}\left\{\frac{C_\gamma \rho N_A g(E_{ph}-h\gamma)N_0}{A_\alpha}e^{-\frac{x}{\lambda(E_{ph},h\gamma)}}\left[1-\left(\frac{E_F+\Phi}{E_{ph}}\right)^{0.5}\right]\right\}dx$$

$$=\frac{C_\gamma \rho N_A g(E_{ph}-h\gamma)N_0\lambda(E_{ph},h\gamma)}{A_\alpha}(1-e^{-5})\left[1-\left(\frac{E_F+\Phi}{E_{ph}}\right)^{0.5}\right]$$

$$[0\leq(E_{ph}-\Phi-E_F)\leq 1.5E_F] \tag{9.41}$$

The $f(E_{ph}, h\gamma)$ of Au are calculated with corresponding parameters (E_{ph}, $h\gamma$, $g(E_{ph}-h\gamma)$ [25], E_F, Φ) and Eqs. (9.26) and (9.41); The comparison between these calculated $f(E_{ph}, h\gamma)$ of Au and experimental ones are shown in Figure 9.7. The $f(E_{ph}, h\gamma)$ (in arbitrary units) of Cu are calculated with corresponding parameters (E_{ph}, $h\gamma$, $g(E_{ph}-h\gamma)$, E_F, Φ) and Eqs. (9.26) and (9.41); the comparison between these calculated $f(E_{ph}, h\gamma)$ of Cu and corresponding experimental ones [27] is shown in Figure 9.8.

In the case that N_0 photons at $\gamma<(E_F+E_g)/h$ enter into metals the electrons with E in the range of $[(E_F+\Phi-h\gamma), (E_F+\Phi)]$ absorbing one $h\gamma$ photon have E_{ph} in the range of $[(E_F+\Phi), (E_F+\Phi+h\gamma)]$, and the photo-emitted electrons with E_{ph} in the range of $[(E_F+\Phi), (E_F+\Phi+h\gamma)]$ may escape into vacuum. Thus, from Eq. (9.41), it is known that in the case that N_0 photons at $\gamma<(E_F+E_g)/h$ enter perpendicularly into metals the number of the photo-emitted electrons can be written as:

$$N_{electrons1}=\frac{C_\gamma \rho N_A}{A_\alpha}\int_{(E_F+\Phi)}^{(E_F+\Phi+h\gamma)}\left\{N_0 g(E_{ph}-h\gamma)\lambda(E_{ph},h\gamma)(1-e^{-5})\left[1-\left(\frac{E_F+\Phi}{E_{ph}}\right)^{0.5}\right]\right\}dE_{ph} \tag{9.42}$$

Eqs. (9.41) and (9.42) are deduced on the basis of the assumption that the number of photons at any x equals N_0, but in fact the number of photons at x decreases with increasing x. Thus, from the characteristics of electron emission and the courses of deducing Eq. (9.42), it is known that in the case that N_0 photons at $\gamma<(E_F+E_g)/h$ enter perpendicularly into metals, the real number of the photo-emitted electrons is < the value of Eq. (9.42) and should be written as:

$$N_{electrons2}=\frac{K(\gamma)C_\gamma \rho N_A}{A_\alpha}\int_{(E_F+\Phi)}^{(E_F+\Phi+h\gamma)}\left\{N_0 g(E_{ph}-h\gamma)\lambda(E_{ph},h\gamma)(1-e^{-5})\left[1-\left(\frac{E_F+\Phi}{E_{ph}}\right)^{0.5}\right]\right\}dE_{ph} \tag{9.43}$$

For given γ and metal, $K(\gamma)$ is a constant which is < 1.0.

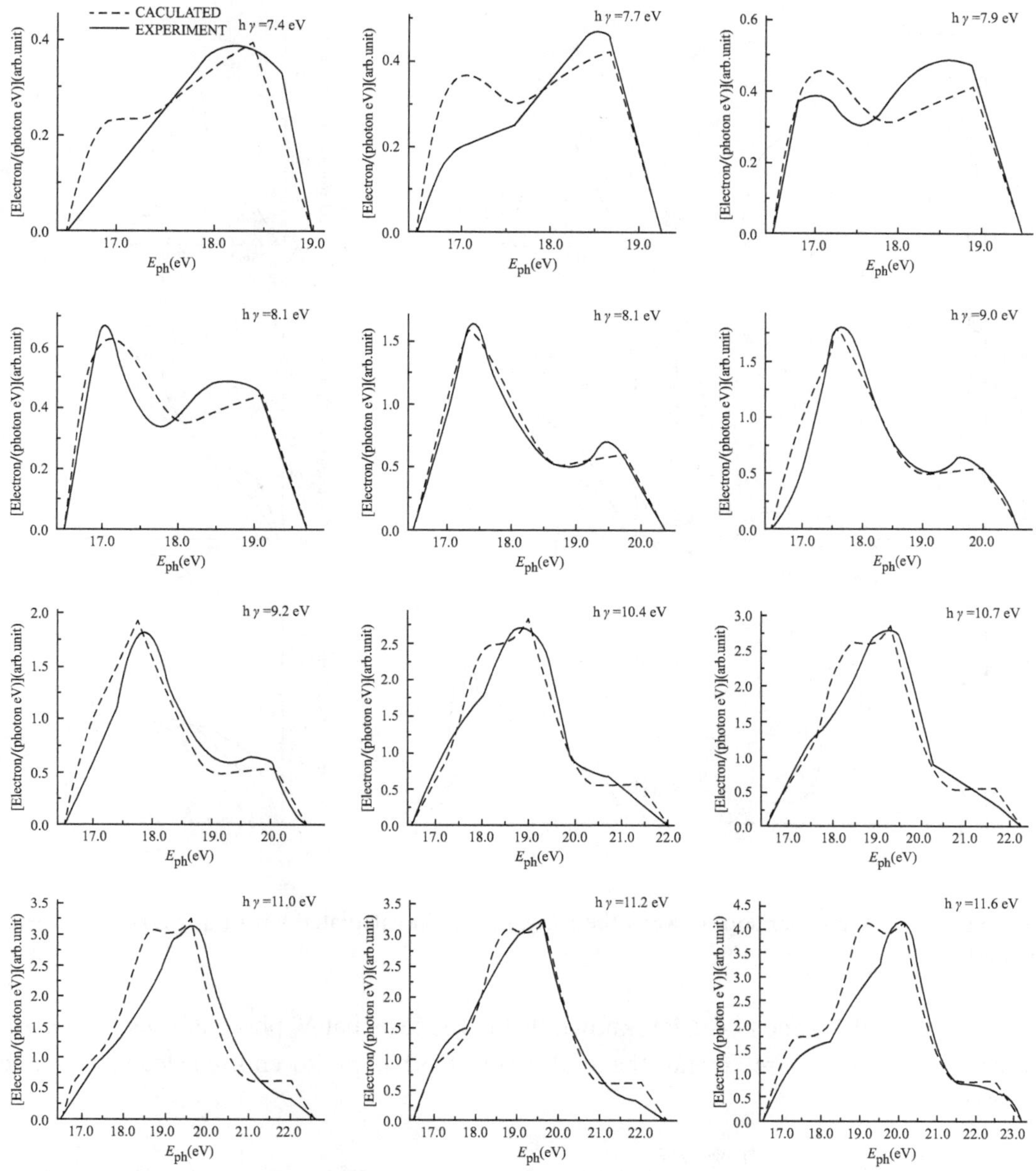

FIGURE 9.7 The comparison between the $f(E_{ph}, h\gamma)$ of Au calculated with Eq. (9.41) and experimental ones.

According to Eq. (9.43) and the three-step processes of photoemission, it is known that in the case that N_0 photons at $\gamma < (E_F + E_g)/h$ enter perpendicularly into metals, the number of internal photo-excited electrons reaching the emission surface can be written as:

$$N_{reach2} = \frac{K(\gamma)C_\gamma \rho N_A}{A_\alpha}\int_{(E_F+\Phi)}^{(E_F+\Phi+h\gamma)} [N_0 g(E_{ph}-h\gamma)\lambda(E_{ph},h\gamma)(1-e^{-5})]\,dE_{ph} \tag{9.44}$$

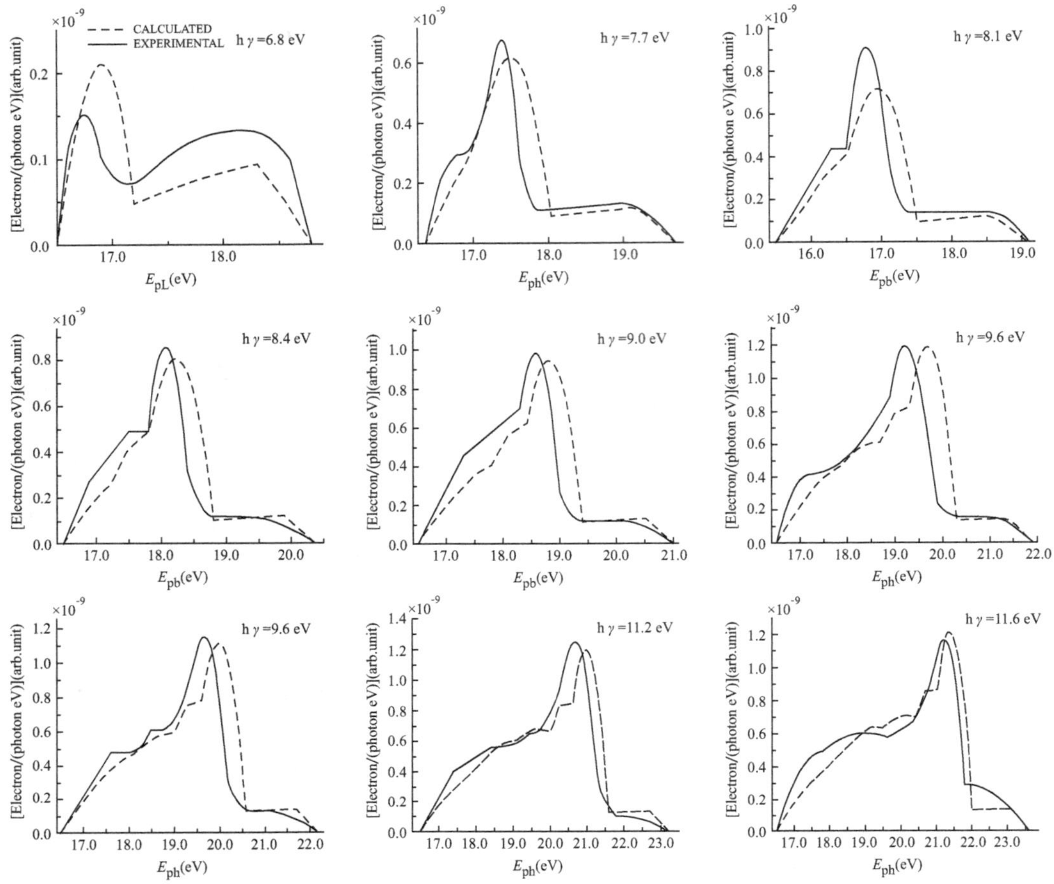

FIGURE 9.8 The comparison between the $f(E_{ph}, h\gamma)$ of Cu calculated with Eq. (9.41) and experimental ones.

Based on Eqs. (9.26) and (9.43), it is known that in the case that N_0 photons at $\gamma < (E_F + E_g)/h$ enter perpendicularly into metals the total escape depth of photo-emitted electrons can be written as:

$$N_{\text{depth2}} = \frac{K(\gamma)C_\gamma \rho N_A}{A_\alpha} \int_{(E_F+\Phi)}^{(E_F+\Phi+h\gamma)} \left\{ N_0 g(E_{ph} - h\gamma)(\lambda(E_{ph}, h\gamma))^2 (1-e^{-5}) \left[1 - \left(\frac{E_F+\Phi}{E_{ph}}\right)^{0.5}\right] \right\} dE_{ph} \tag{9.45}$$

According to the fact that the photo-emitted electrons with E have $E_{vac} = (E_{ph} - E_F - \Phi)$ and Eq. (9.43), it is known that in the case that N_0 photons at $\gamma < (E_F + E_g)/h$ enter perpendicularly into metals, the total E_{vac} of photo-emitted electrons can be written as:

$$N_{\text{energy2}} = \frac{K(\gamma)C_\gamma \rho N_A}{A_\alpha} \int_{(E_F+\Phi)}^{(E_F+\Phi+h\gamma)} \left\{ \begin{array}{l} N_0 g(E_{ph} - h\gamma)(E_{ph} - E_F - \Phi)\lambda(E_{ph}, h\gamma)(1-e^{-5}) \\ \left[1 - \left(\frac{E_F+\Phi}{E_{ph}}\right)^{0.5}\right] \end{array} \right\} dE_{ph} \tag{9.46}$$

Based on the definition of B and Eqs. (9.43) and (9.44), B can be expressed as:

$$B = N_{\text{electrons2}} / N_{\text{reach2}} \tag{9.47}$$

Based on the definition of λ and Eqs. (9.43) and (9.45), λ can be written as:

$$\lambda = N_{\text{depth2}} / N_{\text{electrons2}} \tag{9.48}$$

Based on the definition of λ and Eqs. (9.43) and (9.46), E_{mean} can be expressed as:

$$E_{\text{mean}} = N_{\text{energy2}} / N_{\text{electrons2}} \tag{9.49}$$

The B of Au calculated with parameters of Au (hγ, $g(E_{\text{ph}} - \text{h}\gamma)$, E_F, and Φ) and Eqs. (9.26), (9.43), (9.44), and (9.47) are shown in Table 9.7; The λ of Au calculated with parameters of Au (hγ, $g(E_{\text{ph}} - \text{h}\gamma)$, E_F and Φ) and Eqs. (9.26), (9.43), (9.45), and (9.48) are shown in Table 9.7; The E_{mean} of Au calculated with parameters of Au (hγ, $g(E_{\text{ph}} - \text{h}\gamma)$, E_F, and Φ) and Eqs. (9.26), (9.43), (9.46) and (9.49) are shown in Table 9.7. By the same method, the B, λ, and E_{mean} of Ni and Cu are calculated and shown in Tables 9.8 and 9.9, respectively.

As that we do for the first SEE model, the lower limit of integral in Eqs. (9.43)–(9.46) should be replaced with "11.6 eV" when calculating the corresponding parameters (B, λ, and E_{mean}) at hγ = 11.6 eV of Ni.

According to the simple theories of transport and escape of internal secondary electrons, the probability that an internal secondary electron, which is excited at x, can reach the emission surface and pass over the surface barrier of metal into a vacuum can be written as [33]:

$$f(x)_s = B_s e^{-x/\lambda_s} \tag{9.50}$$

where B_s is the mean probability that an internal secondary electron escapes into vacuum upon reaching emission surface, and λ_s is mean escape depth of secondary electrons.

TABLE 9.7 The Parameters of Au Calculated by the Second SEE Model

M hγ(eV)	C_γ (10^{-23}m^2)	λ_γ (10^{-9}m)	λ (10^{-9}m)	B	E_{mean} (eV)
7.4	45	4.24	2.61	0.0231	1.35
7.7	45.5	4.01	2.58	0.0224	1.42
7.9	48	3.70	2.58	0.0219	1.45
8.1	48.4	3.59	2.58	0.0223	1.46
8.7	49.1	3.34	2.53	0.0262	1.58
9.0	51.2	3.15	2.48	0.0285	1.67
9.2	55.1	2.89	2.45	0.0303	1.74
10.4	53.0	2.92	2.19	0.0459	2.35
10.7	59.4	2.61	2.13	0.0485	2.52
11.0	64.9	2.38	2.07	0.0503	2.69
11.2	66.3	2.33	2.04	0.0515	2.79
11.5	67.6	2.28	2.0	0.0533	2.95

TABLE 9.8 The Parameters of Ni Calculated by the Second SEE Model

M hγ(eV)	C_γ (10^{-23}m²)	λ_γ (10^{-9}m)	λ (10^{-9}m)	B	E_{mean} (eV)
7.6	12.2	9.01	1.74	0.0512	1.81
8.0	18.3	2.99	1.68	0.0569	2.03
8.6	22.2	4.95	1.59	0.0648	2.35
9.2	27.8	3.94	1.52	0.0683	2.65
9.8	30.8	3.56	1.48	0.0676	2.88
10.4	45.1	2.43	1.46	0.0720	3.06
11.0	46.1	2.38	1.4	0.0847	3.31
11.6	36.2	3.03	1.32	0.106	3.69

TABLE 9.9 The Parameters of Cu Calculated by the Second SEE Model

M hγ(eV)	Experimental AQE(hγ)	n%	m%	C_γ (10^{-23}m²)	λ_γ (10^{-9}m)	λ (10^{-9}m)	B	E_{mean} (eV)
6.8	1.6×10^{-3}	0.694	0.128	43.1	3.58	2.92	0.0193	1.32
7.4	2.6×10^{-3}	0.763	0.237	28.9	4.86	2.92	0.0223	1.35
7.7	3.0×10^{-3}	0.788	0.277	22.8	5.95	2.86	0.0263	1.45
8.1	3.9×10^{-3}	0.817	0.324	21.5	6.09	2.74	0.0317	1.65
8.4	4.5×10^{-3}	0.836	0.361	20.4	6.28	2.65	0.0351	1.81
9.0	5.8×10^{-3}	0.873	0.434	19.1	6.43	2.49	0.0418	2.15
9.6	6.8×10^{-3}	0.906	0.506	17.2	6.87	2.34	0.0479	2.49
10.2	7.8×10^{-3}	0.936	0.570	15.9	7.21	2.21	0.0546	2.84
11.2	8.8×10^{-3}	0.976	0.681	13.5	8.13	2.02	0.0634	3.43
11.6	9.2×10^{-3}	0.989	0.731	12.8	8.43	1.96	0.066	3.65

Based on Eq. (9.50) and the fact that SEE and photoemission have common mechanisms of transport and escape, similarly, the probability that an internal photo-emitted electron, which is excited at x, can reach emission surface and pass over surface barrier of metal can be written as:

$$f(x)_{\text{ph}} = Be^{-x/\lambda} \tag{9.51}$$

Thus, according to the three-step processes of photoemission and Eqs. (9.53) and (9.51), it is known that in the case that N_0 photons at $\gamma < (E_F + E_g)/\text{h}$ enter perpendicularly into metals, the number of the photo-emitted electrons can be written as [38]:

$$N_{\text{electrons}} = \int_0^\infty (n(x)f(x)_{\text{ph}})\text{d}x = N_0\alpha_{\text{ph}\gamma}B\int_0^\infty \left[e^{\left[-\left(\alpha_\gamma+\frac{1}{\lambda}\right)x\right]}\right]\text{d}x = \frac{N_0B\lambda\alpha_{\text{ph}\gamma}}{1+\lambda\alpha_\gamma} \tag{9.52}$$

Therefore, from Eq. (9.52) and the definition of $AQE(\text{h}\gamma)$, the $AQE(\text{h}\gamma)$ can be written as [38]:

$$AQE(\text{h}\gamma) = \frac{B\lambda\alpha_{\text{ph}\gamma}}{1+\lambda\alpha_\gamma} \tag{9.53}$$

Based on the energy band structures of metal shown in Figure 9.1, calculated $g(E)$, the values of E_F and Φ, law of energy conservation and the fact that γ is $<(E_F+E_g)/h$, we can calculate the relative number of electrons in the conduction band of metal which may absorb one photon and become internal photo-emitted electrons. For example, when the photons at $h\gamma=7.4$ eV enter Au with $E_F=11.6$ eV and $\Phi=4.9$ eV, it is known that the electrons absorbing one photon and having $E \geqq [E_F-(7.4\text{ eV}-\Phi)]$ (i.e. 9.1 eV) may have enough energy to become internal photo-emitted electrons. Thus, based on the $g(E)$ of Au, the fact that the electrons with $E \geqq 9.1$ eV may become internal photo-emitted electrons and the facts that E_F and Φ of Au equal 11.6 and 4.9 eV, we can calculate the $m\%$ electrons in the conduction band of Au which may absorb one photon and become internal photo-emitted electrons by calculating, and the calculated $m\%$ at $h\gamma=7.4$ eV of Au equals 0.0734; By the same method, the $m\%$ at different $h\gamma$ of Au, Ni and Cu can be calculated and shown in Tables 9.5, 9.6 and 9.9, respectively. Therefore, photons at $\gamma<(E_F+E_g)/h$ enter perpendicularly into metals $\alpha_{ph\gamma}$ of Eq. (9.53) can be expressed as [38]:

$$\alpha_{ph\gamma}=(s\rho N_A C_\gamma m\%)/A_\alpha \tag{9.54}$$

Note that in the case that photons at $\gamma<(E_F+E_g)/h$ enter perpendicularly into metals α_γ of Eq. (9.53) is expressed as Eq. (9.51).

α_γ of NEAS equals $\alpha_{ph\gamma}$. Thus, from Eq. (9.53), $AQE(h\gamma)$ from NEAS can be written as [38]:

$$AQE(h\gamma)=\frac{B\lambda\alpha_\gamma}{1+\lambda\alpha_\gamma} \tag{9.55}$$

Excited electrons, such as electron-induced, ion-induced, and photo-excited, have varying mean escape depth and probability based on their energy (E). Therefore, using Equations (9.50) and (9.51) to express the probability ($f(x)$) of an excited electron escaping into a vacuum during formula derivation is an approximate process due to the physical mechanisms of transport and escape. Specifically, Eq. (9.51) is used as an approximation for $f(x)_{ph}$ in deducing Eq. (9.53). Eqs. (9.55), (9.28), and (9.51) are derived from their respective correct Equations (Eqs. (9.27), (9.50), and (9.53)) under specific conditions, making them valid. Although an approximation is made in the derivation of Eq. (9.53) and Eqs. (9.33) and (9.53), it can be concluded that Eqs. (9.33) and (9.53) are theoretically correct, with Eq. (9.33) being more accurate when E is much greater than 1.0 eV. Additionally, Eq. (9.55) has experimentally shown to describe negative electron affinity photoemission through a diffusion model, successfully analyzing relevant parameters. Thus, the validity of Eq. (9.53) is further supported through its relation to Eq. (9.55).

9.3 CONCLUSIONS

This section is a comprehensive guide to the photoemission properties of metal surfaces and the research methods used to study them. It provides a detailed overview of the importance of photoemission from metal photocathodes in fields such as synthetic metals and high-energy physics, where photoelectric sensitivity (PS) and quantum efficiency (QE) are crucial topics. This section also delves into the widely used classical models for PS and QE of metals, including the three-step model and the Fowler-DuBridge model. These models are explained in detail, along with their applications in research. Furthermore, the secondary electron emission model for photoemission from metals in the vacuum ultraviolet is discussed, providing readers with a deeper understanding of this phenomenon. Overall, Chapter 6 is an invaluable resource for anyone interested in learning about metal's photoemission based on secondary electron emission. It covers a wide range of topics and provides detailed explanations that are easy to understand.

REFERENCES

1. Fowler R H. The analysis of photoelectric sensitivity curves for clean metals at various temperatures. *Physical Review*, 1931, 38(1): 45.
2. DuBridge L A. A further experimental test of Fowler's theory of photoelectric emission. *Physical Review*, 1932, 39(1): 108.
3. Dowell D H, King F K, Kirby R E, et al. In situ cleaning of metal cathodes using a hydrogen ion beam. *Physical Review Special Topics—Accelerators and Beams*, 2006, 9(6): 063502.
4. Dowell D H, Schmerge J F. Quantum efficiency and thermal emittance of metal photocathodes. *Physical Review Special Topics—Accelerators and Beams*, 2009, 12(7): 074201.
5. Xie A G, Wang L, Mu L H. Formula for maximum secondary electron yield from metals. *Surface Review and Letters*, 2015, 22(02): 1550019.
6. Llacer J, Garwin E L. Electron-phonon interaction in alkali halides. I. The transport of secondary electrons with energies between 0.25 and 7.5 eV. *Journal of Applied Physics*, 1969, 40(7): 2766–2775.
7. Xie A G, Dong H J, Liu Y F. Secondary electron emission and photoemission from negative electron affinity semiconductor with large mean escape depth of excited electrons. *Chinese Physics B*, 2022, 32(4): 048102.
8. Matskevich T L. Secondary electron emission of a NaBr single crystal. *Soviet Physics-Technical Physics*, 1956, 1(10): 2320–2320.
9. Shih A, Yater J, Hor C, et al. Secondary electron emission properties of oxidized beryllium CFA cathodes. *IEEE Transactions on Electron Devices*, 1994, 41(12): 2448–2454.
10. Petzel B. Zur Temperaturabhängigkeit der Sekundärelektronenemission von Ionenkristallen. *Annalen der Physik*, 1961, 464(1–2): 44–47.
11. Wolff P A. Theory of secondary electron cascade in metals. *Physical Review*, 1954, 95(1): 56.
12. Winch R P. The photoelectric properties of silver. *Physical Review*, 1931, 37(10): 1269.
13. DuBridge L A, Roehr W. Photoelectric and thermionic properties of palladium. *Physical Review*, 1932, 39(1): 99.
14. Morris L W. Certain photoelectric properties of gold. *Physical Review*, 1931, 37(10): 1263.
15. Lawrence E O, Linford L B. The effect of intense electric fields on the photo-electric properties of metals. *Physical Review*, 1930, 36(3): 482.
16. Warner A H. Variation of the photoelectric effect with temperature and determination of the long wave-length limit for Tungsten. *Physical Review*, 1929, 33(5): 815.
17. Kazda C B. The photo-electric threshold for mercury. *Physical Review*, 1925, 26(5): 643.
18. Goetz A. The photoelectric effect of molten tin and two of its allotropic modifications. *Physical Review*, 1929, 33(3): 373.
19. Xie A G, Dong H J, Pan Z. An electron-induced secondary electron model for photoelectric sensitivity and quantum efficiency of metal surfaces. *Results in Physics*, 2021, 26: 104350.
20. Cazaux J. About the secondary electron yield and the sign of charging of electron irradiated insulators. *The European Physical Journal Applied Physics*, 2001, 15(3): 167–172.
21. Cazaux J. Electron-and X-ray-induced electron emissions from insulators. *Polymer International*, 2001, 50(7): 748–755.

22. Li G C (ed.) *Photon Optics*, 1st edition. National Defense Industry Press, 2010, p. 147.
23. Sommer A H (ed.) *Photoemissive Materials: Preparation, Properties, Uses*, 1st edition. John Wiley and Sons, Inc., 1968, pp. 38–58.
24. Dowell D H, King F K, Kirby R E, et al. In situ cleaning of metal cathodes using a hydrogen ion beam. *Physical Review Special Topics—Accelerators and Beams*, 2006, 9(6): 063502.
25. Kanter H. Slow-electron mean free paths in aluminum, silver, and gold. *Physical Review B*, 1970, 1(2): 522.
26. Blodgett Jr A J, Spicer W E. Experimental determination of the density of states in nickel. *Physical Review*, 1966, 146(2): 390.
27. Krolikowski W F, Spicer W E. Photoemission studies of the noble metals. I. Copper. *Physical Review*, 1969, 185(3): 882.
28. Fong C Y, Cohen M L. Energy band structure of copper by the empirical pseudopotential method. *Physical Review Letters*, 1970, 24(7): 306.
29. Snow E C. Self-consistent energy bands of metallic copper by the augmented-plane-wave method. II. *Physical Review*, 1968, 171(3): 785.
30. ZJ D, HM L, Shimizu R, et al. On the energy distribution of secondary electrons emitted from metals. *Journal of Surface Analysis*, 2008, 15(2): 186–194.
31. Chung M S. Improved calculations of secondary electron energy distributions of metals. *Journal of Applied Physics*, 1975, 46(1): 465–466.
32. Xie A G, Wang L, Mu L H. Formula for maximum secondary electron yield from metals. *Surface Review and Letters*, 2015, 22(02): 1550019.
33. Xie A G, Uhm H S, Chen Y, et al. Maximum secondary electron yield and parameters of secondary electron yield of metals. *Surface Review and Letters*, 2016, 23(05): 1650039.
34. Wolff P A. Theory of secondary electron cascade in metals. *Physical Review*, 1954, 95(1): 56.
35. Xie A G, Dong H J, Pan Z. An electron-induced secondary electron model for photoelectric sensitivity and quantum efficiency of metal surfaces. *Results in Physics*, 2021, 26: 104350.
36. Williams J O D, Lapington J S, Leach S A, et al. Using quantum entangled photons to measure the absolute photon detection efficiency of a multi-pixel SiPM array. *Nuclear Instruments and Methods in Physics Research Section A: Accelerators, Spectrometers, Detectors and Associated Equipment*, 2020, 958: 162167.
37. Xu W, Hou X, Meng Y, et al. Deciphering charging status, absolute quantum efficiency, and absorption cross section of multicarrier states in single colloidal quantum dots. *Nano Letters*, 2017, 17(12): 7487–7493.
48. Xie A G, Liu Y F, Dong H J. Secondary electron emission model for photo-emission from metals in the vacuum ultraviolet. *Nuclear Science and Techniques*, 2022, 33(8): 103.

FSC
www.fsc.org
MIX
Papier | Fördert
gute Waldnutzung
FSC® C083411

Kolibri 360 GmbH
Ferdinand-Jühlke-Straße 7
99095 Erfurt, Deutschland
produktsicherheit@kolibri360.de